Sustainable Built Environment and Urban Growth Management

Sustainable Built Environment and Urban Growth Management

Special Issue Editor
Wann-Ming Wey

MDPI • Basel • Beijing • Wuhan • Barcelona • Belgrade

Special Issue Editor
Wann-Ming Wey
Department of Real Estate
and Built Environment,
National Taipei University
Taiwan

Editorial Office
MDPI
St. Alban-Anlage 66
4052 Basel, Switzerland

This is a reprint of articles from the Special Issue published online in the open access journal *Sustainability* (ISSN 2071-1050) from 2017 to 2018 (available at: https://www.mdpi.com/journal/sustainability/special_issues/Sustainable_Built_Environment_Urban_Growth).

For citation purposes, cite each article independently as indicated on the article page online and as indicated below:

LastName, A.A.; LastName, B.B.; LastName, C.C. Article Title. *Journal Name* **Year**, *Article Number*, Page Range.

ISBN 978-3-03928-186-2 (Pbk)
ISBN 978-3-03928-187-9 (PDF)

© 2020 by the authors. Articles in this book are Open Access and distributed under the Creative Commons Attribution (CC BY) license, which allows users to download, copy and build upon published articles, as long as the author and publisher are properly credited, which ensures maximum dissemination and a wider impact of our publications.

The book as a whole is distributed by MDPI under the terms and conditions of the Creative Commons license CC BY-NC-ND.

Contents

About the Special Issue Editor vii

Wann-Ming Wey
A Commentary on Sustainably Built Environments and Urban Growth Management
Reprinted from: *Sustainability* **2018**, *10*, 3898, doi:10.3390/su10113898 1

Seongwoo Jeon, Hyunjung Hong and Sungdae Kang
Simulation of Urban Growth and Urban Living Environment with Release of the Green Belt
Reprinted from: *Sustainability* **2018**, *10*, 3260, doi:10.3390/su10093260 6

Yu-Hui Chen, Chun-Lin Lee, Guan-Rui Chen, Chiung-Hsin Wang and Ya-Hui Chen
Factors Causing Farmland Price-Value Distortion and Their Implications for Peri-Urban Growth Management
Reprinted from: *Sustainability* **2018**, *10*, 2701, doi:10.3390/su10082701 27

Xiaer Xiahou, Yuchun Tang, Jingfeng Yuan, Tengyuan Chang, Ping Liu and Qiming Li
Evaluating Social Performance of Construction Projects: An Empirical Study
Reprinted from: *Sustainability* **2018**, *10*, 2329, doi:10.3390/su10072329 45

Mattias Höjer and Kristina Mjörnell
Measures and Steps for More Efficient Use of Buildings
Reprinted from: *Sustainability* **2018**, *10*, 1949, doi:10.3390/su10061949 61

Judith Chen-Hsuan Cheng, Ai-Hsuan Chiang, Yulan Yuan and Ming-Yuan Huang
Exploring Antecedents of Green Tourism Behaviors: A Case Study in Suburban Areas of Taipei, Taiwan
Reprinted from: *Sustainability* **2018**, *10*, 1928, doi:10.3390/su10061928 72

Armando Avalos Jiménez, Fernando Flores Vilchez, Oyolsi Nájera González and Susana M. L. Marceleño Flores
Analysis of the Land Use and Cover Changes in the Metropolitan Area of Tepic-Xalisco (1973–2015) through Landsat Images
Reprinted from: *Sustainability* **2018**, *10*, 1860, doi:10.3390/su10061860 89

Kai Fischer, Stefan Hiermaier, Werner Riedel and Ivo Häring
Morphology Dependent Assessment of Resilience for Urban Areas
Reprinted from: *Sustainability* **2018**, *10*, 1800, doi:10.3390/su10061800 104

Byungsuk Kim and Jina Park
Effects of Commercial Activities by Type on Social Bonding and Place Attachment in Neighborhoods
Reprinted from: *Sustainability* **2018**, *10*, 1771, doi:10.3390/su10061771 118

Seungjun Roh, Sungho Tae and Rakhyun Kim
Analysis of Embodied Environmental Impacts of Korean Apartment Buildings Considering Major Building Materials
Reprinted from: *Sustainability* **2018**, *10*, 1693, doi:10.3390/su10061693 132

Mikael Mangold, Magnus Österbring, Conny Overland, Tim Johansson and Holger Wallbaum
Building Ownership, Renovation Investments, and Energy Performance—A Study of Multi-Family Dwellings in Gothenburg
Reprinted from: *Sustainability* **2018**, *10*, 1684, doi:10.3390/su10051684 149

Min Song, Lynn Huntsinger and Manman Han
How does the Ecological Well-Being of Urban and Rural Residents Change with Rural-Urban Land Conversion? The Case of Hubei, China
Reprinted from: *Sustainability* **2018**, *10*, 527, doi:10.3390/su10020527 165

Chou-Tsang Chang and Tzu-Ping Lin
Estimation of Carbon Dioxide Emissions Generated by Building and Traffic in Taichung City
Reprinted from: *Sustainability* **2018**, *10*, 112, doi:10.3390/su10010112 187

Liming Zhang, Wei Yang, Yuan Yuan and Rui Zhou
An Integrated Carbon Policy-Based Interactive Strategy for Carbon Reduction and Economic Development in a Construction Material Supply Chain
Reprinted from: *Sustainability* **2017**, *9*, 2107, doi:10.3390/su9112107 205

Magdalena Celadyn
Environmental Activation of Inner Space Components in Sustainable Interior Design
Reprinted from: *Sustainability* **2018**, *10*, 1945, doi:10.3390/su10061945 220

About the Special Issue Editor

Wann-Ming Wey is currently serving as the Dean of Office of Research and Development as well as a Distinguished Professor of the Department of Real Estate & Built Environment, College of Public Affairs, National Taipei University, Taiwan. His current academic research interests include urban and regional planning and design as well as the applications of the network analysis method and multiple objective programming to the area of urban built environment planning and design topics. He was the 2008–2010 Distinguished Young Scholar Award of the Ministry of Science and Technology (MOST) in Taiwan. His professional memberships include INFORMS, IEEE, Taiwan Institute of Urban Planning and Chinese Institute of Engineers.

Editorial

A Commentary on Sustainably Built Environments and Urban Growth Management

Wann-Ming Wey

Department of Real Estate and Built Environment, National Taipei University, 151, University Road, San Shia District, New Taipei City 23741, Taiwan; wmwey@mail.ntpu.edu.tw

Received: 22 October 2018; Accepted: 23 October 2018; Published: 26 October 2018

Abstract: The concept of urban growth management first emerged in the United States in the 1950s. Its goal was to solve problems stemming from urban sprawl by applying integrated planning, management, and regulation, and to adjust to different development trends in different spaces and times. From the viewpoint of the studies on the link between sustainably built environments, urban growth management, and their interactions, this special issue includes theoretical and empirical studies on sustainable built environment planning and design, sustainable growth management strategies, and other related emerging topics, such as intelligent use of information and communication technologies (ICT) to sustainably build environments, as well as smart cities research with big data, data mining, cloud computing, and internet of things (IOT) ideas.

Keywords: urban growth management; sustainable built environment; quality of life (QoL); smart city & big data

---

1. Background and Literature Review

During the 21st century, numerous urban developments have gradually deviated from their original urban plans, resulting in the spread of urban sprawl and uncontrolled land development. These effects have produced a number of urban problems, such as serious air pollution, poor environment quality, congestion, and inappropriate land development with low urban density, all of which have negatively affected the quality of life (QoL). The growth management concept was first adopted in the United States to solve problems associated with urban sprawl. Contemporary city planners can use multi-objective growth management methods to simultaneously achieve urban development and QoL. In 2010, the EU failed to meet its Lisbon Strategy goals and subsequently adopted a new strategy called Europe 2020, in which it introduced three essential growth constructs: smart, sustainable, and inclusive growth. The EU also asserts that urban development can improve QoL if the concept of development is replaced with the concept of growth [1–3].

Urban development has deviated from its original planned scope, and the resulting urban sprawl and land overuse have generated various problems. The Lisbon Strategy, initiated in 2000, was intended to make the EU the most competitive economy in the world by 2010. Its main targets were (1) labor participation above 70%, (2) research and development (R&D) expenditure above 3% of the GDP, and (3) an average economic growth rate above 3%. However, the EU did not achieve these three targets. After the failure of the Lisbon Strategy, the then-president of the EU, Manuel Barroso, proposed the Europe 2020 strategy based on essential concepts like smart, sustainable, and inclusive growth [2,3]. How to guide urban development toward growth and increase QoL is thus the central question of the present study. Past studies have argued that growth management is achieved by imposing rigorous guidance and control on regional development in order to ensure improved QoL. As described by Timothy Chapin, four emerging waves of growth management policy in the United States have been evolving since the 1950s [4]. Chapin asserted that growth management policy has focused on control in the past, but should focus on smart and sustainable growth in the future. However, cities are complex systems affected by society, as well as the economy, environment, and culture. Traditional urban development in the decades since World War II has faced numerous problems, such as long-distance transportation, inefficient services, air quality deterioration, land use fragmentation, and inadequate city images. These problems have had negative effects on QoL [5]. As to the livability and sustainability of cities, urban planning and its relevant transportation deploying have a particularly profound and positive effect [6], and are crucial to the quality of life to urban residents at the same time.

Although rapid socioeconomic development has accelerated urbanization and suburbanization, it has also placed a sizeable burden on the sustainability of resource use and city livability. To create a humanized and sustainable living environment, Jane Jacobs advocated the concept of human-scale, livable communities in the 1960s, and the World Commission on Environment and Development proposed the concept of sustainable development in *Our Common Future* in 1987. Urban planning and activities have a profound effect on the livability and sustainability of a city [7].

It is worth noting that various scientific innovations in the urban development field are flourishing, especially between cutting-edge technologies and their emerging requirements. Often these cutting-edge technologies claim that they could promote the sustainability and livability of the city, or even make the city, society, or environment smarter. For example, Ahmad and Mehmood [8] suggested that future cities will be further driven by developments in information and communication technology, and that logistics will play a critical role in future cities, due to the increasingly micro-dynamic nature of socioeconomic and globalised production and consumption patterns. Naim and Rashid [9] also reflected that the ICT-based and industry-driven approaches maybe are not the only solutions for urbanization and future city designs. Based on the above insightful rethinking and reflecting, we thus realize that the common ground and means of ICT-based or industry-driven approaches often tend to directly achieve sustainability or livability through innovative technologies, mechanisms, or intelligent systems [7].

2. Important Issues for Sustainably Built Environments and Urban Growth Management

During the past decade, there has been an increasing interest in the link between the sustainably built environment and urban growth management in the field of urban studies. This interest is motivated by the possibility that urban development planning and design principles associated with the built environment can be used to manage individual activities and improve the quality of urban life. Sustainably built environments are relatively important to urban growth management that deals with environmental problems, housing issues, and community well-being. Nowadays, the sustainable built environment planning in most cities has come to a turning point, as the traffic and population growing has become a serious concern and put tremendous pressure on both the environment and people in these cities. It is therefore important to find ways or new lifestyles, such as compact transit-oriented development (TOD) formulations for cities that are more flexible, inclusive, and sustainable. Furthermore, for sustainable built environment and urban growth management, not only the growth management principles (including smart growth, sustainable growth, and inclusive growth) should be taken into account, but the innovative/smart planning strategies like mixed use design, green transport, and new urbanism are utilized in planning a sustainably built environment to prevent urban sprawl. On the other hand, a number of built environment attributes, measured both objectively and subjectively, were related to levels of physical activity, including walking, cycling, driving, etc. The priorities for built environment planning were arranged according to their different weights. Thus, the priorities in resource allocation were clearly defined, thereby preventing poor resource management and waste, which will be importantly addressed in this special issue.

Basic standards for the living environment have long been established, excepting adjustments responding to the rapidly aging society. In 1961, the World Health Organization (WHO) recommended that the fundamentals of a healthy residential environment include effective fulfillment of human needs and harmony with local factors, such as climate, geography, and social practice, as well as customs and traditions, which may be summarized into four key characteristics: safety, health, convenience, and comfort [10].

It is regrettable that administrative subdivisions established by the relevant government units seem not to employ evidence-based governance or consult any objective analytical results based on urban big data; these units appear particularly concerned with whether governance conforms to the agenda and regulations promulgated by the central government. Therefore, it is necessary for us to assist the relevant government agencies in constructing a scientific, quantitative, and objective planning framework to replace the existing outdated planning philosophy, in order to correct the prominent shortcomings of past operations planned solely in accordance with the qualitative judgment and decision-making of official units or planners. The primary goal is to explore or extract big data to improve final decision-making, as well as future strategies, especially in the situation of new urban datasets generating rapidly in the near future [11]. Based on the previous discussion, the numerous datasets and design techniques generated by the decision matrix computing and the grey prediction model involved in this study may also be considered big data [12].

Nowadays people are aware that cities around the globe are being redesigned to become more smart and sustainable. Despite fruitful research progress in sustainability for cities individually, not too much research has been made by integrating the two themes of "smart" and "sustainable" together. Against this background, it is possible to state that there has been growing research that has been systematically investigating sustainable and smart cities, as well as the specific roles planning, development, and management play in those cities' sustained success in the near future. In this special issue, we hope to aim to gather diverse views and report progress towards both smart and sustainable cities under the consideration of urban growth management principles. The main objective of this special issue is to compile and present the cutting edge work of researchers who focus on joined-up thinking regarding studying a sustainably built environment within the urban growth management concept. By doing so, we believe this special issue on "Sustainable Built Environment and Urban Growth Management" contributes to the knowledge pool in this important area, as well as provides new evidences driven from state-of-the-art and state-of-the-practice research.

The issues about the sustainably built environment and urban growth management raise the common paradigm of the future urban development framework. For example, the urban green belt release has led to urban growth within the associated regions and cities, resulting in an increase in the temperature and the accumulation of pollutants in the atmosphere [13]. Also, the distortion and key factors for specific sites to assess urban sprawl and propose a preliminary course of action for peri-urban growth management is discussed [14]. Furthermore, monitoring and quantifying land use and cover changes (LUCC) have been identified as one of the main causes of biodiversity loss and deforestation in the world. LUCC are also essential to achieve proper land management [15]. Achieving sustainability requires the strengthening of resilience. One paper tries to quantify the susceptibility, vulnerability, and the response and recovery behavior of complex systems for multiple threat scenarios. That approach allows the evaluation of complete urban surroundings and enables a quantitative comparison with other development plans or cities [16]. In addition, one research forecasts the development scale of various buildings in future urban blocks, in order to provide an effective approach to estimate the carbon dioxide generated by the traffic volume [17].

3. Concluding Remarks and Research Directions

The special issue generates new insights by investigating the growth management principles that have emerged firstly to confront negative urban influence. Numerous developed countries failed to achieve their initial development strategy goals and soon afterwards implemented more specific and essential strategies of three innovative growth constructions: smart growth, sustainable growth, and inclusive growth. Following various innovative growth constructions or management principles, contemporary urban agents, including practical planners or official decision units, will be able to pursue urban sustainable development and growth, as well as QoL simultaneously.

In light of sustainable built environment and urban growth management-related matters discussed by the contributors of the special issue, we believe that research frameworks, technical requirements, and findings here not only provide new insights into the functioning of current growth constructions or management principles, but will be used as feasible and important directions for practical planners or official decision agents, in order to facilitate urban sustainability and QoL in the end.

Funding: This research received no external funding.

Acknowledgments: The editor wishes to thank the contributed authors of the special issue papers for accepting my cordial invitation and submitting, as well as revising their manuscripts within a short time frame. The editor also thanks the reviewers for their thorough and timely valuable review comments, and the journal's Assistant Editor, Elaine Li, for inviting me to serve as the Guest Editor of this Special Issue.

Conflicts of Interest: The author declares no conflicts of interest.

References

1. European Union. A Strategy for Smart, Sustainable and Inclusive Growth. 2010. Available online: http://ec.europa.eu/eu2020/pdf/COMPLET%20EN%20BARROSO%20%20%20007%20-%20Europe%202020%20-%20EN%20version.pdf (accessed on 25 October 2013).
2. European Union. Regional Policy for Smart Growth in Europe 2020. 2011. Available online: http://ec.europa.eu/regional_policy/information/pdf/brochures/rfec/2011_smart_growth_en.pdf (accessed on 31 October 2013).
3. European Union. Contributing to Sustainable Growth in Europe 2020. 2011. Available online: http://ec.europa.eu/regional_policy/information/pdf/brochures/rfec/2011_sustainable_growth_en.pdf (accessed on 31 October 2013).
4. Chapin, T.S. From Growth Controls, to Comprehensive Planning, to Smart Growth: Planning's Emerging Fourth Wave. *J. Am. Plan. Assoc.* **2012**, *78*, 5–15. [CrossRef]
5. Din, H.S.E.; Shalaby, A.; Farouh, H.E.; Elariane, S.A. Principles of urban quality of life for a neighborhood. *HBRC J.* **2013**, *9*, 86–92.
6. Wey, W.-M. Smart growth and transit-oriented development planning in site selection for a new metro transit station in Taipei, Taiwan. *Habitat Int.* **2015**, *47*, 158–168. [CrossRef]
7. Wey, W.-M.; Huang, J.-Y. Urban Sustainable Transportation Planning Strategies for Livable City's Quality of Life. *Habitat Int.* **2018**, *82*, 9–27. [CrossRef]
8. Ahmad, N.; Mehmood, R. Enterprise systems and performance of future city logistics. *Prod. Plan. Control* **2016**, *27*, 500–513. [CrossRef]
9. Naim, A.; Rashid, M. Enterprise systems: Are we ready for future sustainable cities. *Supply Chain Manag.* **2015**, *20*, 264–283.
10. Zhang, H.; Wey, W.-M.; Chen, S.-J. Demand-Oriented Design Strategies for Low Environmental Impact Housing in the Tropics. *Sustainability* **2017**, *9*, 1614. [CrossRef]
11. Batty, M. Building a science of cities. *Cities* **2012**, *29*, S9–S16. [CrossRef]
12. Huang, J.Y.; Wey, W.-M. Application of Big Data and Analytic Network Process for the Adaptive Reuse Strategies of School Land. *Soc. Indic. Res.* **2018**. [CrossRef]
13. Jeon, S.; Hong, H.; Kang, S. Simulation of Urban Growth and Urban Living Environment with Release of the Green Belt. *Sustainability* **2018**, *10*, 3260. [CrossRef]
14. Chen, Y.-H.; Lee, C.-L.; Chen, G.-R.; Wang, C.-H.; Chen, Y.-H. Factors Causing Farmland Price-Value Distortion and Their Implications for Peri-Urban Growth Management. *Sustainability* **2018**, *10*, 2701. [CrossRef]
15. Jiménez, A.A.; Vilchez, F.F.; González, O.N.; Flores, S.M.L.M. Analysis of the Land Use and Cover Changes in the Metropolitan Area of Tepic-Xalisco (1973–2015) through Landsat Images. *Sustainability* **2018**, *10*, 1860. [CrossRef]

16. Fischer, K.; Hiermaier, S.; Riedel, W.; Häring, I. Morphology Dependent Assessment of Resilience for Urban Areas. *Sustainability* **2018**, *10*, 1800. [CrossRef]
17. Chang, C.-T.; Lin, T.-P. Estimation of Carbon Dioxide Emissions Generated by Building and Traffic in Taichung City. *Sustainability* **2018**, *10*, 112. [CrossRef]

© 2018 by the author. Licensee MDPI, Basel, Switzerland. This article is an open access article distributed under the terms and conditions of the Creative Commons Attribution (CC BY) license (http://creativecommons.org/licenses/by/4.0/).

Article

Simulation of Urban Growth and Urban Living Environment with Release of the Green Belt

Seongwoo Jeon [1], Hyunjung Hong [1,2,*] and Sungdae Kang [3]

1. Department of Environmental Science and Ecological Engineering, Korea University, 145 Anam-ro, Seongbuk-gu, Seoul 02841, Korea; eepps_korea@korea.ac.kr
2. Korea Environment Institute, 370 Sicheong-daero, Sejong 30147, Korea
3. Green Simulation Co., Ltd., 1523 Jungang-daero, Dongrae-gu, Busan 47710, Korea; green.simulation.korea@gmail.com
* Correspondence: hjhong@kei.re.kr or honghj@korea.ac.kr

Received: 29 June 2018; Accepted: 11 September 2018; Published: 12 September 2018

Abstract: Green belts in developing or developed countries have been released because city-center development has reached a saturation point, and the strict protections and restrictions within green belts has led to an increase in opposition from local residents and property owners. However, green belt release has led to urban growth within the associated regions and cities, resulting in an increase in the temperature and the accumulation of pollutants in the atmosphere. We intend to prove quantitatively the effect of the release of green belts at the local level based on the interactions among land cover, climate, and air quality and to provide information for policy decisions accordingly. Our simulation results show that the urban areas of Jeju and Chuncheon, in South Korea, where green belts have been completely released, will increase by up to 21.83% by 2025 and 123.93% by 2020, respectively, compared to areas that have retained green belts. The simulations also show that the surface temperature within the released region of Jeju and Chuncheon will increase by up to 0.83% by 2025 and 0.03% by 2020, respectively. The average atmospheric concentrations within the released region of Jeju and Chuncheon were modelled to increase by up to 256.93% by 2025 and 337.29% by 2020, respectively.

Keywords: green belt; urban growth; land cover; urban living environment; climate change; surface temperature; air quality; atmospheric concentration; conservation; sustainable use

1. Introduction

Rapid urban growth and land development following industrialization and urbanization have led to overpopulation, housing shortages, traffic jams, and damage to the environment [1]. To address these issues, the green belt system was introduced. However, green belts have been released, especially in developing or developed countries [2,3], because (1) development of the city centers has reached a saturation point and (2) the strict protection and development restrictions within green belts has led to an increase in opposition from local residents and property owners [4,5].

Developing or developed countries have been attempting to rationalize the change in the type of land use (land cover) by releasing green belts in locations with low conservation value, or where the purpose of the original designation has been achieved. Portions of green belts have been released, and such adjustments are expected to continue in the near future, based on applicable regional urban master planning. In combination with increasing demand and pressure for community development, land use changes rapidly with the release of green belts. Urban regions released from their green belts have spread in connection with existing downtowns. The expansion of urban regions that have been released from green belts is threatening ecosystems and their associated natural services, owing to the acceleration of production activities and an energy-consumptive lifestyle within them.

Therefore, we should evaluate and predict the impact of green belt releases quantitatively in various fields and deal with these threats to nature conservation and human well-being. Although there have been numerous studies globally on the impact of green belt releases, including in the food, resource, water, biodiversity, real estate, community service and health fields [6–9], there has been a lack of studies aiming to demonstrate its impact on the living environment, especially concerning the climate and air. As awareness increases regarding the fact that land cover change can amplify or mitigate climate change, recently, the number of studies on the evaluation of the interaction between land cover change and climate change has also increased [10–12]. However, there is a shortage of studies evaluating the entire system of green belt release, urban growth, climate change, and air pollution.

In this study, we intended to quantify green belt release effects at the local level, based on the interactions among land cover, climate, and air quality. As, however, the effects of the release of green belts becomes more apparent over a longer period of time, it is difficult to prove them conclusively through short-term, empirical observations. Therefore, we deduced long-term effects indirectly, by setting a future target year, defining scenarios, and simulating urban growth and urban environmental changes that would be expected by that year. This study will support policy decision makers in approving or disapproving the release of green belts, by providing more information regarding the changes to the urban environment caused by green belt release.

2. Materials and Methods

2.1. Study Area

The South Korean Ministry of Land, Infrastructure and Transport (MOLIT) released the entire green belt (82.6 km^2) within Jeju Island, which had been designated and maintained for 30 years, on 4 August 2001. On 13 December 2001, all green belts (291.8 km^2) within Chuncheon and some (2.6 km^2) in Hongcheon (the administrative district next to Chuncheon) were also released.

Using our case studies, we examined urban growth following the release of green belts and its impact on the urban living environment. Jeju and Chuncheon were used as sites for this research (Figure 1) as they have the following characteristics: (1) Due to their geographic and environmental isolation, it was possible to perform an accurate impact assessment of green belt releases, as the possibilities of external urban development pressure, or conurbation were low; (2) these regions comprise medium-sized cities, which allowed spatial analysis at a medium-scale resolution.

2.2. Assessment of Effects from the Release of Green Belts

The study flow is shown in Figure 2, and includes the following six steps: (1) we set up scenarios in which green belts were maintained or released; and (2) estimated the urban region area for the target year, based on increasing trends in the urban area for a certain period of time. (3) Using spatial and statistical techniques, we then analyzed the location characteristics of regions where the land cover was converted into urban land. (4) The probability of the change from non-urban lands to urban lands was estimated based on those characteristics. (5) The estimated probability allowed us to predict changes in the spatial structure within regions released from the green belts by expanding them to the extent of the estimated urban area of the target year. (6) Based on changes in land cover, the ground surface temperature and atmospheric concentrations within those regions were predicted by applying the urban meteorology and climate prediction model.

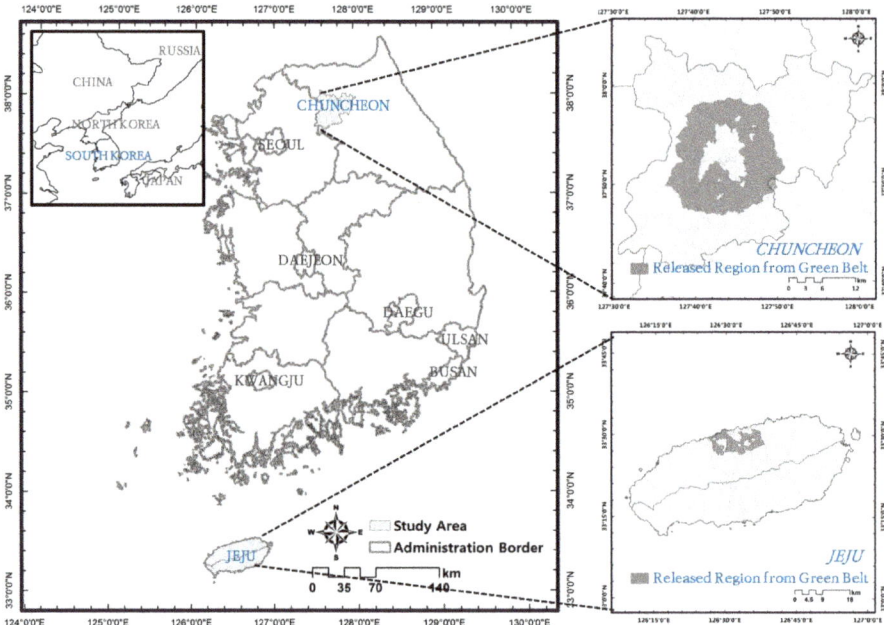

Figure 1. Study area: Jeju and Chuncheon are medium-sized isolated regions, and green belts within these regions which had been maintained for 30 years were released totally in the early 2000s. These regions were used as study sites to prove quantitatively the effect of the release of green belts.

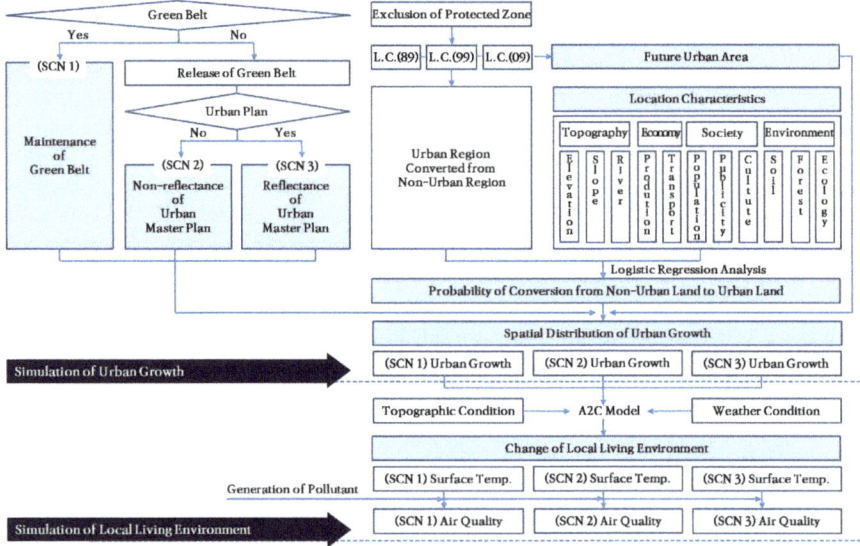

Figure 2. Flow chart: the main flow of this study is as follows: (1) the set of scenarios, (2) the estimation of urban area for the target year, (3) the identification of location characteristics to determine land cover change, (4) the estimation of the probability of conversion from non-urban land into urban land, (5) the simulation of urban growth by green belt release, and (6) the simulation of urban living environment by urban growth.

2.2.1. Scenario Development

We set up scenarios to determine to what extent urban sprawl would continue after release of the green belts, compared to the case in which the green belts were maintained (Table 1). Scenario 1 (SCN 1) was defined as BAU (Business As Usual), and in SCN 1, it was assumed that (1) the sites maintained their green belt boundary as originally designated, and (2) the fluctuation trend of urban areas over the maintenance of the green belt continued until the target year (the target year for simulating urban growth and living environment change following the maintenance or release of the green belts was set as the year to reach an objective in each urban master plan (Chuncheon Urban Master Plan, Jeju Metropolitan City Region Plan). The target year for the simulation of the Chuncheon site was set as 2020, and that of the Jeju site as 2025).

Table 1. Scenarios for analyzing the effect of green belt release: we set up scenarios to determine to what extent urban sprawl would continue after releasing of the green belts compared to that in the case in which the green belts were maintained (Table 1).

	SCN 1	SCN 2	SCN 3
Green Belt	Maintained	Released	Released
Development	Restricted	Non-restricted	Non-restricted
Urban Expansion	Restricted	General Expansion	Intentional Expansion
	Based on Linear Trend before Green Belt Release (1989–1999) *	Based on Linear Trend of Most Recent 20 years of available data (1989–1999–2009) *	Based on Reported Trend in Urban Master Plan

* When estimating the urban area of the target year, we used land cover maps prepared by the South Korean Ministry of Environment using LANDSAT satellite imagery from 1989, 1999, and 2009.

In South Korea, urban growth at the local level has generally progressed via the following process: (1) measuring the demand for development, (2) selecting green belt sites for release, (3) releasing green belts within the total releasable quantity, (4) centrally developing released regions, and (5) expanding built-up zones in connection with existing downtowns. In comparison with BAU, we set up Scenario 2 (SCN 2) to reflect this general progression of urban expansion within South Korea. SCN 2 assumed that an entire green belt was released and that the site maintained its urban growth trend, as shown over the most recently analyzed 20 years (1989–1999–2009) up to the target year.

For urban development to facilitate vitalization of the regional economy, the local government follows three steps in a forward planning process. (1) It estimates the future population of the target year, including the resident population, utilization population, and working population, by reviewing the overall demand for development, the currently available land, and development funds. Based on the estimated forward population, the local government (2) forecasts land demand for additional urban sites, including residential, commercial, and industrial sites, for the long term, and then (3) commences development. We set up Scenario 3 (SCN 3) to reflect the intentional aspect of urban expansion considering the land demand of the local government in comparison with that of BAU. In SCN 3, it was assumed that the entire green belt of the site was released and that the site would be urbanized to the extent of the values published in the urban master plan up to the target year.

2.2.2. Estimation of the Urban Area for the Target Year

A protected zone is the legally controlled geographic space for the purpose of preserving and improving ecological and cultural values [13,14]. It is difficult to alter a land category within the protected zone because property rights and specific actions within it should be reported and permitted in advance based on the relevant law. Therefore, we excluded protected zones when estimating urban areas for target years and the probability of conversion into urban lands.

We excluded the protected zones listed in Table 2 from land cover maps, and confirmed the fluctuation trends of urban areas by considering residential, industrial, commercial, cultural, sports, leisure, transportation, and public areas. Future urban areas based on SCNs 1 and 2 were estimated as follows: we determined linear equations using time-series analysis of urban areas to be derived from land cover maps and estimated future urban areas by inserting target years (2020, 2025) into the equations. In addition, the future urban area based on SCN 3 was substituted with the value of urban areas expected in the urban master plan.

Table 2. Protected zones excluded from the simulation of urban growth from green belt release: we excluded ecological, water and other protected zones when estimating future urban areas and the probability of conversion into urban lands, because it is difficult to alter a land category within protected zones due to the restriction on property rights and activities based on the relevant law.

Ecological Environment	Water Environment	Other Environment
• Ecological and Scenery Conservation Zone • DMZ (Demilitarized Zone) • Wetlands Protection Zone • Wildlife Protection Districts • Natural Park (Nature Conservation District) • Baekdu-daegan * Protection Zone (Core District)	• Riparian Zone • River Zone (Grades I, II, III) • Water Source Protection Zone	• Green Zone for Conservation • Ecosystem Preservation Zone • Cultural Preservation Zone • Green Belt • Absolute Preservation Zone • Natural Reserve • Natural Monument Designation Zone

* Vital Mountain Range of South Korea.

2.2.3. Identification of the Location Characteristics to Determine the Land Cover Change

Simulation of the change in land cover is based on understanding the statistical relationship between spatial patterns of land cover and the driving factors of land cover change [15]. Lim and Choi [16], Allen and Lu [17], Oh et al. [18], Lee et al. [19], and others identified elevation, slope, population density, soil, distance from a river/city/road, and presence of protected zones as factors for determining land cover change. Based on literature review and internal circumstances (in South Korea, the most significant factor leading to the government's policy decision-making on the release of green belts is the determination to retain the growing population within the urban region. Projects to construct public housing, infrastructure, and industrial complexes are implemented when green belts are released. To reflect this phenomenon, we selected production amount, distance from public transport, population density, distance from public facilities, and distance from cultural facilities as the economic and social factors affecting conversion of non-urban land into urban land), we selected factors affecting the conversion of non-urban land into urban land in terms of the overall (1) topography, (2) economy, (3) society, and (4) environment. From these factors, the elements that we selected as affecting the conversion of non-urban land into urban land were: (1a) elevation, (1b) slope, and (1c) distance from rivers; (2a) production amount and (2b) distance from transportation facilities; (3a) population density, (3b) distance from public facilities, and (3c) distance from cultural facilities; and (4a) soil, (4b) forest, and (4c) ecology. We constructed thematic maps with a spatial resolution of 30 m using these factors as raster data (Table 3).

Table 3. Construction of thematic maps via driving factors to determine the urbanization: based on the literature review, we selected factors affecting the conversion of non-urban land into urban land in terms of topography, economy, society, and environment, and constructed thematic maps as raster data.

	Driving Factor	Raw Data	Method
Topography	Elevation	DEM	Construction of Elevation from DEM
	Slope	DEM	Construction of Slope from Elevation
	Distance from River	Stream Order Map	Calculation of Euclidean Distance from Rivers
Economy	Production Amount	Statistics Annual Report	Assignment of 1st, 2nd, and 3rd Industrial Production Amounts to Agricultural/Industrial/Commercial lands
		Land Cover Map (Middle Level)	
	Distance from Transportation Facility	Land Cover Map (Middle Level)	Calculation of Euclidean Distance from Transportation Lands
Society	Population Density	Statistics Annual Report	Assignment of Population Density to Residential Lands
		Land Cover Map (Middle Level)	
	Distance from Public Facility	Land Cover Map (Middle Level)	Calculation of Euclidean Distance from Public Lands
	Distance from Cultural Facility	Land Cover Map (Middle Level)	Calculation of Euclidean Distance from Cultural, Sport, and Recreational Lands
Environment	Soil	Soil Map	Reclassification by Soil Drainage Class *
	Forest	Forest Map	Reclassification by Forest Age Class **
	Ecology	Ecological and Natural Map	Reclassification by Ecological and Natural Class ***

* Soil Drainage Class: 1st class (very well), 2nd class ((very) well), 3rd class (well), 4th class (normal), 5th class (poor), 6th class (very poor), 7th class (exposed rock). ** Forest Age Class: 1st class (1–10 years old), 2nd class (11–20), 3rd class (21–30), 4th class (31–40), 5th class (41–50), 6th class (51–60), 7th class (61–70), 8th class (71–80), 9th class (81–90), 10th class (91–100). *** Ecological and Natural Class: 1st class (major habitat; major ecological axis; spectacular scenery; abundant biodiversity), 2nd class (worthy of conservation in the future; buffer for 1st class zone for the protection of a 1st class zone), 3rd class (all others; suitable for development or utilization).

2.2.4. Estimation of the Conversion Probability from Non-Urban Land into Urban Land

Logistic regression analysis is useful in situations in which the predictor variable is determined by the independent variable or judged by its presence and is suitable when the dependent variable is in the binary mode. It is the statistically applicable method for analysis and prediction of urban growth using several causative factors related to urban growth [11,17,20]. We calculated the conversion probability of transformation into urban regions by using the basic logistic regression model equation—Equation (1).

$$\ln\left(\frac{p}{1-p}\right) = \alpha + \sum \beta_i X_i \tag{1}$$

where p is the conversion probability for transformation into urban regions, X_i is the independent variable, α is constant, and β_i is the regression coefficient.

A sample of the independent variable was extracted via systematic random sampling. We extracted non-urban regions from the land cover map for 1989 and divided them into two categories using time-series analysis: (1) the regions that had been changed from non-urban lands to urban lands since 1989, and (2) all other regions. An equivalent number of pixels was extracted in regions that had been converted from non-urban land to urban land to that in all other regions that had remained non-urban. We conducted regression analysis on these pixels using the driving factors. At this time, logistic regression analysis was performed through phased selection of the variable (forward procedure). We established the logistic regression equation by obtaining the regression coefficient and constant and calculated the probability of the conversion of non-urban land into urban land by applying the regression equations based on the different scenarios, to each thematic map.

2.2.5. Simulation of the Urban Growth by Green Belt Release

The estimated probability allowed us to predict the changes in the spatial structure of the urban regions in which the green belt had been released by expanding them to the extent of the estimated urban area (refer to Section 2.2.2). Based on the conversion probability, we extended the urban region until it converged with the value of the area for the target year derived from the linear equations or urban master plan.

2.2.6. Simulation of the Urban Living Environment by Urban Growth

Land cover change affects the heat balance between the land surface and atmosphere above. The accumulation of heat balance changes would result in local changes of the climate and the air quality (the rise in surface temperature from land cover change induces an increase in the height of the atmospheric mixing layer. This leads to the degradation of air quality by the intensification of the greenhouse effect, because the mixing layer height determines the volume available for dispersion of emitted pollutants [21–24]). We performed simulations of the urban living environment, including the climatic and atmospheric changes, based on the results of the urban expansion prediction. As the Atmosphere to CFD (Computational Fluid Dynamics) (A2C) model is the forecasting model for urban climate and weather based on the 3D non-static primitive equation, it is appropriate for predicting the urban living environment within the complex topography of South Korea. We predicted change to the local climate and air quality within regions released from green belts by inserting land cover and topographic data (digital elevation model) into this model. At that time, the physical properties of the model—such as absorption, reflection, roughness length, maximum moisture, and surface thermal inertia—were reflected as fixed values based on measurements from the early summer when the atmospheric conditions were stable. Areas designated as manufacturing land within urban regions were assumed to be pollutant emission spots (the main cause of air pollution is vehicle operation (0.43 vehicles per capita in South Korea), but it is difficult to predict these for the study areas in the target year. Therefore, we decided to simulate air pollution concentration based on land cover change, because the larger the area of urban lands, the more vehicles there are, leading to more air pollution). In consideration of the overall conditions, such as the atmospheric model, handling devices, and time required, we degraded the spatial resolution of the land cover maps to 200 m and reduced the extent of the target regions to the land surrounding the released green belt area.

3. Results

3.1. Short-Term Effects of the Release of Green Belts Based on Observation

Prior to simulating urban change caused by the release of green belts, we analyzed actual observed short-term changes in land cover and air quality occurring before and after green belt releases.

The changes that appeared within the released regions of the Jeju green belt were as follows (Figures 3 and 4): the urban area within the released regions in 1999 (the 26th year after the designation of the green belt and 2 years before its release) was 3.90 km^2, a decrease of 26.44% compared with 1989 (the 16th year post designation). The agricultural area was 17.90 km^2, which was an increase of 21.00%, and the forest area was 55.10 km^2, increasing by 11.18% compared with 1989. The urban area within the released regions in 2009 (the 8th year after release) was 5.07 km^2, an increase of 30.14% compared with 1999. The agricultural area was 56.29 km^2, which was an increase of 214.50%, and the forest area was 15.34 km^2, a decline of 72.16% compared with 1999. The annual average concentrations of SO_2, NO_2, O_3, CO, and PM_{10} within the released regions (Ido-dong) in 1999 were 0.005 ppm, 0.018 ppm, 0.031 ppm, 0.700 ppm, and 33.00 µg/m^3, respectively. In 1999, the average concentrations of NO_2 and CO had increased by 16.67% and 75.50%, but the concentrations of SO_2, O_3, and PM_{10} had decreased by 40.63%, 0.54%, and 15.38%, respectively, compared with 1995, which was the 1st year of air quality observations for Ido-dong, and the 22nd year after green belt designation. In 2009, the annual average concentrations of SO_2, NO_2, O_3, CO, and PM_{10} were 0.002 ppm, 0.013 ppm, 0.041 ppm, 0.400 ppm, and 40.00 µg/m^3, respectively. In that year, the concentrations of SO_2, NO_2, and CO had decreased by 57.89%, 25.71%, and 42.86%, but the concentrations of O_3 and PM_{10} had increased by 32.97% and 42.86%, respectively, compared with 1999.

Changes appeared within the released regions of Chuncheon as follows (Figures 3 and 4): in 1999 (the 26th year after designation of the green belt, and 2 years before its release), the urban area within the released regions was 2.17 km^2, an increase of 87.32% compared with 1989, the 16th year after its designation. Compared with 1989, the agricultural area was 47.70 km^2, a decrease of 3.22%, and the forest area was 210.18 km^2, a decrease of 1.12%. The urban area within the released regions in 2009 (8 years after release) was 4.75 km^2, an increase of 119.48% compared with 1999. Agricultural area was 46.40 km^2, a decrease of 2.73%, and forest area was 213.53 km^2, an increase of 1.59% compared with 1999. However, grassland coverage was 5.13 km^2—a decrease of 44.33% compared with 1999. In 2009, the annual average concentrations of SO_2, NO_2, O_3, CO, and PM_{10} within the released regions (Seoksa-dong) were 0.004 ppm, 0.018 ppm, 0.025 ppm, 0.600 ppm, and 58.00 µg/m^3, respectively. By that year, the concentrations of NO_2, CO, and PM_{10} had increased by 100.00%, 50.00%, and 13.73%, respectively; SO_2 remained unchanged, and O_3 had decreased by 10.71% compared with 2004, which was the 3rd year after the release of the green belt, and the year that air quality monitoring commenced at Seoksa-dong.

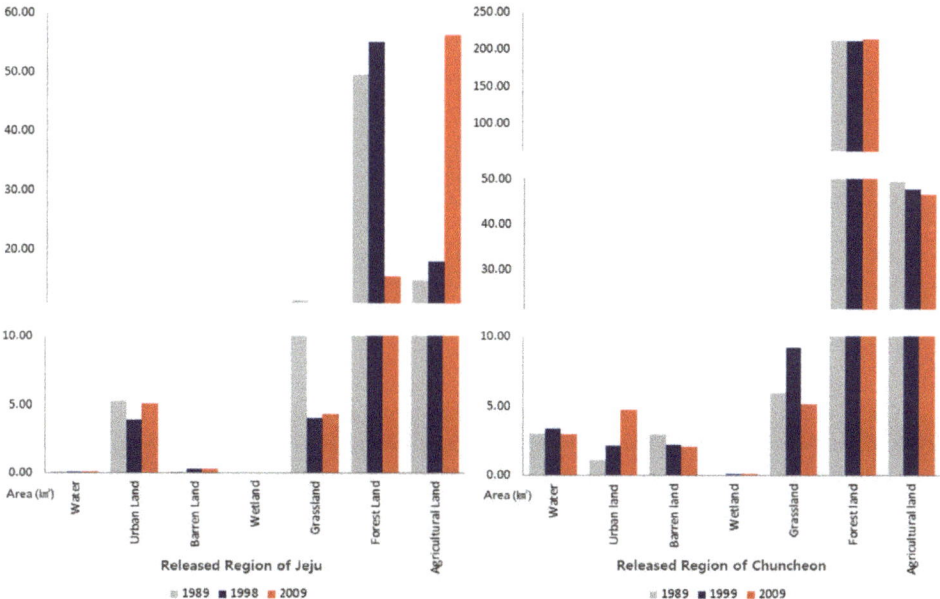

Figure 3. Land cover change before and after the release of green belts, based on observation; we conducted the time series (1989, 1999, 2009) analysis on the observed change of the area based on land cover type within the regions released from green belts.

Figure 4. Air quality change before and after the release of green belts, based on observation: we conducted the time series (1999, 2009) analysis based on the observed change of air quality (SO_2, NO_2, O_3, CO, PM_{10}) within the regions released from green belts.

We intended to demonstrate the long-term effect of the release of green belts indirectly, by setting the future target year, defining scenarios, and simulating urban growth and environmental change based on the interactions between land cover, climate, and air quality. We decided to do this for the following reasons:

(1) As the effect of the release of green belts appeared to accumulate over a long period of time, it was difficult to demonstrate effects from just short-term empirical observations.
(2) Atmospheric gas concentrations are monitored continually, in order to track the status of air pollution and climate-changing substances in real time and at the national level. There are approximately 300 monitoring stations, acting as representative sites for air quality in South Korea, with just one of these in each of the released regions of Jeju and Chuncheon. Numerous external influences contribute to influence the monitored air quality values, not just effects from release of the green belts, and these other air quality influences cannot be separated from any actual air quality change caused by the release of the green belts. Therefore, there was a limit to our ability to validate the effect of the green belt release accurately, using observational data alone.

3.2. Long-Term Effects of the Release of Green Belts Based on the Simulation

3.2.1. Urban Area of the Target Year

The urban area of Jeju for the target year (2025) was 146.20 km² in SCN 1, 125.81 km² in SCN 2, and 178.11 km² in SCN 3 (Table 4). The urban area of Chuncheon for the target year (2020) was 19.64 km² in SCN 1, 28.99 km² in SCN 2, and 43.98 km² in SCN 3 (Table 4).

Table 4. Linear equation and estimated urban area for each scenario: we calculated the future urban areas by adding urban areas in protected zones to values to be estimated from the linear equations or to be cited from the urban master plan.

Scenario		Linear Trend Equation	Urban Area (Including Urban Area of Protected Zones)			
			1989	1999	2009	2020(25)
Jeju	SCN 1	y = 2.19x − 4290.39 (R^2 = 1.00 *)	69.39	88.86	98.89	146.20
	SCN 2	y = 1.44x − 2788.77 (R^2 = 0.95)				125.81
	SCN 3	-				178.11
Chuncheon	SCN 1	y = 0.68x − 1345.12 (R^2 = 1.00 *)	11.87	15.86	26.04	19.64
	SCN 2	y = 0.44x − 861.58 (R^2 = 0.93)				28.99
	SCN 3	-				43.98

* SCN 1 is the scenario considering the trend of urban expansion at the time of urban growth control via the designation and maintenance of the green belts, and the urban area of SCN 1 was derived by the following process: (1) calculating areas of urban lands to be extracted from land cover maps of 1989 and 1998, (2) deriving the linear equation between areas of two periods, (3) inputting the target year into the linear equation and deducting the urban area at the target year. Therefore, R^2 of the linear equation is 1.00.

Regarding Jeju, the value of the urban area in SCN 1 was higher than that of SCN 2, which assumes green belt release. The cause of this is considered to be that urbanization within South Korea progressed rapidly from the 1970s to the 1990s, but the speed of urban growth has slowed since then, causing the rate of urban growth from 1989 to 1999 to be higher than that from 1999 to 2009. As the equation based on the above-mentioned phenomenon was applied to calculating the future urban area, although the green belt was not released in SCN 1, the urban area in SCN 1 is 1.16 times greater than that in SCN 2. However, these values do not exceed the values reported in the urban master plan.

3.2.2. Location Characteristics and the Probability of Conversion into Urban Land

The analysis confirmed that population density, production amount, and barren land on the land cover map were positively correlated with the conversion of non-urban land into urban land (Table 5). The distance from public facilities, distance from transportation facilities, water on the land cover map, and ecological and natural class are negatively correlated with the conversion of non-urban land into urban land (Table 5).

In SCN 1, ecological and natural class, forest age class, and water on the land cover map primarily affected the conversion of non-urban land into urban land in Jeju, with wetland, barren land, forest age class, ecological and natural class, and soil drainage class the main conversion influencers in Chuncheon. In SCNs 2 and 3, wetland, barren land, grassland, and forest land on the land cover map, forest age class, and ecological and natural class affected the conversion of non-urban land into urban land in Jeju, with forest age class, soil drainage class, and barren land affecting the conversion in Chuncheon. The coefficients, constants, and thematic maps based on driving factors were applied to each logistic regression equation according to the scenarios, and the conversion probabilities were computed.

Table 5. Coefficients and constants based on the logistic regression analysis of the conversion into urban land; we obtained coefficients and constants for each driving factor contributing to the conversion of non-urban land into urban land by performing binary logistic regression analysis on systematic random sampling data ($p < 0.05$).

			Jeju		Chuncheon	
	Item		SCN 1	SCN 2, 3	SCN 1	SCN 2, 3
Coefficient	Elevation		−0.0003	0.0004	−0.0079	−0.0027
	Slope		0.0031	−0.0005	-	−0.0089
	River		−0.0001	−0.0001	0.0009	−0.0004
	Population Density		0.0187	0.0313	0.0117	0.0109
	Public Facilities		−0.0006	−0.0004	−0.0004	−0.0002
	Cultural Facilities		−0.0001	−0.0001	-	−0.0001
	Production Amount		0.0003	0.0004	0.0004	0.0005
	Traffic Facilities		−0.0013	−0.0021	−0.0014	−0.0017
	Land Cover	Water	−1.2475	−0.4934	−0.6369	−0.7078
		Urban Land	-	-	-	-
		Barren Land	0.7086	1.2794	1.4004	1.7318
		Wetland	0.2250	3.0034	21.5742	-
		Grassland	−0.8611	−1.0632	0.9838	0.2797
		Forest Land	−0.8811	−1.0400	−0.1193	−0.1090
		Agricultural Land	-	-	-	-
	Soil Drainage Class	Very Well	-	-	0.9563	−0.9589
		(Very) Well	0.2890	0.4902	0.4139	−1.9050
		Well	0.6968	0.0033	0.9169	−0.7887
		Normal	-	-	1.4265	−0.7833
		Poor	0.3731	0.3299	0.9357	−0.6466
		Very Poor	0.3328	−0.0479	-	-
		Exposed Rock	-	-	-	-
	Forest Age Class	1st Class	−2.5156	2.2282	16.8759	−21.2623
		2nd Class	−0.5642	0.8666	15.1851	−0.6532
		3rd Class	−0.2923	0.1544	15.7588	−0.4955
		4th Class	0.0600	0.0748	16.5151	−0.9146
		5th Class	0.3716	0.1450	15.2279	−0.5156
		6th Class	0.3387	−0.2757	15.4118	−0.6132
		7th Class	-	-	-	-
	Ecological and Natural Class	1st Class	−2.6659	−1.4699	−2.3240	−0.7857
		2nd Class	−0.7273	−0.4247	−1.8046	−0.3780
		3rd Class	−1.0542	−0.3806	−1.8881	−0.4549
	Constant		1.2022	1.1127	−15.1472	3.6239

3.2.3. Change in Urban Growth, Climate, and Air Quality via Green Belt Release within Jeju

Results are shown in Table 6 and Figure 5. In SCN 1 (green belt maintained), the urban area of Jeju in 2025, including protected zones, was 146.20 km^2, an increase of 64.53% compared to the urban area in 1999 (the last year prior to the green belt release in which the actual urban area was known). The agricultural area was 357.19 km^2, which is a decrease of 9.37%, and the forest area was 1018.29 km^2, a decrease of 1.66%, compared to the area in 1999. Of the regions that were converted into urban land, 64.41% was utilized as agricultural land, 29.94% as forest land, and 4.24% as barren land in 1999. In SCN 2 (release of the green belt and development reflecting the past and current trends), the urban area of Jeju in 2025, including protected zones, was 125.81 km^2, an increase of 27.22% compared to that of 2009 (the last year after green belt release in which the actual urban area was known). The agricultural area was 799.28 km^2 (2.50% decrease since 2009), and the forest area was 646.96 km^2 (0.36% decrease since 2009). Of the regions that were converted into urban land, 76.08%

was used as agricultural land, 11.18% as barren land, and 8.66% as forest land in 2009. In SCN 3 (release of the green belt and development reflecting the local demand), the urban area of Jeju in 2025 was 178.11 km^2, reflecting an 80.10% increase since 2009. The agricultural area was 762.07 km^2 (7.04% decrease since 2009), and the forest area was 637.81 km^2 (1.77% decrease since 2009). Of the regions that were converted into urban land, 72.82% was classified as agricultural land, 14.49% as forest land, and 6.82% as grassland in 2009.

Based on SCN 1, the urban area within the released regions was 3.90 km^2 in 2025. The area based on SCN 2 was 7.11 km^2, an increase of 82.51%, and the area based on SCN 3 was 10.86 km^2, reflecting an increase by 178.70% since 1999, the year before the green belt release. The forest area within the released regions decreased rapidly owing to the expansion of urban areas. In 2025, the forest area of the released regions was 55.00 km^2 in SCN 1, 15.21 km^2 in SCN 2 (72.39% decrease), and 14.60 km^2 in SCN 3 (73.51% decrease). The more green belt released, and the higher the demand for local development, the more urban areas within the released regions and downtown expanded.

Table 6. Land cover change and living environment change in Jeju; we proved quantitatively the effect of the release of green belts within Jeju through the prediction of the urban expansion and the simulation of the urban living environment based on the interactions among land cover, climate, and air quality.

		Land Cover (km^2)						Surface Temperature (°C)		Atmospheric Concentration (g/m^2)	
		Urban Land	Agri. Land	Forest Land	Grass Land	Wet Land	Barren Land	Water	Released Region	Down Town	Released Region	Down Town
1989		69.39	379.32	924.93	466.55	0.59	7.83	37.12	-	-	-	-
1999		88.86	394.12	1035.46	322.08	0.39	15.10	29.17	-	-	-	-
2009		98.89	819.76	649.29	271.38	0.38	15.34	31.29	-	-	-	-
2025	SCN 1	146.20	357.19	1018.29	321.28	0.39	12.67	29.16	31.45	32.47	2.67×10^{-07}	4.42×10^{-06}
	SCN 2	125.81	799.28	646.96	270.40	0.38	12.33	31.17	31.71	32.63	8.70×10^{-07}	4.33×10^{-06}
	SCN 3	178.11	762.07	637.81	265.98	0.38	10.81	31.17	31.45	32.36	9.53×10^{-07}	4.51×10^{-06}

The average surface temperature in the released region in 2025 was 31.45 °C in SCN 1, 31.71 °C in SCN 2, and 31.45 °C in SCN 3. The surface temperature in the existing downtown (classified as downtown on the land cover map of 2009) was 32.47 °C in SCN 1, 32.63 °C in SCN 2, and 32.36 °C in SCN 3. The surface temperature in regions released from the green belt and the existing downtown in SCN 2 increased by 0.83% and 0.49%, respectively, compared to those of SCN 1. However, the temperatures in SCN 3 decreased by 0.01% and 0.34%, respectively, compared to those of SCN 1, because the spatial resolution of the land cover had been degraded from 30 m to 200 m, resulting in the decrease of urban expansion (we tried to improve the spatial resolution of the input and result data from 200 m to 90 m and simulated the local climate change. At 90 m spatial resolution, the surface temperature in the released regions was 29.21 °C in SCN 1, 29.38 °C in SCN 2, and 29.49 °C in SCN 3. The surface temperature in the downtown was 29.81 °C in SCN 1, 30.13 °C in SCN 2, and 29.89 °C in SCN 3. The surface temperature in regions released from the green belt and in the existing downtown in SCN 3 increased by 0.96% and 0.27%, respectively, compared to those of SCN 1. This shows that the release of the green belt led to the increase in the urban surface temperature compared with the maintenance of the green belt). However, the difference between the average surface temperature within downtown and the released region decreased by 1.02 °C in Scenario 1, 0.92 °C in SCN 2, and 0.91 °C in SCN 3, indicating that higher-temperature regions expanded after the green belt release.

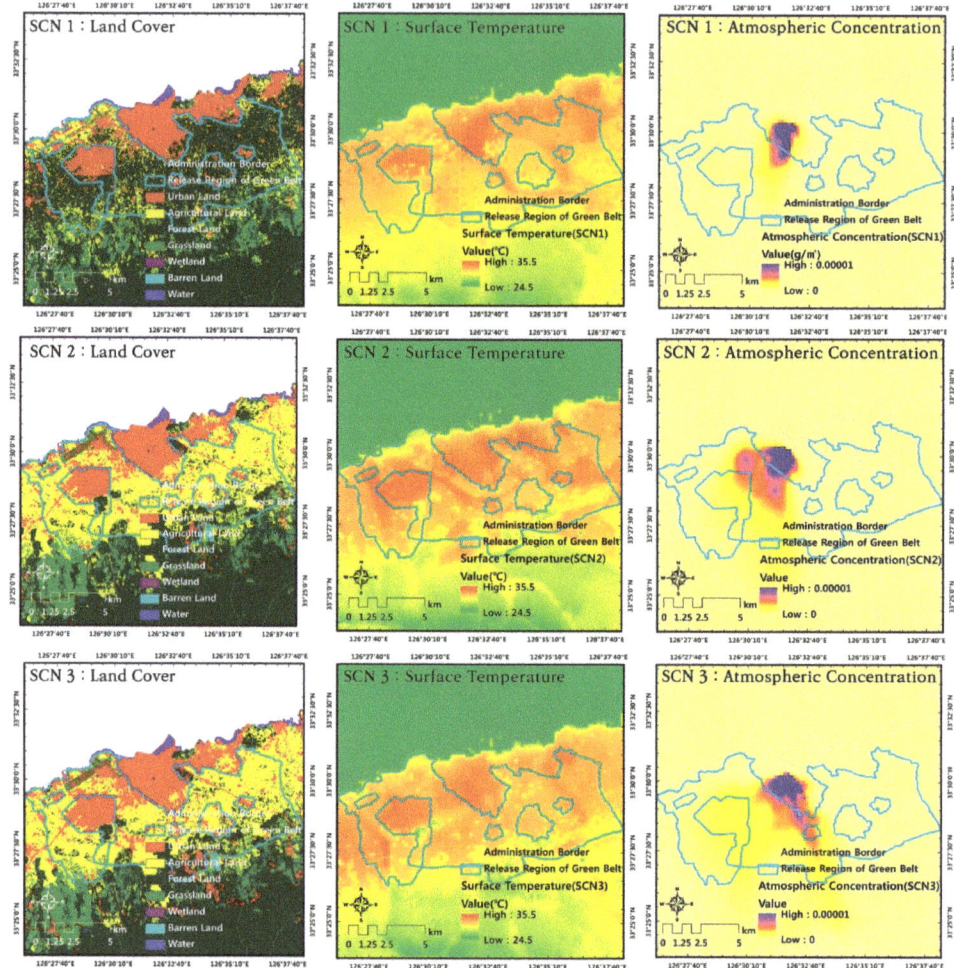

Figure 5. Spatial distribution of urban growth, climate, and air quality in the region released from the green belt within Jeju; we represented spatially the effect of the release of green belts within Jeju through the prediction of the urban expansion and the simulation of the urban living environment based on the interactions among land cover, climate, and air quality.

The average atmospheric concentrations within the released regions in 2025 were 2.67×10^{-07} g/m^2, 8.70×10^{-07} g/m^2, and 9.53×10^{-07} g/m^2 in SCNs 1, 2, and 3, respectively. The average atmospheric concentrations in SCNs 2 and 3 rapidly increased by 225.84% and 256.93%, respectively, compared to that of SCN 1. In 2025, the atmospheric concentrations in the existing downtown in each scenario were 4.42×10^{-06} g/m^2, 4.33×10^{-06} g/m^2, and 4.51×10^{-06} g/m^2, respectively. The average atmospheric concentration of the downtown in SCN 2 decreased by 2.04% compared to that of SCN 1. However, the average atmospheric concentration of the downtown in SCN 3 increased by 2.04% compared to that of SCN 1.

3.2.4. Changes in Urban Growth, Climate, and Air Quality via Green Belt Release within Chuncheon

Results are presented in Table 7 and shown in Figure 6. In SCN 1, the urban area of Chuncheon in 2020, including the protected zone, was 19.64 km^2, reflecting an increase of 28.83% since 1999. The agricultural area was 105.03 km^2 (1.22% decrease since 1999), and the forest area was 906.10 km^2 (0.15% decrease since 1999). Of the regions that were converted into urban land, 36.24% was utilized as forest land, 34.39% as agricultural land, and 17.20% as barren land in 1999. In SCN 2, the urban area of Chuncheon was 28.99 km^2 in 2020, an increase of 11.33% since 2009. The agricultural area was 101.01 km^2 (1.73% decrease since 2009), and the forest area was 915.29 km^2 (0.05% decrease since 2009). Of the regions that were converted into urban land, 60.34% was classified as agricultural land, 16.61% as forest land, and 12.54% as barren land in 2009. In SCN 3, the urban area of Chuncheon was 43.98 km^2 in 2020, an increase by 68.89% since 2009; the agricultural area was 90.05 km^2 (12.39% decrease since 2009); and the forest area was 912.64 km^2 (0.34% decrease since 2009). Of the regions that were converted into urban land, 71.01% was used as agricultural land, 17.50% as forest land, and 5.74% as barren land in 2009.

In 2020, the urban area within the released regions in SCN 1 was 2.09 km^2. It was 5.63 km^2 in SCN 2, reflecting an increase of 170.21% since 1999, before the release of the green belt. In SCN 3, it was 14.03 km^2, reflecting an increase of 573.84% since 1999. The agricultural area decreased as the urban area within the released region expanded. In 2020, the agricultural area within the released region was 47.47 km^2 in SCN 1, 45.76 km^2 in SCN 2 (3.62% decrease since 1999, before the release of the green belt), and 39.45 km^2 in SCN 3 (decrease of 16.90% since 1999).

Table 7. Land cover change and living environment change in Chuncheon; we proved quantitatively the effect of the release of green belts within Chuncheon through the prediction of the urban expansion and the simulation of the urban living environment based on the interactions among land cover, climate, and air quality.

		Land Cover (km^2)							Surface Temperature (°C)		Atmospheric Concentration (g/m^2)	
		Urban Land	Agri. Land	Forest Land	Grass Land	Wet Land	Barren Land	Water	Released Region	Down Town	Released Region	Down Town
	1989	11.87	114.04	908.59	16.28	0.05	8.66	56.04	-	-	-	-
	1999	15.86	106.33	907.47	23.97	0.12	10.19	51.58	-	-	-	-
	2009	26.04	102.79	915.78	13.81	0.14	7.09	49.85	-	-	-	-
2020	SCN 1	19.64	105.03	906.10	23.51	0.12	9.54	51.58	31.02	33.37	2.95×10^{-08}	3.50×10^{-05}
	SCN 2	28.99	101.01	915.29	13.51	0.14	6.72	49.84	31.03	34.34	1.28×10^{-07}	3.71×10^{-06}
	SCN 3	43.98	90.05	912.64	12.79	0.14	6.06	49.84	31.03	34.41	1.29×10^{-07}	4.64×10^{-06}

The average surface temperatures within the released regions in 2020 were 31.02 °C, 31.03 °C, and 31.03 °C in SCNs 1, 2, and 3, respectively, and the average surface temperatures within the existing downtown were 33.37 °C, 34.34 °C, and 34.41 °C. These values indicated that the surface temperatures of the downtown in all scenarios increased after the green belt was released. As a result of releasing the green belt, the surface temperatures within the released region increased by 0.03%, on average, and those within the downtown increased by 2.91–3.12% compared to those in SCN 1.

In 2020, the average atmospheric concentrations within the released region were 2.95×10^{-08} g/m^2, 1.28×10^{-07} g/m^2, and 1.29×10^{-07} g/m^2 in SCNs 1, 2, and 3, respectively, and the average atmospheric concentrations within the downtown were 3.50×10^{-05} g/m^2, 3.71×10^{-06} g/m^2, and 4.64×10^{-06} g/m^2. The atmospheric concentration within the downtown was improved by releasing the green belt. However, the average atmospheric concentrations within the released regions in SCNs 2 and 3 increased by 333.90% and 337.29%, respectively, compared to that in SCN 1.

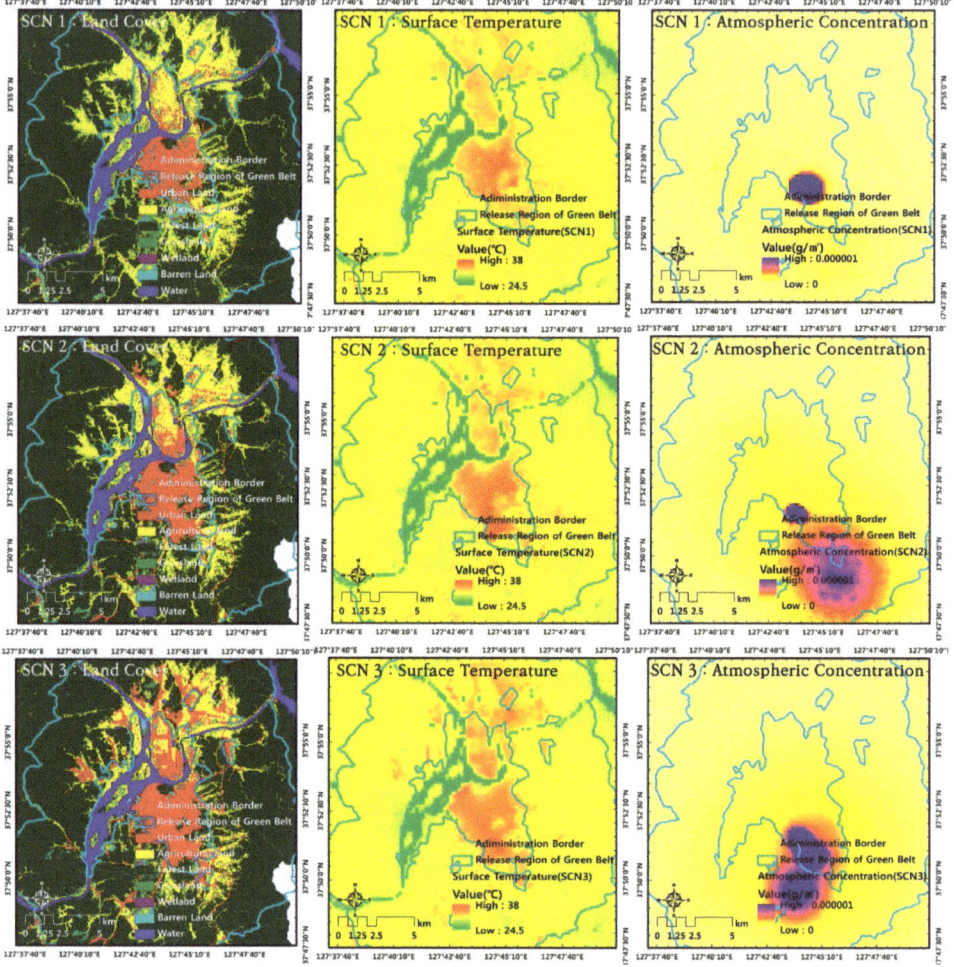

Figure 6. Spatial distribution of urban growth, climate, and air quality in the region released from the green belt within Chuncheon; we represented spatially the effect of the release of green belts within Chuncheon through the prediction of the urban expansion and the simulation of the urban living environment based on the interactions among land cover, climate, and air quality.

4. Discussion

4.1. Effect of Green Belt on the Restriction of the Urban Expansion

The green belt system was introduced to prevent urban sprawl through the designation of development restriction zones and the limitation of activities performed within these zones. We identified the effect of green belts on urban growth control by simulating the future urban area based on the maintenance or the release of the green belts. This simulation demonstrated that green belts promoted and integrated the development district into the center of the city, preventing large-scale development and the spread of development outside of the city. This was shown in our projections, whereby future urban areas of Jeju and Chuncheon decreased by up to 17.92% and 55.34% in the scenario in which the green belt was maintained compared with those in the scenario in which the green belt was released.

However, green belts within developing or developed countries that had been maintained for several decades have been released, and their release is expected to continue for economic vitalization reasons [2,3]. Based on this study, green belt release could lead to expansion of existing urban regions and even acceleration of leapfrogging development (the logistic regression analysis of conversion into urban land showed that population density, production, and barren land on the land cover map were positively correlated with conversion of non-urban land into urban land. On this basis, it can be confirmed that urban land expanded around existing urban lands, as shown in Figures 5 and 6). This goes against the purpose of the green belt, that is, to restrict urban expansion, conserve the natural environment, and secure the living environment of urban residents.

It is expected that urban development will continue to expand as a result of urban master planning focusing on urban growth and targeting agricultural land surrounding cities, which is vulnerable to conversion into urban land. It is possible that the urban spatial structure will be changed in the following sequence: urbanization of regions released from green belts, urban spread connected with existing downtowns, and urban sprawl. Broad urban sprawl negatively affects forest land, which produces resources and energy, improves air quality, mitigates climate change, conserves ecosystems, and improves health [25,26]. There is concern that urban expansion will lead to encroachment onto urban farmland and will damage forest land. Protected zones, as well as forest land, might be released, which would reduce the quality of the urban environment if development demand increases continuously and development capacity exceeds the availability of developable land within non-protected zones of the city.

4.2. Effect of the Green Belt on Conservation of the Urban Living Environment

The green belt ensures a healthy living environment for citizens through the prevention of urban sprawl and conservation of the natural environment that surrounds cities. However, urban growth that results from green belt release leads to a higher temperature in downtown centers, and a lower temperature outside the city. The high-temperature region of >32 °C will expand across the existing downtown and new urban regions as a result of green belt release. The average surface temperature within the released region in the scenario that was set for green belt release increased by up to 0.83% in Jeju and 0.03% in Chuncheon, compared to that of the scenario in which the green belt was maintained. Surface temperature rise due to land cover change induces an increase in the atmospheric mixing layer height which in turn determines the volume of the air shed into which emitted pollutants disperse. This enhances degradation of the air quality through intensification of the greenhouse effect [21–24], because the higher the mixing layer height, the less it is influenced by the external weather conditions that dilute pollutants.

In the short term, this phenomenon will cause pollutants to accumulate in the urban canopy directly above their emission source and the atmospheric concentrations will increase. The average atmospheric concentrations within the released region in the scenario of green belt release increased by up to 256.93% in Jeju and 337.29% in Chuncheon compared to that of the scenario set for green belt maintenance. If entire green belts are released and development is accomplished to accommodate all local development demand, more pollutants will be accumulated, and higher atmospheric concentrations will appear. In the medium/long term, this could result in the more frequent occurrence of severe weather phenomena, including tropical nights and localized heavy rain. This could deplete the output of primary and secondary industries and accelerate energy consumption. In addition, it is postulated that this, in turn, could create obstacles to the supply of energy within the urban region.

4.3. Sustainable Management of Green Belts for Coexistence between Nature and Humans

In the initial stage of urbanization, the primary purpose of the national policy within the developing or developed countries is economic development. Existing downtowns that have already been equipped with infrastructure are generally selected as the development sites for economic growth because the effectiveness and productivity of the national land use are considered important. Because of significant investments in existing downtowns, population growth is remarkable and social infrastructure is created accordingly. This results in the expansion of urban lands.

The green belt system was introduced as a means for controlling urban growth because development expanded from urban regions to suburban regions. However, green belts have been released to promote the rationalization of land use change, because the discomfort of residents and landowners worsened owing to standardized and excessive regulations. In South Korea, particularly, the population is concentrated in urban regions—the urban residential population proportion was 91.82% in 2017, and it is necessary to prepare housing and infrastructure for their support of the concentrated population. However, development sites are in short supply within urban regions. Accordingly, South Korea's government releases green belts surrounding urban regions and implements development plans and projects for these released regions from green belts. As a result, horizontal sprawl, and vertical growth within the released regions, have caused cities to expand and to be combined, resulting in changes in the urban spatial structure. A wide range of human activities that consume natural resources have led to changes in the living environment, including increases in the surface and atmospheric temperatures, expansion of high temperature regions, and changes in wind speed. These changes have caused accumulation of pollutants in the atmosphere.

In South Korea, England, Japan, Canada, United States of America, Australia, New Zealand, France, Netherland, and Russia, green belts are being maintained [27]. Even then, parts of green belts have been released, or the related system abolished, in some of these countries—South Korea, England, and Japan [27]. Especially in South Korea, where there has been great environment benefit from retaining green belts over decades, green belts are now being released based on urban master plans, and this phenomenon is expected to continue into the near future. As green belt release occasionally contributes to resolving urban crises, in combination with time and space situations, such as housing shortages and their high cost [9,28], there is debate with regard to the need to adjust green belt borders and to the regulations that apply to their conservation [28–31]. Importantly, it is almost impossible for released green belt areas to be re-designated as green belts, for economic and social reasons, and it is also difficult to reverse the environmental damage; therefore, designation and release of green belts should be carefully decided, in order to promote coexistence between nature and humanity.

First, for the management of sustainable urban growth, green belts should be designated as confined regions, in which it is necessary to restrict development. This is necessary (1) to prevent disorderly urban sprawl, (2) to prevent neighboring cities from being linked by continuous urban corridor areas, (3) to conserve the natural environment and ecosystems surrounding cities, and (4) to ensure a healthy living environment for city dwellers. More effort should be made to preserve green belts with high conservation values, and to restore broken and damaged ecosystems within existing green belts, by restoring ecological corridors (the green (forest), blue (river), yellow (soil), and white (wind) networks) around green belts, connecting with ecosystems inside and outside of the green belts, and building green spaces.

Second, green belts should be used and managed sustainably, as they directly affect the lives of local residents. We should encourage lifestyle activities that have minimal influence on the ecosystem, through transferring technology, providing supporting finance for sustainable agricultural production, diversifying community support programs, and aiding development of compatible facilities. We could render parts of the green belts eco-friendly, by transforming them into public parks and constructing green leisure facilities on land purchased for that purpose. Traffic and movement networks (white belts) around the green belts should be constructed for commuting, while still ensuring low-density or low-impact use and maintaining a pleasant urban environment.

Third, green belts could have their shapes adjusted, or could be released if necessary, if issues arise concerning the adequate supply of urban land, or land-use rationalization, in their vicinity, provided environmental assessment of the green belt shows that the released land embodies low conservation value. Such a review should include the following questions:

- Is the urban development necessary, taking into consideration the decline of the city (that is, that the rate of urbanization has slowed down)?
- Should urban development definitely be accommodated within the green belt?

If the release and development of green belts is inevitable, it should at least be discouraged within more environmentally sensitive regions, in order to conserve the urban ecosystem that provides diverse ecosystem services, generates pleasant urban environments, and responds to climate change. It should be developed in connection with areas in need of urban regeneration, or urban revitalization, not in the context of building a new town or industrial complex.

We should encourage eco-friendly use and development of released areas, for tours or as recreational spaces for citizens, where the environmental burden of use is low, and the environmental impact assimilating capacity can be maintained. We should also ensure that a minimum green zone width is retained within the released region to prevent urban conurbation. We should:

- Select the released and developed locations from the green belt in light of the current situation and potential development impacts,
- Develop and apply techniques to minimize environmental change,
- Prepare measures to mitigate development impacts,
- Arrange solutions to reasonably manage pollutants,
- Establish and apply customized methods to reflect the local characteristics, and
- Periodically monitor whether the above measures are implemented or not.

Basic investigation should be carried out, and environmental monitoring using purpose-built facilities should be enforced. We expect the above measures to create a comfortable urban environment which promotes and maintains a harmonious balance between nature and human beings through the sustainable management of green belts.

5. Conclusions

Green belts are development-restricted zones created with the aim of preventing urban sprawl and ensuring a healthy living environment for citizens through conservation of the natural environment surrounding cities. However, green belts have been released to promote the rationalization of land use change, and the urban land within the released regions has enlarged through existing downtowns in response to increasing demand for community development. The environment of these regions has been degraded and threatened by the acceleration of development activities and a lack of management. Therefore, we conducted a quantitative assessment to prove the impact of green belt releases on urban growth and urban living environment degradation, focusing on the changes in land cover, climate, and air quality.

This study has concluded that the urban areas of Jeju (by 2025), and Chuncheon (by 2020), where green belt land was released, would increase by up to 21.83% and 123.93%, respectively, compared to the BAU scenario, where green belts were maintained intact. The surface temperatures within the released regions of Jeju (by 2025) and Chuncheon (by 2020) are predicted to increase by up to 0.83% and 0.03%, respectively. The average atmospheric concentrations within the released regions of Jeju (by 20205) and Chuncheon (by 2020) would increase by up to 256.93% and 337.29%, respectively.

However, these results have the following limitations. (1) Scenarios were set up based on the simple assumption that the current urban growth trends would continue into the future for the study areas. (2) A linear equation was applied to estimate the future urban area, owing to a lack of input data. (3) This prediction was accomplished at the medial spatial resolution level because of limited simulation models and infrastructure. Environmental change trends following the release of the green belts could be analyzed; however, detailed estimates were impossible. Therefore, a database including historic information on green belts should be generated. In addition, further studies should set and verify scenarios in consideration of future change and potential demands on the environment, economy, and society. Finally, various models based at the local, metropolitan, and national levels should be developed and applied; these models would help researchers and policy makers to predict future change, and the likely influence and demands of the urban environment, and to establish policies for the sustainable management of associated green belts.

Although limitations exist within this study, our results show that the significant benefits of green belts are compromised when they are released, their designated purposes are not followed, and their spatial characteristics are not considered. This causes exaggerated growth and expansion of existing cities, conurbation between existing towns and new towns, degradation of the environmental quality of downtown and released regions, and renders remaining green belt areas non-sustainable. It is impossible for green belts to solve all urban problems, and they in fact contribute occasionally to urban crises, in combination with issues of time and space [9,28]. There is debate with regard to the need to adjust green belt borders, and on the regulations to apply for their conservation [28–31]. This study has shown, however, that green belts are able to contribute positively to controlling urban growth, and to conserving the urban living environment, and that they require a minimal but necessary amount of maintenance. It is almost impossible for released green belt areas to be re-designated as green belts, or for them to be returned to their pre-release environmental state. Therefore, the designation and the release of green belts should be carefully decided in order to promote the coexistence between nature and human beings.

For the management of sustainable urban growth, green belts should be designated as confined regions in which it is necessary to restrict development. This is necessary for the following reasons: (1) to prevent disorderly urban sprawl, (2) to prevent neighboring cities from being linked by continuous urban corridors, (3) to conserve the natural environment and ecosystems surrounding cities, and (4) to ensure a healthy living environment for city dwellers. More effort should be made to conserve green belts with high conservation value, and to restore damaged ecosystems within them. Within green belts, human activities that do not degrade ecosystem functions, and minimally affect natural resources within the green belt zone should be encouraged. Parts of green belts should be dedicated to eco-friendly uses, by transforming them into public parks and constructing environmentally friendly facilities in land purchased for the purpose. Traffic and movement networks (white belts) around the green belts should be constructed for commuters, while ensuring that low-density use and low-impact use are encouraged, maintaining a pleasant urban environment. Green belts could have their shapes adjusted, or could be released if necessary, if issues arise concerning the adequate supply of urban land, or land-use rationalization, in their vicinity, as long as environmental assessment of the green belt shows that the released land embodies low conservation values. However, the release of green belts should be discouraged within environmentally sensitive regions in order to conserve the urban ecosystem, and released green belt land should be developed in connection with areas that are in need of urban regeneration or revitalization, and should not be used for building new towns or industrial complexes. We should select sites for release from green belts only after conducting analysis considering geographic, economic, social, and environmental factors, and should encourage eco-friendly use and development of released areas for tourism, or as recreational spaces for citizens, where the environmental burden is low, and the environmental impact assimilating capacity can be maintained. We should contribute to the sustainable management of green belts by introducing eco-friendly or low-impact development techniques, implementing environmental monitoring, and constructing environmental infrastructure within released areas. Finally, we should also ensure that a minimum green zone width is retained within the released region, to prevent urban conurbation.

This study is significant, as we intended to support policy decision-making with respect to green belts by predicting urban growth and the changes that green belt release could cause in the urban living environment using various development scenarios. We anticipate that a balance between conservation and use will be kept by managing green belts reasonably through more studies based on our research, and that healthy urban environments will be promoted and improved.

Author Contributions: Conceptualization, S.J.; Methodology, H.H., and S.K.; Software, H.H., and S.K.; Validation, H.H., and S.K.; Formal Analysis, S.J., H.H., and S.K.; Investigation, H.H., and S.K.; Resources, H.H., and S.K.; Data Curation, S.J.; Writing-Original Draft Preparation, H.H.; Writing-Review & Editing, S.J. and H.H.; Visualization, H.H., and S.K.; Supervision, S.J.; Project Administration, S.J.; Funding Acquisition, S.J.

Funding: This study was funded by Korea Environment Institute project (GP2013-01-32) and Korea University Grant.

Acknowledgments: This study was conducted as part of "Sustainable Management of Green Belt by Changing Land Environment Management" supported by the Korea Environment Institute. For publication, this study was supported by Korea University Grant.

Conflicts of Interest: The authors declare no conflict of interest.

References

1. Bhatta, B. Analysis of urban growth and sprawl from remote sensing data. *Adv. Geogr. Inf. Sci.* **2010**, 17–36. [CrossRef]
2. Yamamoto, K. Comparison of the garden city concept and green belt concept in major Asian and Oceanic cities. *J. Hum. Soc. Sci.* **2009**, *3*, 1055–1064.
3. Department for Communities and Local Government. Local Planning Authority Green Belt: England 2014/15. Available online: https://www.gov.uk/government/collections/green-belt-statistics (accessed on 3 August 2018).
4. Kim, J.H.; Jung, S.H.; Hwang, K.M.; Lee, J.M.; Kim, S.M.; Hwang, S.K.; Lee, J.Y. *Policy Development for Quality Innovation of Forest Resources*; Korea Forest Service: Seoul, Korea, 2012.
5. Yang, J.C.; Kim, S.W. Policy Agenda Change and Perspective for the Restricted Development Zone, Green Belt under Socio-Economic Paradigm Change. *J. Korea Reg. Sci. Assoc.* **2013**, *29*, 123–143.
6. Choi, J.W.; Lee, K.J.; Noh, T.H.; Min, S.H. Changes of landuse for eight years (1999–2007) in greenbelt area. *Seoul Environ. Sci.* **2010**, *19*, 1025–1034.
7. Oh, J.K.; Yang, S.W. The development activity control in the German outer-area. *Seoul City Res.* **2011**, *12*, 65–82.
8. Steffen, W.; Sanderson, A.; Tyson, P.D.; Jäger, J.; Matson, P.A.; Moore, B., III; Oldfield, F.; Richardson, K.; Schellnhuber, H.-J.; Turner, B.L.; et al. *Global Changes and the Earth System: A Planet under Pressure*; Springer: Heidelberg, Germany, 2004.
9. Han, A.T. Effects of relaxing the urban growth management policy: Greenbelt policy of Seoul metropolitan area, South Korea. *J. Plan. Educ. Res.* **2017**. [CrossRef]
10. Civerolo, K.; Hogrefe, C.; Lynn, B.; Rosenthal, J.; Ku, J.-Y.; Solecki, W.; Cox, J.; Small, C.; Rosenzweig, C.; Goldberg, R.; et al. Estimating the effects of increased urbanization on surface meteorology and ozone concentrations in the New York City metropolitan region. *Atmos. Environ.* **2007**, *41*, 1803–1818. [CrossRef]
11. Rosenthal, J.K.; Sclar, E.D.; Kinney, P.L.; Knowlton, K.; Crauderueff, R.; Brandt-Rauf, P.W. Links between the built environment, climate and population health: Interdisciplinary environmental change research in New York City. *Ann. Acad. Med. Singap.* **2007**, *36*, 834–846. [PubMed]
12. Ward, P.J.; Balen, R.T.; Verstraeten, G.; Renssen, H.; Vandenberghe, J. The impact of land use and climate change on late Holocene and future suspended sediment yield of the Meuse catchment. *Geomorphology* **2009**, *103*, 389–400. [CrossRef]
13. Dudley, N. *Guidelines for Applying Protected Area Management Categories*; World Conservation Union: Gland, Switzerland, 2008; Available online: http://www.iucn.org/dbtw-wpd/edocs/paps-016/pdfShare (accessed on 3 September 2017).
14. Park, Y.H.; Jeon, S.W.; Eum, J.H.; Hong, H.J.; Choi, H.A.; Byun, B.S. *Sustainable Use and Management of Protected Areas: Site-Based Approaches Considering Ecological and Socio-Economic Factors*; Korea Environment Institute: Seoul, Korea, 2012.
15. Verburg, P.H.; Soepboer, W.; Veldkamp, A.; Limpiada, R.; Espaldon, V.; Mastura, S.S. Modeling the spatial dynamics of regional land use the CLUE-S model. *Environ. Manag.* **2002**, *30*, 391–405. [CrossRef] [PubMed]
16. Lim, C.H.; Choi, D.S. Predicting micro land use dynamics: A cellular automata modelling approach. *Korea Plan. Assoc.* **2002**, *37*, 229–239.
17. Allen, J.; Lu, K. Modeling and prediction of future urban growth in the Charleston region of South Carolina: A GIS-based integrated approach. *Conserv. Ecol.* **2003**, *8*, 2. [CrossRef]

18. Oh, Y.G.; Choi, J.Y.; Bae, S.J.; Yoo, S.H.; Lee, S.H. A probability mapping for land cover change prediction using CLUE model. *Korea Soc. Rural Plan.* **2010**, *16*, 47–55.
19. Lee, D.G.; Ryu, D.H.; Kim, H.G.; Lee, S.H. Analyzing the future land use change and its effects for the region of Yangpyeong-gun and Yeoju-gun in Korea with the Dyna-CLUE model. *Korea Soc. Environ. Restor. Technol.* **2011**, *14*, 19–310.
20. Kwon, H.C.; Lee, B.G.; Lee, C.S.; Ko, J.W. The landslide probability analysis using logistic regression analysis and artificial neural network methods in Jeju. *Korean Soc. GIS* **2011**, *19*, 33–40.
21. Russell, P.B.; Uthe, E.E.; Ludwig, F.L.; Shaw, N.A. A comparison of atmospheric structure as observed with monostatic acoustic sounder and lidar techniques. *J. Geophys. Res.* **1974**, *79*, 5555–5566. [CrossRef]
22. Menut, L.; Flamant, C.; Pelon, J. Urban boundary-layer height determination from lidar measurements over the Paris area. *Appl. Opt.* **1999**, *38*, 945–954. [CrossRef] [PubMed]
23. Kim, S.W.; Yoon, S.C.; Won, J.G.; Choi, S.C. Ground-based remote sensing measurements of aerosol and ozone in an urban area: A case study of mixing height evolution and its effect on ground-level ozone concentrations. *Atmos. Environ.* **2007**, *41*, 7069–7081. [CrossRef]
24. Wang, X.Y.; Wang, K.C. Estimation of atmospheric mixing layer height from radiosonde data. *Atmos. Meas. Tech.* **2014**, *7*, 1701–1709. [CrossRef]
25. Jeon, K.W. Role and direction of forest business for the promotion of water recharge function. *Ann. Acad. Korean For. Soc. Korea* **2002**, *0*, 25–44.
26. Kim, J.I.; Yeo, C.H. Measuring greenbelt policy effects through the urban growth prediction model. *Korea Plan. Assoc.* **2008**, *43*, 211–223.
27. Kwon, Y.W. *City and Green Belt*; Purungil: Seoul, Korea, 2006.
28. Zhao, Y. Space as method: Field sites and encounters in Beijing's green belts. *City* **2017**, *21*, 190–206. [CrossRef]
29. Thomas, K.; Littlewood, S. From green belts to green infrastructure? the evolution of a new concept in the emerging soft governance of spatial strategies. *J. Plan. Pract. Res.* **2010**, *25*, 203–222. [CrossRef]
30. Morrison, N. A Green Belt under Pressure: The Case of Cambridge, England. *J. Plan. Pract. Res.* **2010**, *25*, 157–181. [CrossRef]
31. Mace, A. The Metropolitan Green Belt, changing an institution. *Prog. Plan.* **2018**, *121*, 1–28. [CrossRef]

© 2018 by the authors. Licensee MDPI, Basel, Switzerland. This article is an open access article distributed under the terms and conditions of the Creative Commons Attribution (CC BY) license (http://creativecommons.org/licenses/by/4.0/).

Article

Factors Causing Farmland Price-Value Distortion and Their Implications for Peri-Urban Growth Management

Yu-Hui Chen [1], Chun-Lin Lee [2,*], Guan-Rui Chen [1], Chiung-Hsin Wang [3]  **and Ya-Hui Chen [4]**

[1] Department of Agricultural Economics, National Taiwan University, Taipei City 106, Taiwan (R.O.C.); yhc@ntu.edu.tw (Y.-H.C.); r04627009@ntu.edu.tw (C.-R.C.)
[2] Department of Landscape Architecture, Chinese Culture University, Taipei City 11114, Taiwan (R.O.C.)
[3] Department of Natural Resources and Environmental Studies, National Dong Hwa University, Hualien 97401, Taiwan (R.O.C.); s24105.wang@gmail.com
[4] Department of Business Administration, Hsuan Chuang University, Hsinchu City 300, Taiwan (R.O.C.); yhchen558@gmail.com
* Correspondence: chunlin1977@gmail.com

Received: 20 June 2018; Accepted: 26 July 2018; Published: 1 August 2018

Abstract: Taiwan's Agricultural Development Act (ADA) of 2000 relaxed farmland ownership criteria and allowed non-farmers to own farms. Although this opened up the market and induced a growth in farmland trading, relaxing these criteria without proper monitoring resulted in rapid development of farmhouses that fragmented farmlands, adversely affecting agricultural production and the quality of peri-urban environments, and increased management difficulties. Relaxing farmland ownership criteria also provided opportunities for speculation, which pushed up farmland prices, causing farmland price to deviate from its production value. We used a price:value ratio as an index of price-value distortion to explore farmland price-value distortion spatially using a geographical information system (GIS). Yilan County was used as a case study since its agricultural lands suffer high development pressure due to ready accessibility from the Taipei metropolitan area. Ordinary least square and quantile regression were used to identify factors driving distortion in Yilan County. Finally, we discuss the distortion and key factors for specific sites in Yilan to assess the urban sprawl and propose a preliminary course of action for peri-urban growth management. Our findings suggest that residential activities stimulate farmland price-value distortion but do not enhance farmland value. Designation of a land parcel as agricultural within an urban area allows for speculation and increases distortion. The land parcel's association with infrastructure such as road and irrigation systems, and the price of agricultural products, are significantly correlated with distortion. Most of these identified factors increased farmland price because of the potential for non-agricultural land-use. We propose that to resolve farmland price-value distortion in Yilan, multi-functional values, in addition to agriculture, must be envisioned.

Keywords: driving factors; farmland price-value distortion; GIS; price:value ratio; quantile regression; spatial spectrum

1. Introduction

A number of environmental impacts within farmlands and their surrounding areas are the result of current land-use planning and policies, which emphasize economic development over environmental protection, ecological conservation, and control of urban sprawl into farmlands [1,2]. Urban sprawl encompasses a complex pattern of land-use, transportation, and socioeconomic development [3], depending on the type of urban and rural planning and farmland policies [4]. Factories in rural areas, regarded as industrial land-use, are inextricably linked with urbanization and the urban sprawl pattern in Taiwan [5–7]. Moreover, urban sprawl into farmlands in Yilan shows a unique pattern due to the presence of farmhouses occurring both along roadsides and in the middle of farmlands, a result of amendments to the Agricultural Development Act (ADA) in 2000 that allowed for residential activities on farmland. The resultant pattern differs from the "Frog Jump" pattern observed in western countries. The urban sprawl developing along the roadside and in the middle of farmlands caused adverse impacts on the peri-urban environment and complicated sprawl management.

While agricultural production is the most conventional way of gaining revenue from farmlands currently, it is anticipated that there will be greater returns from farmlands through farmland readjustment in the future [8]. Urban sprawl onto farmlands in Yilan is a result of the comparative advantage of farmland owners and developers. The findings of Hardie et al. [9] suggested that in general, farmland and housing prices are determined by income, population, and accessibility variables. This implies that option values associated with irreversible and uncertain land development are capitalized into current farmland values. Plantinga et al. [8] decomposed farmland value into two components: (1) the quasi-rents from agricultural production and (2) the gains from potential land development at the nation's county level. Kostov [10] applied spatial quantile regression and hedonic land price to model agricultural land sales in Northern Ireland. Later, McMillen [11] applied quantile regression to spatial grid modelling in the Chicago area and in this way predicted the change in land values as one moves closer to the central business district of the city. Yoo and Frederick [12] used quantile regression to statistically explore the effects of land subsidence and earth fissures to residential property values in a study from Arizona. Few studies, however, have concentrated on the spatial pattern of both farmland price and value and their distribution difference at site scale with geographical information system (GIS) spatial analyses.

Current farmland prices are mostly driven by the option values of potential development, which often surpass the land's production value. Farmland prices are thus frequently higher than would be expected from their production value. With a lack of proper monitoring, speculators entered the farmland market, and a scattering of farmhouses emerged on the Yilan Plain. The agricultural production zone and surrounding areas continue to suffer from an array of environmental impacts caused by the increase of residential housing and over-development that occurred once the ADA was amended. Therefore, current farmland prices, representing the option values of potential farmland development, are a paradoxical excuse for farmland speculation in Taiwan. Because the agricultural development policy no longer stipulated that farmland had to be utilized for agricultural production and allowed development of residential buildings on farmlands, building companies and land owners triggered speculation of real estate on farmland. This caused a substantial farmland price increase and thereby distorted its price-value ratio. In Yilan, the price-value distortion of farmlands resulted in decreased agricultural production and affected farmland ecosystem services.

We used Yilan County as a case study to investigate spatial patterns of this price-value distortion. We used a GIS platform to integrate a farmland transaction database (price) with the value of crop production (using a "relative value from production" approach). A quantile regression (QR) was used to identify factors potentially impacting the price-value distortion for each farmland transaction. These factors were explored with spatial analyses to review the spatial pattern of the distortion relative to planning and management policies. This was used to propose potential measures to manage the price-value distortion and enhance ecosystem services of farmlands in Yilan.

2. Research Background

2.1. Farmland and Farmhouse Development in Yilan

Yilan County, located in the northeast of Taiwan, is adjacent to the Taipei metropolitan area (Figure 1). It has an area of 2144 km^2. 14.93% of the land is a plain, known as Lanyang Plain (24°37′–24°50′ N, 121°37′–121°50′ E). Its soil quality, sufficient precipitation, and efficient irrigation system make it an important agricultural production region in Taiwan. After Taiwan's Agricultural Development Act revised farmland ownership criteria to allow non-farmers to own farmlands in 2000, many urban civilians became farm owners and are now hobby farmers or simply build farmhouses in the production area. The completion of National Freeway No. 5 in 2006 shortened travel time, and commuting between Yilan County and Taipei city became feasible. This provided further incentives for urban residents to pursue agri-tourism in Yilan, or even to purchase farmland there. These factors all increased the transformation of Yilan from a conventional farming area into an agri-tourism industry. As a result, a large number of farmhouses were built on farmlands which caused a rapid urban sprawl pattern on the Lanyang Plain, hindering production as well as the development of agricultural industries in Yilan County [13].

To mitigate the problem, the "Regulations for Constructing Farmhouses on Agricultural Land in Yilan" (hereafter "Farmhouse Regulations") were implemented by the Yilan County Government in 1994. They provided guidance for building farmhouses that suited the landscape and environment. Subsidies were offered for compliance, which attracted speculators and led to more farmhouses appearing on the Lanyang Plain. The demand for farmland was strong, and elevated farmland prices and a unique pattern of urban sprawl emerged on the Lanyang Plain. As Huang [14] pointed out, when it is anticipated that farmland prices will be inflated in the future, speculation prevails. This is exactly what occurred in Yilan County. It is important to realize that once farmland has been converted into buildings, both cultivation area and agricultural production value are decreased, and farmland ecosystem services, which are socioeconomically and environmentally important, will suffer from the loss of farmland [15].

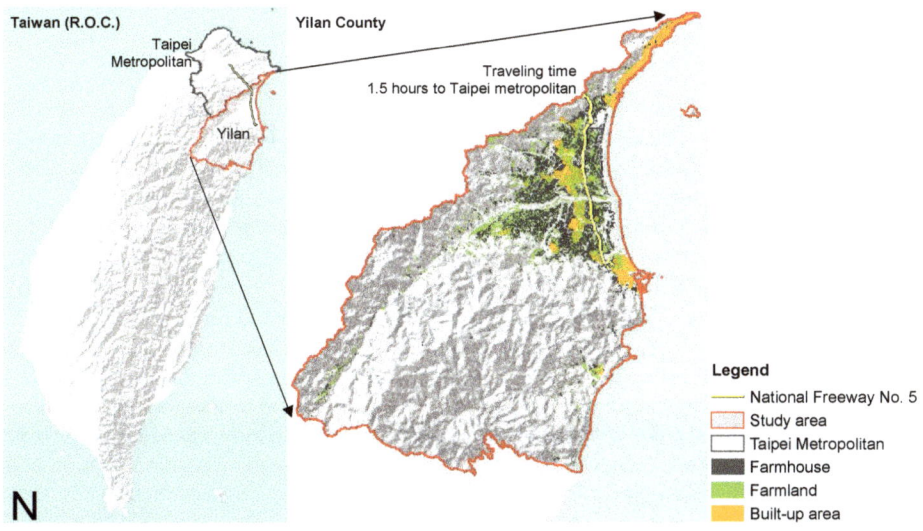

Figure 1. Study area.

2.2. Farmland Price and Value

When only farmers owned farmland, agricultural production was the most common source of revenue. As a result, farmland value was determined based on the output level or profitability of agricultural production, and transaction prices of farmland usually reflected its production value [16], implying little deviation between farmland price and value. Since the criteria for farmland ownership were relaxed and farmland policy shifted from "farmland should be owned by farmers" to "farmland should be used for agricultural production", non-farmers have been allowed to own farmland. This caused a loophole allowing for speculation. It is anticipated that farmland readjustment (consolidation) will generate high returns and that farmland price will begin to deviate from its original purpose (i.e., production [8]). Rather than agricultural production value, the development potential of farmlands will play a vital role in determining farmland value and price. Hardie et al. [9] suggested that income, population, and other variables determine average farmland and housing prices. Their research suggests that optional values associated with irreversible and uncertain land development are capitalized into present farmland values.

The value and benefits of farmlands are worth assessing from a conservation and ecosystem services perspective. The multi-functionality of farmlands in providing such benefits as production value [16], water retention [17], wildlife habitats [18], air quality improvement [19], recreational and landscape aesthetics [20], and ecosystem services [21] has been the subject of many studies. However, this multi-functionality is based on the premise that farmlands are utilized for agricultural production rather than speculation. The capital gain from the potential development of farmlands cannot be combined with production value and other ecosystem services. Although taking ecosystem services into policy consideration is important, due to the difficulties of evaluating the value of ecosystem services of farmlands, its impacts on farmland price are not included in this study. Farmland prices in the following study therefore reflect the expected value of potential development (i.e., the possibility of building farmhouses) and agricultural production value. Discrepancies between farmland production value and farmland prices can be treated as proxies of "distortion". Few studies have investigated the difference between prices and production values of farmlands, without which it is impossible to assess the significance of the role played by speculation.

3. Research Design and Data

3.1. Definition

Most previous studies have focused on factors influencing the price of farmlands, such as soil condition, production environment, and farmland productivity [22–24]. Since the advent of urbanization, additional factors have begun to influence land value. Arbitrage that hopes for appreciation may play a vital role in price decision. Therefore, analyzing the deviation between land price and its production value is helpful in understanding the current situation of farmland price distortion. The ratio of farmland price to farmland production value adjusted for the farmland price index, as defined in the present and earlier studies, reflects the level of distortion as follows:

$$\text{price} - \text{value distortion} = \frac{\text{price of farmland}}{\text{value of farmland}} \times \text{farmland price index}$$

The higher the ratio, the greater the distortion; a high distortion indicates a considerable discrepancy between farmland price and production value. Farmland price data were downloaded and summarized from the real estate transaction database provided by the Ministry of Internal Affairs. Farmland value was generated from the raw data of Taiwan's agricultural census in 2010. Farmland price indices were collected from the Platform of Real Estate Information from the Ministry of Internal Affairs.

Before 2000, most Yilan farmlands were utilized for agricultural production because of the limitations of the ADA and zoning control for agricultural areas (Figure 2). During this period, farmland value was usually influenced by agricultural production and transportation cost from the perspective of land economics and location theory. Enhancing production conditions by irrigation system and soil improvement are examples of key methods for increasing farmland value and price. In this period, there was no obvious difference between farmland price and value. The farmland price increased substantially after the amendments of the ADA. Since farmland production value did not change, the farmland price:value ratio increased, resulting in a clear upward trend of price-value distortion. Subsequent studies revealed that farmland buildings along roads usually dominate the unique urban sprawl pattern in the surrounding environment of urban areas.

We adopt the price:value ratio firstly to generate a single index to analyze the phenomenon of farmland price-value distortion, and secondly to organize various scenarios for analyzing development impacts on good quality farmland. We then use this to discuss potential issues arising from farmland protection and demand for urban development (Figure 3). For example, farmland with both a high transaction price and a high productive value is identified as a good quality farmland in Yilan. Its price:value ratio (price-value distortion) is usually a result of serious urban development impact. Therefore, besides distortion, relative and standardized measures based on crop production value and farmland price as determined from land transactions were also used to explore farmland value. The farmland price:value ratio shows an interesting distortion spectrum which highlights the conflict between land speculation, farmland preservation, and urban growth management.

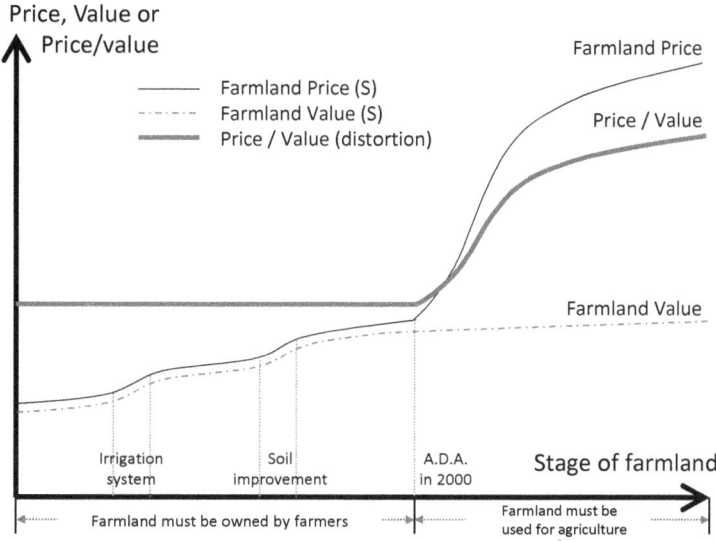

Figure 2. Change in farmland price-value distortion.

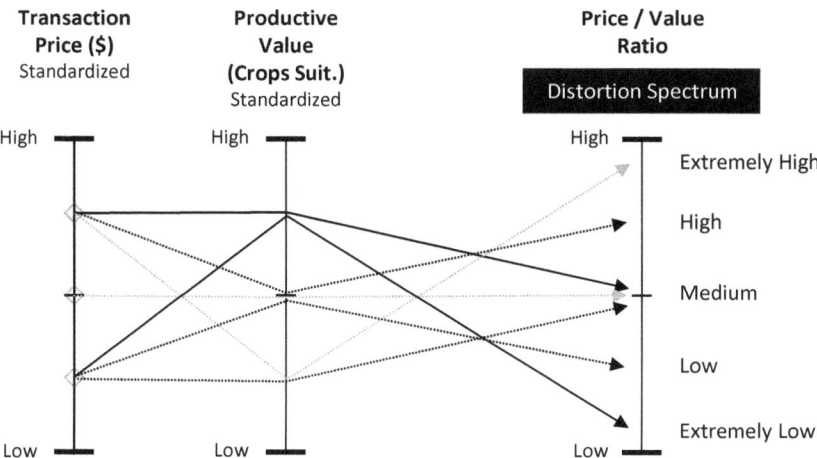

Figure 3. The farmland price-value distortion spectrum.

3.2. Research Design

A geographical information system (GIS) was used to integrate farmland transaction, to spatially illustrate the distortion spectrum (Figure 4), and to identify and estimate potential factors influencing the distortion. Quantile regression (QR) was also adopted to estimate the impact of factors on the distortion. The spatial pattern of the distortion relative to planning and management policies and their implications were discussed as well. Suggestions were provided to manage distortion, preservation, and urban growth management in Yilan.

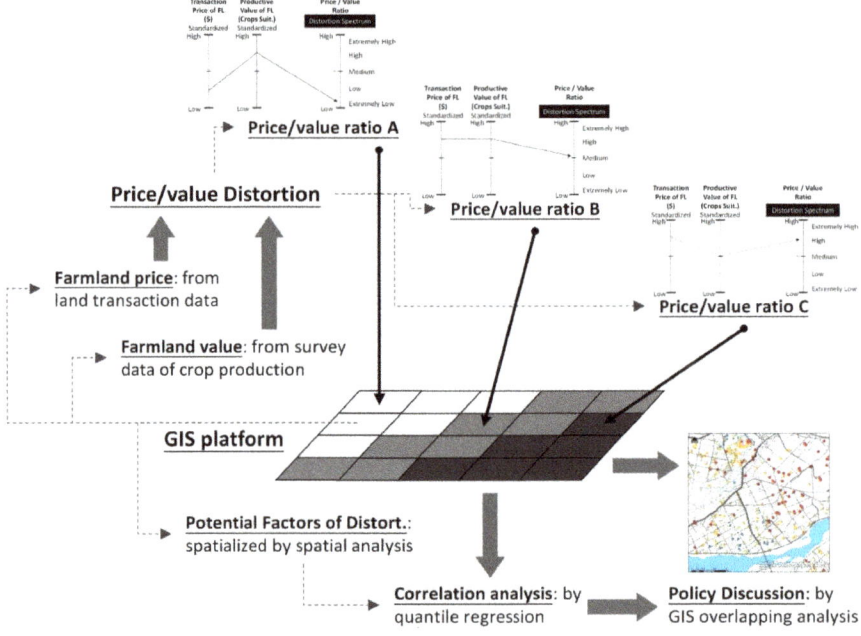

Figure 4. Research design.

3.3. Data

We selected eight potential factors influencing farmland price and production value: soil condition (a proxy for crop suitability); irrigation system; crop price variation; farmland use regulation; distance of farmland from urban area; width of the nearest road; complexity of land count and transaction; and real estate loan index [13,22–28]. Calculating farmland price and value in this study focuses on farmland itself without considering the price of farmhouses. In addition to calculating distortion and these factors (Tables 1 and 2), QR and ESRI ArcGIS were used to analyze the farmland price determination in Yilan County. Our findings indicate that the factors (or explained variables) do affect distortion.

Table 1. The definitions and sources of variables.

Variable	Definition and Data Description	Data Source
Price	Real transaction price of farmland (NT$/m^2) (transaction price of farmland from 2012 to 2015 is adjusted by consumer price index to have a common comparative base in 2016)	Actual real estate transaction cases, Ministry of the Interior (August 2012–September 2016)

Table 1. Cont.

Variable	Definition and Data Description	Data Source
Value	Production value of farmland (NT\$/m^2/year) (average production value of crops per year is calculated from the original database based on the location for each farmland transaction)	Agricultural, Forestry, Fishery and Husbandry Census, Directorate General of Budget, Accounting and Statistics (2016)
Price_value	Price-value distortion (ratio of the real transaction price to production value, %)	(Same as above)
Avgattr	Soil conditions (0 is the poorest, 4 is the best)	NGIS Ecological Resources Database
near_water	Distance to the nearest canal (m)	Platform for the National Land Use Inventory
Agriprice	Annual crop prices variation (%)	Statistical database of Yilan County; Agricultural Statistics Yearbook, Council of Agriculture
use_urban	Agricultural zone in urban area (1 yes, 0 no)	Actual information of real estate transaction case, Ministry of the Interior
use_regular	General agricultural zone in non-urban area? (1 yes, 0 no)	(Same as above)
use_special	Special agricultural zone in non-urban area? (1 yes, 0 no)	(Same as above)
use_mt	Sloping and conservation zone in non-urban area? (1 yes, 0 no)	(Same as above)
near_urban	Distance to the nearest urban area (m)	Layers of use in urban planning, Construction and Planning Agency, Ministry of the Interior
Width	The width of the nearest road (m)	Taiwan Electronic Map, Ministry of the Interior
Landcount	Complexity of land counts (1, 2, 3, etc.) (many approaches could be used to define complexity of land. To simplify the problem, as well as to cope with the real situation of Yilan, the complexity of farmland ownership was applied in this research instead of common landscape metrics, Shannon index, or landscape complexity [29].)	Agricultural cadastral maps, Yilan County
Interestrate	Mortgage rate (%)	Deposit and loan rate of five leading domestic banks, Central Bank

Table 2. Descriptive statistics of variables.

Variable	Mean	SD	Maximum	Minimum
Price_value	306.14	360.11	5223.48	0.19
Avgattr	3.16	1.27	4	0
near_water	258.94	507.44	4921.98	0
Agriprice	2.73	7.22	29.84	−50.34
use_urban	0.08	0.28	1	0
use_regular	0.16	0.37	1	0
use_special	0.68	0.47	1	0
use_mt	0.08	0.27	1	0
near_urban	2047.44	2200.73	38,770.77	0
Width	5.77	3.24	40	1
Landcount	8.74	6.66	75	1
Interestrate	1.37	0.04	1.38	1.12

3.4. Results

3.4.1. Illustration of Price-Value Distortion in Yilan County

The categories of distortion, from low to high, are shown as blue, green, orange, and red (Figure 5). To facilitate comprehension, quartiles were used to define the bounds of each distortion category. The calculated distortion categories of Yilan County indicate that areas in the plain were the most heavily traded. Among these, areas with a high density of high distortion instances were Yilan City (Figure 5(b),A), Luodong Township (Figure 5(b),B), Zhuangwei Township, Wujie Township, Dongshan Township, and Jiaosi Township. Transactions decreased with distance from the city or towns (as illustrated by the color distribution of the symbols changing from orange to green and then to blue in Figure 5). Figure 5 shows that these areas are east of Yuanshan Township, Sansing Township, Datong Township (Figure 5(b),C), and Nan'ao Township (Figure 5(b),D).

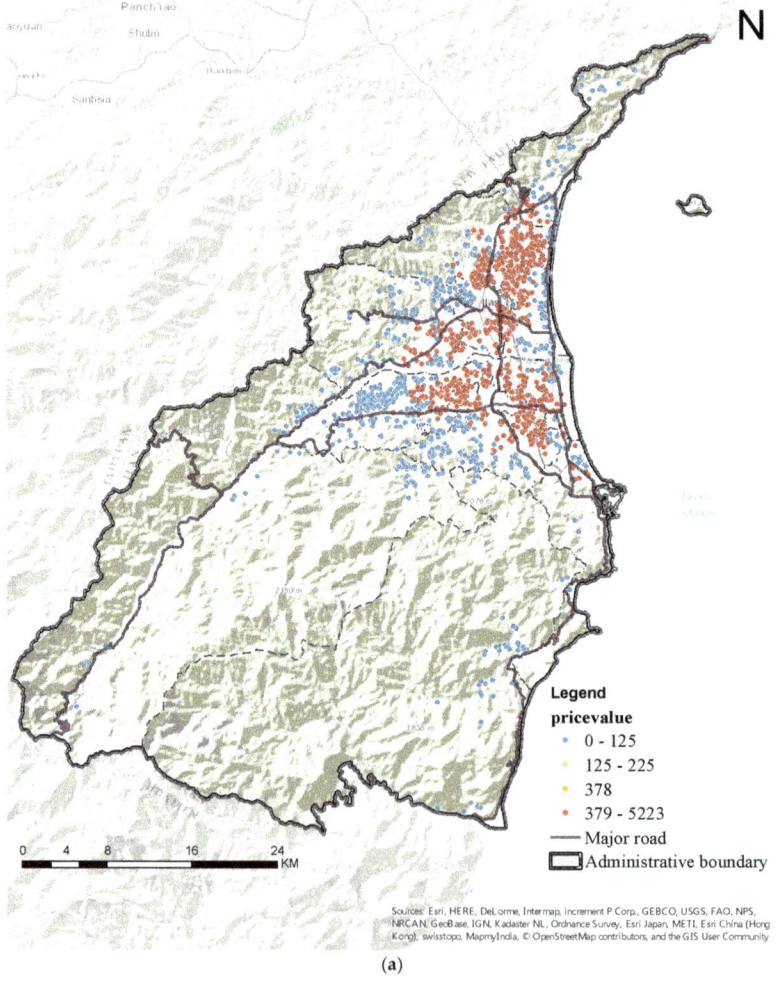

(a)

Figure 5. *Cont.*

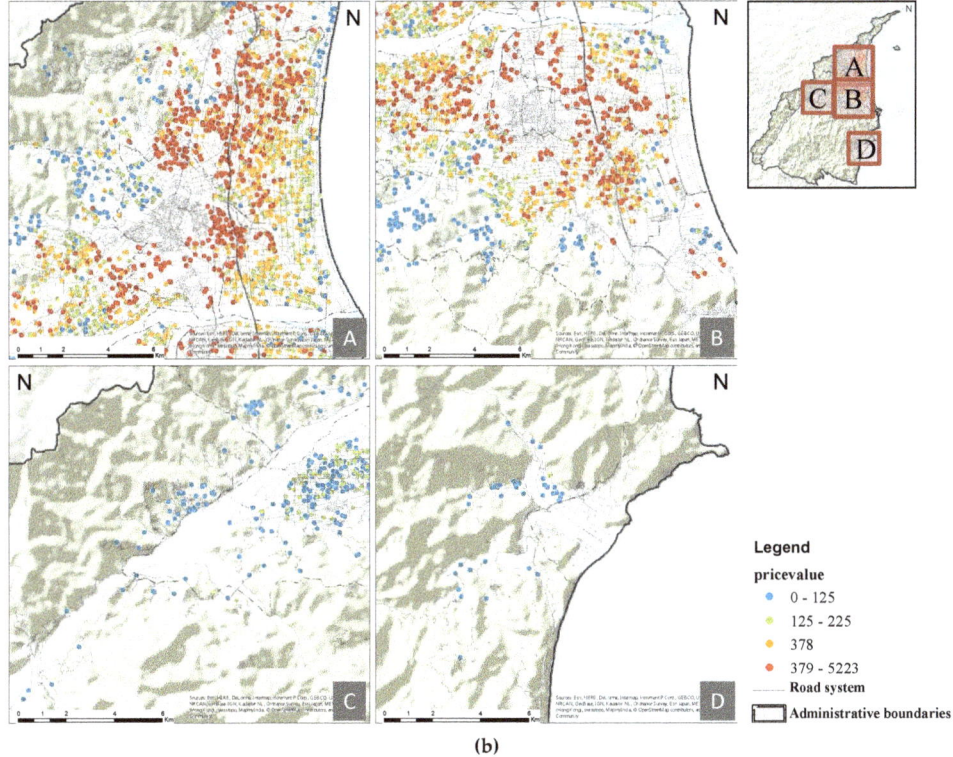

Figure 5. Price-value distortion of farmlands in Yilan County. (**a**) Distribution of high and low distortion categories in Yilan County; (**b**) Distribution of distortion categories in specific areas. A and B present the areas with high distortion, in which A instances Yilan City and B is Luodong Township; C and D present the areas with low distortion, in which C instances the eastern of Yuanshan Township, Sansing Township, Datong Township and D is Nanao Township.

3.4.2. Ordinary Least Square (OLS) Regression and Quantile Regression (QR) Analyses for Price-Value Distortion of Farmlands in Yilan County

In addition to visualizing distortion by area with ArcGIS, quantitative analysis was used to examine the factors influencing distortion. In an ordinary least square regression model, the calculated distortion was treated as the dependent variable, and suitability for crop plantation, the distance to the nearest canal, crop price variation, land usage regulation, distance to the nearest urban planning areas, width of the nearest road, the complexity of land counts, and mortgage rates were treated as explanatory (independent) variables. The full regression model can be expressed as:

$$\text{price\_value}_i = \beta_0 + \beta_1 \text{avgattr} + \beta_2 \text{near\_water} + \beta_3 \text{agriprice} + \beta_4 \text{use\_urban} + \beta_5 \text{use\_special} + \beta_6 \text{use\_mt} + \beta_7 \text{near\_urban} + \beta_8 \text{width} + \beta_9 \text{landcount} + \beta_{10} \text{interestrate} + \varepsilon_i, \quad i = 1, 2, 3, \ldots, 4641$$

The results of the OLS model are shown in Table 3. All factors except for land count complexity significantly affected price distortion.

We also applied quantile regression, an approach initially introduced by Koenker and Bassett [30]. The model is used to examine the dynamic movement in each quantile and its main concept focuses on minimizing an asymmetrically weighted sum of absolute errors [31]. The relationships between dependent and independent variables under various circumstances at each quantile were discussed. In comparison to OLS models, a QR model can process samples with a non-normal distribution. Furthermore, the QR model was used to overcome a sample selection bias and outlier problems that were encountered when running the OLS model [31]. To conduct the QR, five quantiles (θ = 0.1, 0.25, 0.5, 0.75, 0.9) were used in our analysis to evaluate the impact of each factor on different distortion categories. The results (Table 3) show that the nearest distance to the canal, crop price variation, land use regulation, width of the nearest road, and the complexity of land count have positive impacts on distortion. The higher θ, the greater the distortion and when θ is higher, the impact of factors influencing distortion are higher. The confidence intervals calculated from the results of the OLS and QR are shown in Figure 6.

Table 3. OLS (ordinary least square) and QR (quantile regression) results for the price-value distortion of farmland in Yilan County.

Variable	OLS	QR (θ)				
		0.1	0.25	0.5	0.75	0.9
constant	430.334 ***	10.264 **	119.658 *	266.637 ***	249.612	1291.145 ***
	(3.02)	(3.11)	(1.88)	(3.13)	(1.29)	(3.31)
avgattr	2.196	1.807	0.736	4.619 *	15.066 ***	17.748
	(0.43)	(1.38)	(0.46)	(1.73)	(3.98)	(1.47)
near_water	−0.0284 **	−0.004	−0.012 ***	−0.018 ***	−0.017 ***	−0.017 *
	(−2.33)	(−1.44)	(−3.50)	(−4.17)	(−3.79)	(−1.73)
agriprice	1.779 ***	0.125	0.577 ***	1.080 ***	1.170 ***	1.719
	(2.79)	(0.87)	(2.1)	(3.81)	(2.54)	(1.45)
use_urban	600.541 ***	176.976 ***	298.907 ***	382.246 ***	690.964 ***	1335.272 ***
	(29.82)	(4.20)	(18.49)	(12.20)	(9.47)	(16.13)
use_special	82.934 **	63.083 ***	79.733 ***	92.831 ***	120.048 ***	117.442 ***
	(6.34)	(18.25)	(16.82)	(14.00)	(13.54)	(3.57)
use_mt	−64.989 **	−23.172 ***	−48.019 ***	−77.340 ***	−84.817 ***	−157.409 ***
	(−2.37)	(−4.52)	(−7.48)	(−7.55)	(−4.91)	(−3.33)
near_urban	−0.011 ***	−0.001	−0.000107	−0.002 *	−0.005 **	−0.007
	(−4.33)	(−1.30)	(−0.02)	(−1.56)	(−2.03)	(−1.34)
Width	17.831 ***	2.900 ***	4.733 ***	7.828 ***	15.472 ***	22.099 ***
	(12.62)	(3.31)	(6.38)	(6.76)	(7.12)	(5.00)
landcount	1.243 *	−0.329	0.299	1.173 **	3.271 ***	2.968 ***
	(1.84)	(−1.46)	(1.15)	(2.39)	(3.51)	(2.84)
interestrate	−235.224 **	−71.725 **	−54.222	−130.992 *	−111.256	−784.926 ***
	(−2.28)	(−2.44)	(−1.18)	(−2.17)	(−0.81)	(−2.81)
Adj./Pseudo R–Squared	0.283	0.1674	0.1664	0.1652	0.1902	0.2658
F statistic	183.99 ***					
VIF	1.57					

Note: 1. n = 4641; * $p < 0.10$, ** $p < 0.05$, *** $p < 0.01$. 2. The variance inflation factor (VIF) is used as an indicator of multicollinearity.

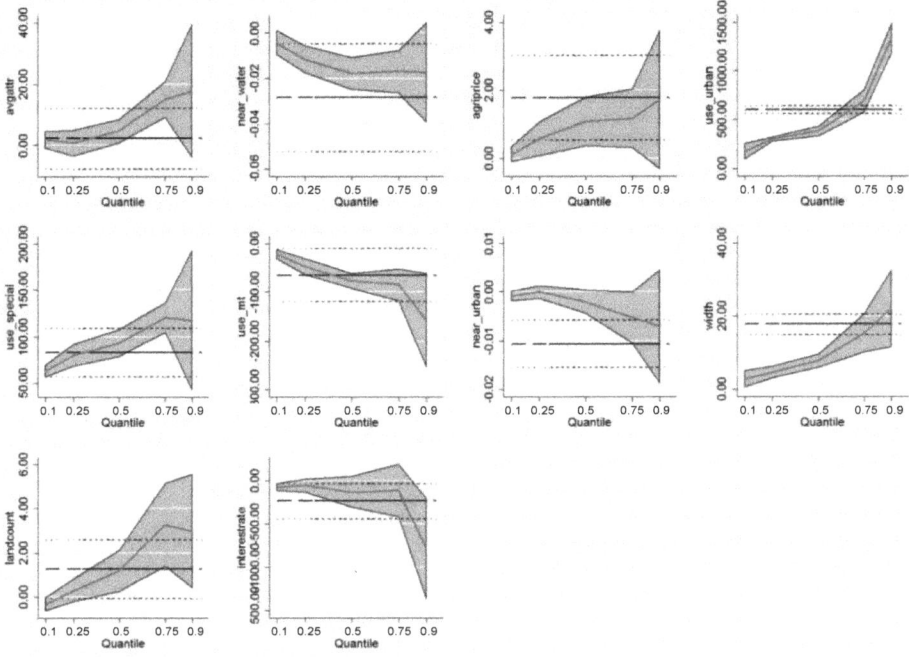

Figure 6. The distribution of OLS (ordinary least square) and QR (quantile regression) 95% confidence interval for each variable. Quantiles range from 0.1 to 0.9 and each quantile gap is 0.1. The shaded area represents the QR 95% confidence interval. The OLS regression line is the solid black line between the upper and lower dotted lines, which show the OLS confidence interval at 0.95.

1. Distance to the main canal

The results suggest that distance to the nearest canal has a significant negative impact on distortion, with $p < 0.01$ in the 0.25, 0.5, and 0.75 quantiles, and $p < 0.1$ in the 0.9 quantile. Our findings suggest that when the farmland is separated by 1 m from the nearest canal, the distortion is reduced by 0.012%, 0.018%, and 0.017%, for the 0.25, 0.5, and 0.75 quantiles, respectively.

2. Crop price variation

Crop price variation has significant impact on distortion with $p < 0.01$ in the 0.25, 0.5, and 0.75 quantiles. Higher crop price variation is associated with greater distortion. The model indicates that when crop price variation increases by 1%, the distortion will increase by 0.577%, 1.080%, and 1.170% for the 0.25, 0.5, and 0.75 quantiles, respectively.

3. Land-use regulation

Land-use regulation has a significant impact on the price distortion with $p < 0.01$ under various quantiles. The price-value distortion is higher when the farmland is located in urban areas. In addition, the higher the quantile, the greater the distortion when farmlands are located in urban regions. The distortion is lower when the farmlands are located on slopes or conservation fields in non-urban zones. Compared to urban areas, distortion decreased at higher quantiles in non-urban zones.

4. Width of the nearest road

Width of the nearest road significantly affects price-value distortion under various quantiles. When the width of the nearest road widens by 1 m, the distortion is increased by 2.900%, 4.733%, 7.828%, 15.472%, and 22.099% for the 0.1, 0.25, 0.5, 0.75, and 0.9 quantiles, respectively. This suggests that the higher the quantile, the greater the impact of road width on distortion.

5. Complexity of land count

The complexity of land count significantly affects farmland distortion with $p < 0.01$ in the 0.5 and 0.9 quantiles. This implies that an increase in the number of farmland transactions in buffer areas causes distortion to increase by 1.173%, 3.271%, and 2.968% in the 0.5, 0.75 and 0.9 quantiles, respectively. If the quantile is less than or equal to 0.75, higher complexity of land count results in a greater distortion. The distortion reached its peak in the 0.75 quantile.

4. Discussion

4.1. Designating an Urban Area Tends to Increase the Price-Value Distortion of Agriculture Zones

Few of Taiwan's urban plans have been designed from the perspective of agricultural development or the multi-functionality of agriculture. These plans often imply that farmlands located in urban areas will eventually be developed for non-agricultural use. Farmlands located in urban areas are targets for speculation (Figure 7). Due to the anticipated capital gain, the prices of these farmlands are increasing, which will hinder development in these areas. It will also increase the price-value distortion ratio. Therefore, designating urban zoning in agricultural zones tends to increase the price-value distortion of farmlands.

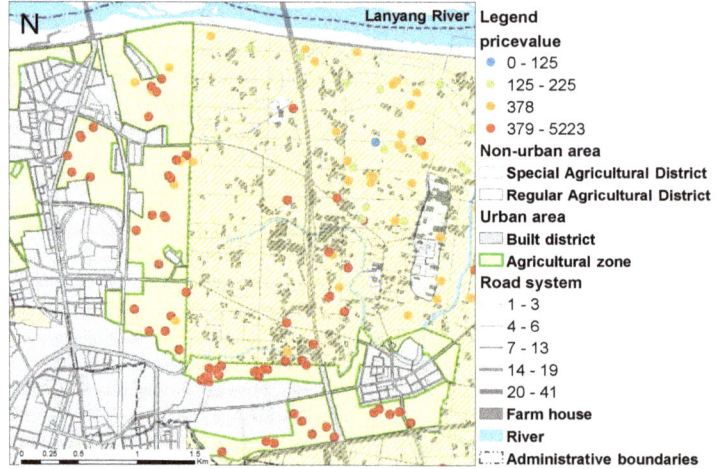

Figure 7. The farmland price-value distortion relative to the urban plan of Wujie Town.

4.2. A Better Road or Transportation System in Agricultural Zones Tends to Increase the Price-Value Distortion

Road or transportation system accessibility is important for agricultural marketing. The width of roads is often regarded as an indicator of transportation conditions. Under the Building Act of Yilan County, the width of the nearest road was set as a minimum requirement for constructing farmhouses. A better road system therefore tends to elevate the price of farmlands. Our model suggests that road condition is highly correlated with price-value distortion, and this is consistent with reality. Figure 8 illustrates clusters of high distortion that correspond to the pattern of road width in Yuanshan Town.

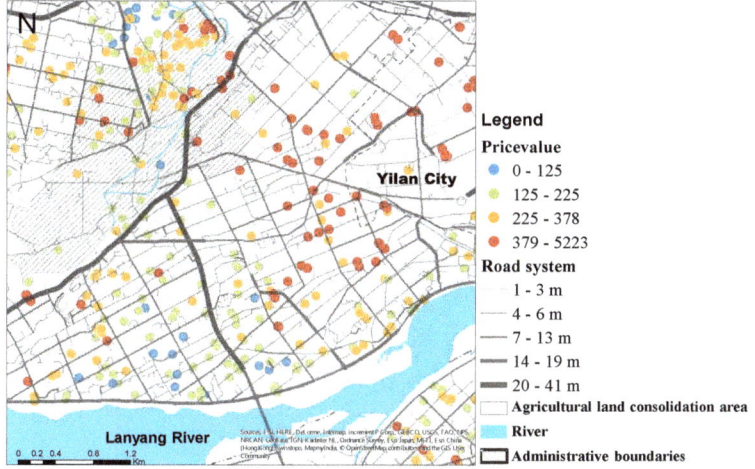

Figure 8. Distribution of price-value distortion in relation to road widths in Yuanshan Town.

4.3. Farmland Reform Policy Significantly Affects Price-Value Distortion

Farmland readjustment is one of the major policy mechanisms for improving the productivity of agriculture. The process aims to produce ordered farmland parcels suitable for mechanization and better irrigation systems. A better agricultural production environment increases land price but enhances the possibility of higher distortion. The regression analysis outcomes suggest that there is an important relationship between distortion and distance between farmlands and irrigation systems in Yilan. Figure 9 shows that in Yuanshan Town, higher distortion occurs in areas with better irrigation systems developed through the farmland readjustment plan. The Agricultural Development Act amendments in force since 2000 shifted the policy goal from "farmland owned by farmers" to "farmland used for agriculture". This policy relaxed the criteria for ownership of farmland, which positively affected farmland price, and this, in turn, elevated the price-value distortion of farmlands. Higher distortion and the appearance of farmhouses on good quality farmland as a result of this agricultural policy have become major challenges for the development of Yilan's agriculture.

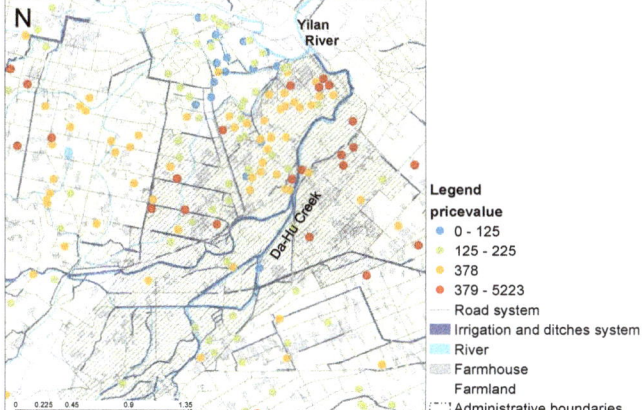

Figure 9. Distribution of price-value distortion relative to irrigation systems in Yuanshan Town.

4.4. Cases of Various Price-Value Distortions and Their Potential Impacts on Agriculture

Cases of high distortion are shown in Figure 10 (A1,A2). Farmlands located in the agricultural zone of an urban area often have higher distortions due to their higher transaction prices and lower production values. This situation suggests a possibility of engendering an expectation for developing farmlands in urban areas. Development would destroy any opportunity of benefiting from the farmlands' ecosystem services in urban areas because development is irreversible. Further, farmlands located in the special agricultural districts in the suburbs but in close proximity to urban areas often have higher transaction prices and production values; here, price-value distortions are moderate (Figure 10 (B1)) and there is serious conflict between farmland development and agricultural production.

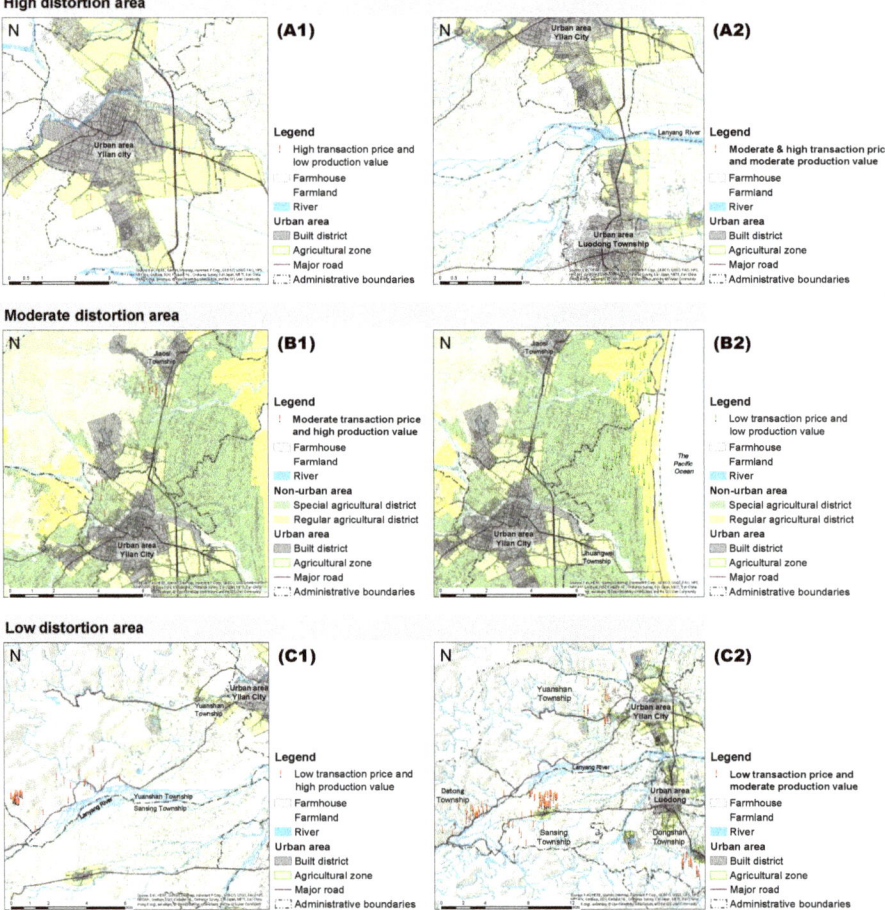

Figure 10. Cases of various price-value distortions. A1 represents the area of high transaction price with low production value; A2 represents the area of moderate and high transaction price with moderate production value; B1 represents the area of moderate transaction price with high production value; B2 represents the area of low transaction price with low production value; C1 represents the area of low transaction price with high production value; C2 represents the area of low transaction price with moderate production value.

Moderate distortions also occur when both the prices and production values of farmlands are low. This occurs in the regular agricultural district of suburbs (Figure 10 (B2)) and leads to a unique and widespread urban sprawl pattern in the rural areas of Yilan. Lower farmland price and higher production value generate low distortions, which occurs in the hilly area of Yilan County (Figure 10 (C1,C2)). Agricultural production in this area does not suffer from serious urbanization impacts. However, agricultural production itself has the potential to affect and pollute the environment through the use of fertilizers and pesticides. Clearly there are different cases of price-value distortion associated with the different farmland situations and characteristics, and these need to be taken into account by agricultural and urban planning departments of local and central governments. Therefore, a systematic approach of land administration is needed to consider the heterogeneity of these cases and inter-disciplinary cooperation from different departments in government is required [32].

5. Conclusions

Our findings indicate that policy reforms have significant impact on the farmland environment of Lanyang Plain and its surrounding regions. Investment in agricultural infrastructure was expected to improve production efficiency and increase production value. However, irrigation and road system improvements increased the price and thus the price-value distortion of farmland, which led to urbanization and impeded agricultural production in the region. The prices of farmland in agricultural zones within urban areas increased substantially due to speculation. This amplified the discrepancy between farmland price and its production value, which in turn increased the price-value ratio or distortion. Price-value distortion can be used as a proxy for the deviation between price and production value of farmlands, and can be mapped onto locations to explore reasons for the distortion and to understand its impact on agricultural development. Allowing non-farming use of farmlands can result in irreversible impacts on agriculture. Therefore, future urban planning has to consider the multi-functionality of farmland, and that the agricultural zone is not just for "reserving land for urban development." Greater consideration of factors such as the multi-functionality of agriculture may provide a healthier balance between economic benefits and environment sustainability, as well as a finer balance between farmland price and value (Figure 11). Moreover, location assessment of farmland readjustment and agricultural investment policy should work with payment policy for the ecosystem services of farmland to decrease the distortion.

Land Administration Systems (LAS) can facilitate the sustainable development of farmland. LAS focuses land on systematic relations among various factors, ecosystem services and interaction between local and central governments [32]. The complex adaptive system can provide insights when deconstructing the complex agricultural landscape system in peri-urban area based on physical flaws and influence [33]. Additionally, Land Value Tax for non-agricultural use, real estate boom and recession may influence farmland price and distortion. It will be inspiring to employ these approaches to explore the spatial pattern of distortion ratio based on adequate and available transaction data in the future. Multifunctional use of farmland is absent from this study, but could be taken into account in future research.

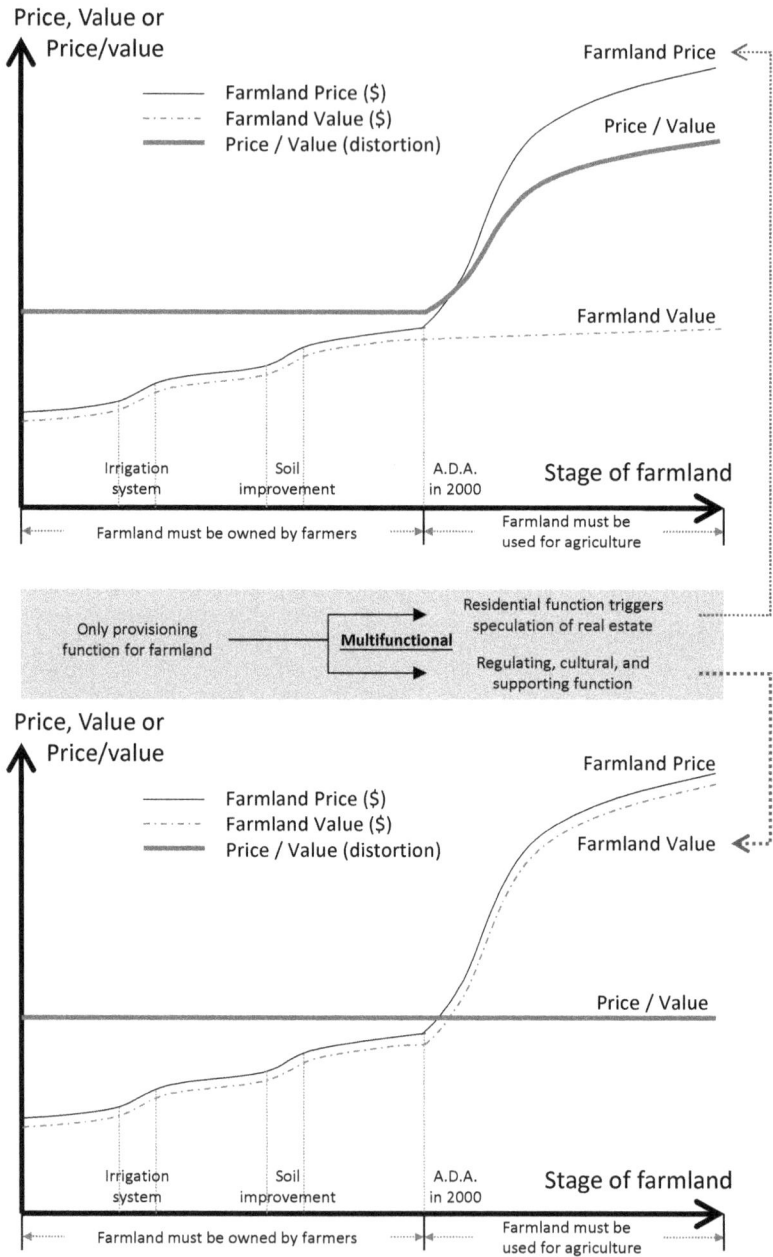

Figure 11. Enhancing farmland value from an ecosystem service perspective to manage farmland price-value distortion.

Author Contributions: Yu-Hui Chen and Chun-Lin Lee conceived and designed the research and decided the methodology applied in the study. Guan-Rui Chen and Ya-Hui Chen collected and analyzed data, and made quantile regression analysis; Chiung-Hsin Wang applied GIS to analyze the data and provided visualized outcomes. Yu-Hui Chen and Chun-Lin Lee wrote the paper.

Funding: This research received no external funding.

Conflicts of Interest: The authors declare no conflict of interest.

References

1. Gilg, A. Perceptions about land use. *Land Use Policy* **2009**, *26*, S76–S82. [CrossRef]
2. Park, S.; Smardon, R.C. Worldview and social amplification of risk framework: Dioxin case in Korea. *Int. J. Appl. Environ. Sci.* **2011**, *6*, 173–191.
3. Frumkin, H. Urban sprawl and public health. *Public Health Rep.* **2002**, *117*, 201–217. [CrossRef]
4. Couch, C.; Petschel-Held, G. *Urban Sprawl in Europe: Landscape Land Use Change and Policy*; Wiley-Blackwell: London, UK, 2008.
5. Chiang, Y.-C.; Tsai, F.-F.; Chang, H.-P.; Chen, C.-F.; Huang, Y.-C. Adaptive society in a changing environment: Insight into the social resilience of a rural region of Taiwan. *Land Use Policy* **2014**, *36*, 510–521. [CrossRef]
6. Wang, C.-H. *Analysis of Agricultural Land Change in Lanyang Plain Using a System Perspective*; Chinese Cultural University: Taipei, Taiwan, 2014. (In Chinese)
7. Wey, W.-M.; Hsu, J. New Urbanism and Smart Growth: Toward achieving a smart National Taipei University District. *Habitat Int.* **2014**, *42*, 164–174. [CrossRef]
8. Plantinga, A.J.; Lubowski, R.N.; Stavins, R.N. The effects of potential land development on agricultural land prices. *J. Urban Econ.* **2002**, *52*, 561–581. [CrossRef]
9. Hardie, I.W.; Narayan, T.A.; Gardner, B.L. The Joint Influence of Agricultural and Nonfarm Factors on Real Estate Values: An Application to the Mid-Atlantic Region. *Am. J. Agric. Econ.* **2001**, *83*, 120–132. [CrossRef]
10. Kostov, P. A Spatial Quantile Regression Hedonic Model of Agricultural Land Prices. *Spat. Econ. Anal.* **2009**, *4*, 53–72. [CrossRef]
11. McMillen, D. Conditionally parametric quantile regression for spatial data: An analysis of land values in early nineteenth century Chicago. *Reg. Sci. Urban Econ.* **2015**, *55*, 28–38. [CrossRef]
12. Yoo, J.; Frederick, T. The varying impact of land subsidence and earth fissures on residential property values in Maricopa County—A quantile regression approach. *Int. J. Urban Sci.* **2017**, *21*, 204–216. [CrossRef]
13. Chen, W.-B.; Lee, C.-L.; Chang, C.-R.; Wang, C.-H. An analysis of agricultural land change factors and spatial planning policies review for Lan-Yang Plain: The system approach. *J. Taiwan Land Res.* **2016**, *19*, 1–35.
14. Huang, W.-S. *The Farmland Price under Urbanization Pressure—A Case Study of I-Lan County*; Department of Real Estate & Built Environment, College of Public Affairs, National Taipei University: Taipei, Taiwan, 2008.
15. MA Synthesis. *Ecosystems and Human Well-Being: Synthesis Report*; Island Press: Washington, DC, USA, 2005.
16. Stewart, P.A.; Libby, L.W. Determinants of Farmland Value: The Case of DeKalb County, Illinois. *Rev. Agric. Econ.* **1998**, *20*, 80–95. [CrossRef]
17. Zhang, W.; Ricketts, T.H.; Kremen, C.; Carney, K.; Swinton, S.M. Ecosystem services and dis-services to agriculture. *Ecol. Econ.* **2007**, *64*, 253–260. [CrossRef]
18. Bergstrom, J.C. Postproductivism and changing rural land use values and preferences. In *Land Use Problems and Conflicts: Causes, Consequence and Solutions*; Goetz, S.J., Shortle, J.S., Bergstrom, J.C., Eds.; Routledge: London, UK, 2005; pp. 64–76.
19. De Groot, R.; Hein, L. Concept and valuation of landscape functions at diff erent scales. In *Multifunctional Land Use: Meeting Future Demands for Landscape Goods and Services*; Mander, Ü., Helming, K., Wiggering, H., Eds.; Springer: Berlin, Germany, 2007; pp. 15–36.
20. Kong, F.; Yin, H.; Nakagoshi, N. Using GIS and landscape metrics in the hedonic price modeling of the amenity value of urban green space: A case study in Jinan City, China. *Landsc. Urban Plann.* **2007**, *79*, 240–252. [CrossRef]
21. Lee, C.-L.; Lee, C.-H. Economic Assessments of Ecosystem Service Impacts of Agricultural Land-use Change (in chinese). *Taiwa. Agric. Econ. Rev.* **2012**, *17*, 111–144.
22. Dehring, C.A.; Lind, M.S. Residential Land-Use Controls and Land Values: Zoning and Covenant Interactions. *Land Econ.* **2007**, *83*, 445–457. [CrossRef]
23. Delbecq, B.A.; Kuethe, T.H.; Borchers, A.M. Identifying the Extent of the Urban Fringe and its Impact on Agricultural Land Values. *Land Econ.* **2014**, *90*, 587–600. [CrossRef]
24. Faux, J.; Perry, G.M. Estimating Irrigation Water Value Using Hedonic Price Analysis: A Case Study in Malheur County, Oregon. *Land Econ.* **1999**, *75*, 440–452. [CrossRef]

25. Beaton, W.P. The Impact of Regional Land-Use Controls on Property Values: The Case of the New Jersey Pinelands. *Land Econ.* **1991**, *67*, 172–194. [CrossRef]
26. Wu, T.-C.; Lin, F.-T.; Lin, S.-T.; Hsu, Y.-L. Changes in Farmhouse Distribution Pattern and Factors Affecting Farmhouse Change in Yilan (in chinese). *J. City Plan.* **2013**, *40*, 31–57.
27. Choumert, J.; Phélinas, P. Determinants of agricultural land values in Argentina. *Ecol. Econ.* **2015**, *110*, 134–140. [CrossRef]
28. Henneberry, D.M.; Barrows, R.L. Capitalization of Exclusive Agricultural Zoning into Farmland Prices. *Land Econ.* **1990**, *66*, 249–258. [CrossRef]
29. Papadimitriou, F. The Algorithmic Complexity of Landscapes. *Landsc. Res.* **2012**, *37*, 591–611. [CrossRef]
30. Koenker, R.; Bassett, G. Regression Quantiles. *Econometrica* **1978**, *46*, 33–50. [CrossRef]
31. Koenker, R.; Hallock, K.F. Quantile Regression. *J. Econ. Perspect.* **2001**, *15*, 143–156. [CrossRef]
32. Williamson, I.; Enemark, S.; Wallace, J.; Rajabifard, A. *Land Administration for Sustainable Development*; Esri Press: California, CA, USA, 2010.
33. Holland, J.H. Studying Complex Adaptive Systems. *J. Syst. Sci. Complex.* **2006**, *19*, 1–8. [CrossRef]

© 2018 by the authors. Licensee MDPI, Basel, Switzerland. This article is an open access article distributed under the terms and conditions of the Creative Commons Attribution (CC BY) license (http://creativecommons.org/licenses/by/4.0/).

Article

Evaluating Social Performance of Construction Projects: An Empirical Study

Xiaer Xiahou [1], Yuchun Tang [1], Jingfeng Yuan [1,*], Tengyuan Chang [1], Ping Liu [1,2] and Qiming Li [1]

1. Department of Construction Management and Real Estate, School of Civil Engineering, Southeast University, Nanjing 211189, China; xh@seu.edu.cn (X.X.); tangyuchunseu@163.com (Y.T.); changty@seu.edu.cn (T.C.); liupvip@foxmail.com (P.L.); seulqming@163.com (Q.L.)
2. School of Civil Engineering, Lanzhou University of Technology, Lanzhou 730050, China
* Correspondence: jingfeng-yuan@seu.edu.cn; Tel.: +86-25-8379-3257

Received: 4 June 2018; Accepted: 4 July 2018; Published: 5 July 2018

Abstract: The concept of sustainable development is gaining increasing popularity in construction industry. Previous studies have prioritized on the sustainable performance of construction projects from perspectives of economy and environment, social performance of construction projects has not drawn much attention. Social performance of construction projects refers to the extent which the projects meet the needs of current and future generations. Therefore, social performance of construction projects is critical for project success as well as social sustainability. However, a systematic framework for evaluating social performance of construction projects is absent. At the same time, existing methods are time-consuming and subject to certain degree of subjectivity. To overcome these limitations, the fuzzy analytical hierarchy process (FAHP) method is introduced in this paper to evaluate social performance of construction projects. A real-world hospital redevelopment project was employed as an empirical study to develop the systematic framework for social performance evaluation using FAHP method. By analyzing previous studies and the hospital redevelopment project, a systematic framework with 18 indicators of five dimensions (i.e., socio-economy development, socio-environment development, social flexibility, public service development, and environment and resource conservation) was developed. Social performance of two proposed schemes for hospital redevelopment project were evaluated using the FAHP approach. Results show Scheme 2 has a relative higher social performance sore than that of Scheme 1 and the hospital redevelopment project would improve socio-economy development, socio-environment development, social flexibility, and public service development, while it brings challenges to environment and resource conservation. More seriously, results indicate the hospital project may threaten healthcare and disease prevention of the local communities. Therefore, more measures should be taken to improve social performance of the hospital redevelopment project. The empirical study shows the proposed framework using FAHP method is viable for conducting social performance evaluation of construction projects, which could be helpful to improve social performance, reduce negative social impacts, and contribute to the social sustainability of construction projects.

Keywords: social performance; social performance evaluation; fuzzy analytical hierarchy process; empirical study

1. Introduction

It has been recognized that construction industry plays significant roles for the sustainable development [1,2]. Activities within lifecycle of construction projects have inherent impacts (i.e., economic impacts, environmental impacts, and social impacts) to the society [3]. Compared to economic and environmental impacts, social impacts associated with construction projects, are the least explicit in the "triple bottom line" principle of sustainable development [4,5]. Social impacts of construction projects refer to certain social consequences to human populations of construction projects that change the ways in which people live, work, play, relate to one another, organize to meet their needs, and generally cope as members of society [5,6]. The development of construction projects may lead to both positive and negative social impacts, e.g., land acquisition and disposal, resettlement of nearby residents, and resource depletion [7,8]. Under certain conditions, social impacts could evolve into social risks and even lead to social conflicts between different stakeholders if they are not dealt with carefully and properly [9]. To mitigate negative social impacts, social impact assessment (SIA) has widely been used within lifecycle of construction projects [10,11]. International Association for Impact Assessment (IAIA) defines social impact assessment (SIA) as processes of analyzing, monitoring and managing the intended and unintended social consequences, both positive and negative, of planned interventions (policies, programs, plans, and projects) and any social change processes invoked by these interventions [12]. Instead of addressing social performance of the project, the primary goal of SIA is to ensure the sustainability and equity of biophysical and human environment [13,14].

Social performance refers to an organization's response to anticipated or existing social demands [15,16]. As a project-oriented industry, the major goal of construction firms/organizations is to provide the society with varieties of construction projects, social performance of construction projects could be aggregated to the construction firms/organizations level and the industry level [17]. Therefore, social performance of construction projects is crucial for the social sustainability, which is to meet the demands of current and future populations and communities [14,18], and improving social performance has been a major concern by all stakeholders [5,19]. Previous studies have measured economic and environmental performance of construction projects, while social performance, as a critical and indispensable dimension for project success and sustainable development [20,21], has not been well studied [21–23]. It is vital to study the social performance of construction projects and evaluate social performance of construction projects, which would not only help the decision makers to make decisions when proposing a construction project but also enable the project managers to audit a project and determine what kind of future improvements could be made [24].

Social performance evaluation of construction projects could be defined as improving social performance by providing information about achievement of social aspects, it allows decision makers to determine its ongoing performance in meeting social criteria, which helps reduce social impacts, prevent social risks, improve the overall performance of the project, and finally contribute to social sustainability [15,25]. While various studies have discussed the SIA of construction projects [7,8,26], social performance evaluation of construction projects are rarely studied. Yuan [20] studied the social performance of construction waste management, which emphasized the management process of construction waste and the introduced research approach could not be transferred to evaluate social performance of construction projects. A previous study conducted by Shen, et al. [2] presented a checklist for evaluating sustainability performance, indicators reflecting social sustainability performance of construction projects were listed as reference for conducting further evaluation. Prior studies indicate that the research gap exists in social performance evaluation of construction projects.

This research developed a systematic framework for social performance evaluation of construction projects. A hospital redevelopment project was introduced as the empirical study. In addition, this study could also help to understand social performance of construction projects. The remainder of this article is as follows: Section 2 illustrates research background and the literature review. Detailed processes of FAHP-based method to evaluate social performance of construction projects are described in Section 3. A systematic framework for social evaluation is developed in Section 4 by analyzing prior studies and the introduced hospital redevelopment project. An empirical study of social performance of two proposed schemes for the hospital redevelopment project are evaluated using the developed approach, which is presented in Section 5. Results of the empirical study are discussed in Section 6. Finally, research findings and limitations are stated.

2. Research Background and Literature Review

2.1. Social Performance of Construction Projects

Social performance of construction projects reflects the extent to which the lifecycle of construction projects meets the demands of anticipated or existing social demands. Therefore, social performance of construction projects could be obtained by analyzing social impacts of construction projects and the requirements for social sustainability by diverse stakeholders. Shen, et al. [2] explored the indicators for social sustainability performance evaluation of different stages. Valdes-Vasquez, et al. [14] identified 50 processes for social sustainability consideration during planning and design phase of construction projects, and these processes were categorized into six categories, namely stakeholder engagement, user considerations, team formation, management considerations, impact assessment, and place context. Zuo, et al. [27] interviewed domain experts and 26 criteria of social sustainability were identified, which were further discussed from three dimensions, i.e., macro level, external stakeholders, and internal stakeholders. Tilt, et al. [7] applied SIA to explore social impacts in a large dam project. These impacts are identified as migration and resettlement of people near the dam sites, changes in the rural economy and employment structure, effects on infrastructure and housing, impacts on non-material or cultural aspects of life, and impacts on community health and gender relations. Almahmoud, et al. [28] studied social core functions (SCFs) of a construction project from perspectives of diverse stakeholders. Capital performance, health and physical comfort, accessibility, integration, usability psychological comfort, and operation health and safety were identified as SCFs of a construction project. Li, et al. [8] studied social impacts of an affordable housing project and indicators reflecting social impacts were discussed from three aspects as socio-economic effects, adaptabilities, and social risks. Wang, et al. [5], Shi, et al. [9], Liu, et al. [19], and Liu, et al. [29] also addressed the social risks of the construction projects. They suggested that the projects should not only be compliant with the regulations but also meet the requirements of diverse stakeholders, especially the end-users, which will improve project social flexibility and thereby contribute to project social sustainability.

2.2. Performance Management and Evaluation

Performance management, which is defined as a closed loop control system that deploys policy and strategy, and obtains feedback from various levels to manage the performance of the system [30], has been widely adopted by a wide range of industries [31,32]. Performance management could help the project managers to continuously improve its project management practices [33]. Construction industry is no exception to the gaining popularity of performance management. Several construction firms have adopted performance management to enhance the performance of construction industry at different levels, e.g., organization level, stakeholders level, and project level [31,32]. As a project-oriented industry, the performance of construction projects is crucial to project success as well as the satisfaction of the organization and diverse stakeholders [34]. However, there are diverse goals of construction projects; accordingly, construction projects performance management should also cover a wide range of themes, e.g., quality, cost, time, safety and health, environment, and client satisfaction [34–37]. Traditionally, the management of construction performance relied on three indicators (i.e., cost, time, and quality), which are lagging and fails to provide a holistic view [24]. Construction projects, e.g., infrastructure projects, are complex engineering systems, which require tremendous investment and would have profound and long-time impacts on the economy, environment, and society [38]. As opined by Atkinson [39], apart from "The Iron Triangle", more success criteria should be accepted in project management. There are increasing studies investigating on the social aspects of construction projects from diverse perspectives including safety and health [34,40], social impacts, social risks, social conflicts [7,9,19,34], etc. Social performance has been perceived as critical aspects for sustainable development and project success.

Performance evaluation is the process of determining the efficiency and/or effectiveness of past action, which has been widely adopted to measure the performance of construction projects [34].

Several frameworks were proposed to help the evaluation of construction projects performance. Among the proposed performance evaluation frameworks, three are the most prevalent in construction industry: European Foundation for Quality Management (EFQM) Excellence Model, Balanced Scorecard (BSC), and Key performance indicators (KPIs) model [32]. EFQM Excellence Model is a quality-based framework [31], which is frequently and more appropriately applied at the organizational level [32,41]. BSC framework consists of a range of "leading and lagging" indicators and the scorecard is divided into four perspectives, namely financial, custom, internal business, and innovation learning [31,34]. Different perspectives interact under certain principles [31]. Even though the BSC is more prevalent in performance management at project-level, four perspectives are far from enough to cover the performance of construction projects, and more perspectives should be supplemented to measure the performance of construction projects. KPI framework, which adopts the method of benchmarking, selects time, cost, quality, client satisfaction, change orders, business performance, and health and safety as the seven key performance indicators to measure the performance of construction projects. In practice, KPI framework is more flexible, indicators of different aspects could be clustered to measure certain aspects of project performance, which makes it widely adopted in construction projects performance measurement, and various studies have been conducted to measure the overall or partial performance of construction projects [34,42]. Based on KPI framework, several tools have been developed to help the evaluation of construction projects social performance such as SIA tool [12], checklist method [2], success factors identification, social network analysis [28], etc. However, current calculation methods for social performance management based on KPI framework are subjected to certain degree of objective [32]. Alternative approaches should be devised to reduce this limitation.

To sum up, performance management has been employed to help project managers continuously improve project outcomes, social performance as one of the key aspects for projects success and sustainability is gaining increasing interests by research scholars and practitioners. Unfortunately, social performance of construction projects has not been systematically studied. In addition, among three most prevalent performance measurement frameworks for performance management in construction industry, KPI framework based on benchmarking method is the most appreciated for developing tools to evaluate construction project social performance. However, since indicators for social performance are intangible, KPI–based approach is time-consuming in identifying indicators and relatively subjective. Therefore, considerations should be given to mitigate the subjectivity of these tools.

3. The FAHP-Based Method for Social Performance Evaluation

To improve the efficiency and reduce the degree of subjectivity, the fuzzy analytical hierarchy process (FAHP) was adopted in this study to develop the approach for the evaluation of construction projects social performance. The method of FAHP is developed from AHP method, which was first established by Saaty [43] in 1981. The AHP method has been widely used by scholars in many fields, including performance measurement [32]. However, the application of AHP method is yield to the degree of uncertainty and subjectivity. Zeng, et al. [44] pointed that experts may find it hard to select a single number in the comparison process. Instead of a definite value, it would be proper to give a range values for comparison, e.g., 2 to 5. To avoid the deficiency, FAHP approach proposed by Zeng, et al. [44] was employed in this research to conduct social performance evaluation. The FAHP approach consists of the following six successive steps.

Step 1: Identification of attributes

In the first step, the FAHP is adopted to identify the common attributes of the given problem. A comprehensive and accurate identification of these attributes is fundamental for the evaluation approach, as the study relies heavily on these attributes while the misconception of these attributes results in a failure of the research model. Similarly, as depicted by Bititci, et al. [30], performance measurement is usually determined by the metric of a number of indicators, and thereby the fundamental step would be the identification of indicators that could be employed to evaluate social performance of construction projects.

Step 2: Pairwise comparisons under fuzzy environment

Once the attributes related to the problem were clearly identified, a pairwise comparison among the common attributes has to be made: one over and another under the fuzzy environment. To set up this fuzzy pairwise comparison, experts are invited to complete a comparative questionnaire. Answers from experts are used to compare each attribute and convert the linguistic comparison into the fuzzy pairwise relation matrix.

$$\widetilde{A} = \begin{bmatrix} 1 & \widetilde{r}_{12} & \widetilde{r}_{13} & \cdots & \widetilde{r}_{1n} \\ \widetilde{r}_{21} & 1 & \widetilde{r}_{23} & \cdots & \widetilde{r}_{2n} \\ \vdots & \vdots & \vdots & \ddots & \vdots \\ \widetilde{r}_{n1} & \widetilde{r}_{n2} & \cdots & \cdots & 1 \end{bmatrix} \qquad (1)$$

Step 3: Defuzzification

The pairwise comparison was composed of fuzzy numbers in Step 2, and then these triangular fuzzy numbers are transformed into crisp numbers. The conversation process is called defuzzification process. For various types of defuzzification methods are buried in the literatures, this study utilizes the centroid method of defuzzification for its wide acceptance [45]. The linguistic variables described by fuzzy numbers are denoted by membership functions [45], which are presented in Table 1 and Figure 1, separately.

Table 1. Scale for relative importance used in the pairwise comparison matrix.

Intensity of Importance	Fuzzy Number	Linguistic Variables	Triangular Fuzzy Numbers (TFNs)	Reciprocal of TFNs
1	$\widetilde{1}$	Equally important	(1,1,3)	(1/3,1,1)
3	$\widetilde{3}$	Weakly important	(1,3,5)	(1/5,1/3,1)
5	$\widetilde{5}$	Important	(3,5,7)	(1/7,1/5,1/3)
7	$\widetilde{7}$	Strongly important	(5,7,9)	(1/9,1/7,1/5)
9	$\widetilde{9}$	Extremely important	(7,9,11)	(1/11,1/9,1/7)

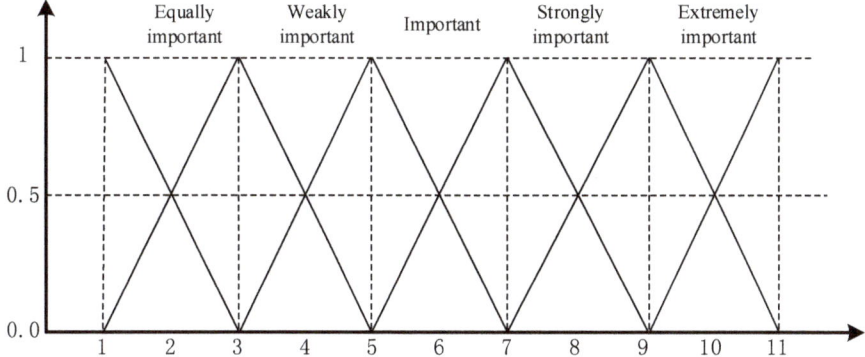

Figure 1. Fuzzy membership function for linguistic expressions for criteria.

Step 4: Estimation of global weights

The defuzzified pairwise comparison gained from the previous steps is processed through various standard arithmetic operations of formal AHP to seek the global weights of each attribute (W). The calculations processes standardize the defuzzified pairwise comparison matrix (all values in the matrix should be kept between 0 and 1) and calculate the eigenvalue (x) with the help of the sum of standardized rows, where the eigenvalue is the global weight of each attribute (W) [46].

Step 5: Consistency check

The collected data based on the experts' judgements are subject to a certain degree of subjectivity, which naturally has errors. Therefore, the consistency of these attributes of the criteria and the relevant steps should be checked. The consistency index (CI) introduced for a pairwise comparison matrix is listed below:

$$CI = \frac{\lambda_{max} - n}{n - 1} \quad (2)$$

where λ_{max} is the largest eigenvalue of the comparison matrix; and n is the dimension of the matrix or the number of decision criteria under consideration.

The consistency ratio (CR) is given as:

$$CR = \frac{CI}{RI(n)} \quad (3)$$

where RI(n) is a random index relying on the size of matrix. The random index values of random matrices introduced are listed in Table 2. If the consistency ratio (CR) is equal to or less than 0.1, the result is acceptable. However, if the value is greater than 0.1, the decision makers have to remake pairwise comparisons to achieve consistency in their responses.

Table 2. Random index values.

N	1	2	3	4	5	6	7	8	9	10
RI(n)	0	0	0.58	0.90	1.12	1.24	1.32	1.41	1.45	1.51

Step 6: Establishment of assessment sets and fuzzy comprehensive assessment of multiple indices

Fuzzy comprehensive assessment method can be used to quantify the fuzzy indices or define the membership degree via building hierarchical fuzzy subsets, which is based on the fuzzy transformation to synthesize each index.

To quantify different attributes of each assessment index, the semantic scales of subjective assessments are used to establish assessment classes, namely, $V = (v_1, v_2, \cdots, v_n)$. Then, according to experts' responses, the score assessment matrix of the entire index system (R) can be acquired, which is denoted as follows:

$$R = (r_{ij})_{m \times n} = \begin{bmatrix} r_{11} & r_{12} & \cdots & r_{1n} \\ r_{21} & r_{22} & \cdots & r_{2n} \\ \cdot & \cdot & & \cdot \\ r_{m1} & r_{m2} & \cdots & r_{mn} \end{bmatrix} \quad (4)$$

where n is the number of the decision schemes; m is the number of indicators of index system; $R_i = (r_{i1}, r_{i2}, r_{i3}, \cdots, r_{im})$, $(i = 1, 2, 3, \cdots, m)$ is the single index fuzzy assessment class of index i; and r_{ij} is the membership degree.

The assessment score matrix (R) and the global weight of each attribute (W) obtained from Steps 4 and 5 are multiplied to get fuzzy evaluation results of the index system. The result of fuzzy comprehensive assessment of multiple indices is as follows:

$$Z = W \cdot R = \{Z_1, Z_2, \ldots, Z_n\} = \{w_1, w_2, \ldots, w_m\} \cdot \begin{bmatrix} r_{11} & r_{12} & \cdots & r_{1n} \\ r_{21} & r_{22} & \cdots & r_{2n} \\ \cdot & \cdot & \cdot & \cdot \\ r_{m1} & r_{m2} & \cdots & r_{mn} \end{bmatrix} \quad (5)$$

where Z_i is the final score of the ith plan after fuzzy comprehensive assessment. The decision scheme with the largest Z_i is the optimal plan.

4. Framework for Social Performance Evaluation of Hospital Redevelopment Project

To develop a systematic framework to evaluate social performance of construction projects, prior studies on social impacts associated with construction projects and requirements for social sustainability indicators for social performance evaluation are beneficial. Besides, to help the understandings of the proposed indicators, a hospital redevelopment project named "South New-Town hospital (Nan Bu Xin Cheng (NBXC) hospital in Chinese)", which locates in the southern part of Nanjing, the capital city of Jiangsu province, China, was introduced in this research as a complement to understand the social performance. The real-world project also helped to conduct the empirical study of the proposed FAHP-based evaluation approach. Social performance of two schemes proposed for the hospital redevelopment project were presented and evaluated using the proposed approach. Figure 2 presents the location and sketch of NBXC hospital projects.

Figure 2. Location and sketch of NBXC hospital project.

4.1. Social Impacts of the Hospital Redevelopment Project

Before NBXC hospital project was developed, a community hospital, which only had basic equipment and provided only outpatient service to the nearby residents, had run for years. However, the rapid urbanization accompanied with fast increase in aging population have led to great challenges to the capacity and service of hospitals [47]. As the community hospital could only provide basic medical service, it could no longer satisfy the demands of the nearby communities due to its outdated equipment and capacity. As shown in the left part of Figure 2, a military airport and several residential communities are around the former hospital. Therefore, this area was quite noisy due to the takeoff and landing of aircraft. In addition, the development of transportation system as well as the local economy were highly restricted by the military zone.

In 2010, the local government decided to redevelop the hospital and two schemes were proposed. The first scheme (Scheme 1) was to redevelop the hospital without demolishing it and just replace the outdated equipment with the latest one. This scheme would not change the land use of nearby area and thereby make little disturbance to the nearby communities. Accordingly, it would require little investment and improve the capacity of the hospital, which is seriously concerned by the local government when the populations are gradually increasing with the rapid urbanization of Nanjing. The second scheme (Scheme 2) suggested building the NBXC hospital project. The former building for the community hospital would be demolished in Scheme 2 to build the NBXC hospital project. Medical experts and advanced medical equipment would be introduced in the new hospital. Finally, Scheme 2 was selected and NBXC hospital project started from January 2012 and would begin operation in June 2018. Table 3 briefly compares the parameters of Schemes 1 and 2.

Table 3. A brief comparison of Schemes 1 and 2.

Parameter	Scheme 1	Scheme 2
Investment	Less than 1 million CNY	3.5 billion CNY
Capacity	No bed available	1500-bed
Service	Outpatient service only	Outpatient and inpatient service
Floor area	Less than 1000 square meter	Over 300,000 square meter
Other function	No other function	Research and development
Nearby facilities	Military airport	Subway station, business centers
Service area	Local community	National

To build NBXC hospital, the government negotiated with the military sector and changed the land use; the airport was relocated to another area of the city. The affected residents were resettled and compensations were also paid. As presented in Figure 2 and Table 3, the government invested 3.5 billion CNY on this project with an ambitious plan to improve the nearby circumstance. Finally, NBXC hospital covers an area of over six acres, which consists of several buildings providing inpatient, outpatient, and emergency healthcare services; buildings for research purpose; etc. The main building is a nineteen-story building with two stories underground, and a total floorage of >300,000 m^2, which could provide 1500-bed capacity for the public. One subway line and several municipal roads were built to make it easier for the public to get to the hospital. In addition, the newly built NBXC hospital was equipped with the latest technologies, and special considerations were also given to the medical waste management to avoid polluting the local environment. As declared by the government, NBXC hospital ranks among the first tier nationwide according to the Chinese medical ranking system. The operation of NBXC hospital would greatly help to enhance healthcare service of this area and satisfy the demands of the local residents for high quality healthcare service. In addition, with the advanced technologies and most famous experts, NBXC is also attractive for patients of all the country, which would drive the development of local economy.

4.2. Framework for Social Performance Evaluation of Hospital Redevelopment Project

Based on the extensive literature review and the case of the major hospital projects, 18 indicators were proposed for the social performance evaluation of the hospital redevelopment project and these indicators could be categorized into five dimensions (i.e., socio-economy development, socio-environment development, social flexibility, public service development, and environment and resource conservation). Eighteen indicators within five dimensions are presented in Table 4.

The first dimension shows that the development of the hospital redevelopment project may enhance the socio-economy. As stated by Shen, et al. [2], the development of construction projects would provide local employment opportunities and regional economic development. The development of a certain type of construction project, e.g., the hospital project, could change the industrial structure of the local economy. In addition, the development of construction projects may change the land use [48]. In this case, to develop the hospital project, the local government changed the land use by relocating the military airport and resettling some nearby residents.

The second dimension indicates that the hospital redevelopment project would change the socio-environment. The development of construction projects would help to deliver more infrastructure and enhance the service to the public [49]. Therefore, it would improve their living standard and the level of social security. As described in the case, the development of the NBXC hospital project changed the situation that people living near the military airport had to endure the noisy warplanes. Instead, people could receive high level of healthcare service. Major projects such as hospitals would be landmarks of the city and improve the regional reputation.

The third dimension indicates that the social performance of construction projects would be influenced by social flexibility of the construction projects, which is the adaptabilities of projects to the society (e.g., government, investor, end-users, etc.). The development of construction projects is to satisfy the appeals of diverse stakeholders within the lifecycle of projects [26] and meet the compliance with the policies (e.g., safety and health regulations). Specifically, the hospital project should give special considerations for the patients.

The fourth dimension is the public service offered by the hospital redevelopment project. Different types of construction projects may enhance specific public services by the development of specific infrastructure. For example, the development of transportation infrastructure would provide a faster, more convenient and economical means of transportation method for the citizens [50]. In the hospital project, NBXC hospital would improve the regional capacity of healthcare and disease prevention, promote the advancement of medical technologies and education, and enhance the ability of response to emergency healthcare service.

The last dimension is the influences of the hospital redevelopment project on environment and resource conservation. Most construction projects have great impacts on the landscape of the city. In most cases, the newly built projects are national, regional, or local landmarks. Therefore, it would help to prompt the image of cities. Construction activities and operation of the infrastructure would consume nature resource and generate waste, which may contaminate the environment [20]. In hospital projects, medical waste and disposals generated in the operation stage may spread viruses and threat the health of the public, which should be carefully dealt with [51].

Table 4. Framework for social performance evaluation of hospital redevelopment projects.

Target Layer	Dimensions	Indicators	Source
Indicators for social performance evaluation of the hospital redevelopment projects A	Socio-economy development B_1	C_{11} Employment rate C_{12} Regional economic development C_{13} Industrial structure C_{14} Land use	[2,50] [14,27,50] [2,50] [7,48]
	Socio-environment development B_2	C_{21} Resident's living standard C_{22} Infrastructure and public service C_{23} Social security C_{24} Regional reputation	[8,27] [50] [2] [9,19]
	Social flexibility B_3	C_{31} Stakeholders satisfaction C_{32} Compliance with policies C_{33} Patients satisfaction	[5,50] [40] Cases
	Public service development B_4	C_{41} Healthcare and disease prevention C_{42} Development of medical technologies and education C_{43} Emergency healthcare service ability	Cases Cases Cases
	Environment and resource conservation B_5	C_{51} Landscape of the city C_{52} Wastage management C_{53} Resource depletion C_{54} Environmental protection	[48] [2,20] [2,21,27] [14,27,52]

5. An Empirical Study

To help evaluate the social performance of hospital projects, ten domain experts were investigated separately to contribute their expertise in this research. The criteria for the selection of experts are working experience and background. These experts should have over ten years of working experience in their working fields, which ensures that the investigated experts have a good knowledge of social impacts associated with the project. The expert team consisted of three from different department of local government, who are responsible for the development of this area; three from research institutions, who are familiar with construction project management and social sustainability; two from the contractor of NBXC hospital projects; and two representatives (i.e., one doctor and one nurse) of the NBXC hospital staff. These experts were visited individually and face-to-face investigations were conducted at their office to ensure the independence of this research. Research background and research approach were introduced to these experts to help them get a quick and comprehensive understandings about the project and this research, which would further ensure the quality of this research. After the introduction, the experts were first required to give remarks on the weight of social performance indicators based on their expertise.

Fuzzy numbers given by the experts determined relative importance of one index over the other, which further helped to build a fuzzy judgment vector. The judgment vectors help to form portions of the fuzzy pairwise comparison matrix, which is then adopted to determine the weight of each criterion. The result of consistency check indicates the objectiveness of the weight of each criterion. Meanwhile, the total sequencing weight sets of social performance assessment index system (W) is built, as shown in Table 5.

To quantify the magnitude of hospital redevelopment project affecting each social performance indicator, the semantic scales of each subjective assessment index was quantified and the assessment class $V = (v_1, v_2, \cdots, v_n)$ was further classified into fivescales (i.e., high, relatively higher, average, relatively lower, and low), as shown in Table 6. As suggested by Table 6, the standard values of social performance evaluation should be over 0.6, which indicates that the construction projects are contributory to the social sustainability.

Table 5. The weight for each indicator in social performance evaluation system.

A	B_1	B_2	B_3	B_4	B_5	Weight
	0.442	0.252	0.063	0.129	0.114	
C_{11}	0.182					0.080
C_{12}	0.512					0.226
C_{13}	0.084					0.037
C_{14}	0.223					0.099
C_{21}		0.533				0.134
C_{22}		0.237				0.060
C_{23}		0.140				0.035
C_{24}		0.098				0.025
C_{31}			0.406			0.026
C_{32}			0.322			0.020
C_{33}			0.271			0.017
C_{41}				0.209		0.027
C_{42}				0.391		0.050
C_{43}				0.401		0.052
C_{51}					0.134	0.015
C_{52}					0.383	0.044
C_{53}					0.197	0.022
C_{54}					0.286	0.033

Table 6. Weighted and standard values of the magnitude of construction redevelopment project.

Influential Degree	High	Relatively High	Average	Relatively Low	Low
Weighted Values	1.0	0.8	0.5	0.2	0
Standard Values	0.900~1.0	0.6~0.899	0.4~0.599	0.2~0.399	0~0.2

Notes: (1) Influential degree of social performance can vary according to the need of actual classifications, which is generally divided into five scales; and (2) weighted values and standard values are all determined according to experiences.

To provide a comprehensive view about social performance of the hospital redevelopment project, a comparative analysis of Schemes 1 and 2 is presented in this paper. Ten invited experts were asked to score the social performance indices of Schemes 1 and 2 based on Table 4. The mean values of these indicators in both scheme were calculated and the membership degree of 18 two-level indices was obtained, as shown in Table 7.

Table 7. Membership degree of social performance evaluation of both scheme.

Indicator	Membership Degree		Indicator	Membership Degree	
	Scheme 2	Scheme 1		Scheme 2	Scheme 1
C_{11}	0.911	0.256	C_{32}	0.867	0.156
C_{12}	0.722	0.244	C_{33}	0.756	0.111
C_{13}	0.600	0.156	C_{41}	0.433	0.533
C_{14}	0.533	0.300	C_{42}	0.778	0.211
C_{21}	0.756	0.211	C_{43}	0.633	0.278
C_{22}	0.689	0.244	C_{51}	0.467	0.300
C_{23}	0.533	0.244	C_{52}	0.622	0.722
C_{24}	0.567	0.089	C_{53}	0.466	0.567
C_{31}	0.911	0.022	C_{54}	0.700	0.811

The synthetized evaluation matrix (R) of both scheme could be developed according to Table 7. By taking the advantage of Table 3, total sequencing weight sets of social performance (*W*) can be obtained. Therefore, social performance of each scheme could be obtained as below.

$$Z = W \cdot R = \begin{bmatrix} 0.080 \\ 0.226 \\ 0.037 \\ 0.099 \\ 0.134 \\ 0.060 \\ 0.035 \\ 0.025 \\ 0.026 \\ 0.020 \\ 0.017 \\ 0.027 \\ 0.050 \\ 0.052 \\ 0.015 \\ 0.044 \\ 0.022 \\ 0.033 \end{bmatrix}^T \cdot \begin{bmatrix} 0.911 & 0.256 \\ 0.722 & 0.244 \\ 0.600 & 0.156 \\ 0.533 & 0.300 \\ 0.756 & 0.211 \\ 0.689 & 0.244 \\ 0.533 & 0.244 \\ 0.567 & 0.089 \\ 0.911 & 0.022 \\ 0.867 & 0.156 \\ 0.756 & 0.111 \\ 0.433 & 0.533 \\ 0.778 & 0.211 \\ 0.633 & 0.278 \\ 0.467 & 0.300 \\ 0.622 & 0.722 \\ 0.466 & 0.567 \\ 0.700 & 0.811 \end{bmatrix} = [0.692, 0.285]$$

The social performance of Schemes 2 and 1 are 0.692 and 0.285, respectively, which shows social performance of the hospital redevelopment project of Scheme 2 is much higher than that of Scheme 1. In addition, the result of Scheme 2 (0.692) is relatively high according to membership degree given by Table 6. It also suggests that the construction of NBXC project (Scheme 2) is a successful project from the perspective of social aspect and would contribute to the social sustainability of the society.

6. Discussion

This research proposed 18 indicators for social performance evaluation of the hospital redevelopment project from five dimensions. Ten experts were asked to contribute their expertise to evaluate the social performance of two proposed schemes based on FAHP method. The result of comparative analysis shows the development of NBXC hospital project has a relative high social performance and it would contribute to project success and social sustainability. A modified K-chart was employed to help the discussion of research results. In Figure 3, the shaded rectangle shows the indicators are positively associated with development of NBXC hospital project, where social performance of Scheme 2 is higher than that of Scheme 1, while the white rectangle indicates the negative relationship between indicators and development of NBXC hospital project, where social performance of Scheme 1 is higher than Scheme 2.

As shown in Figure 3, social performance of NBXC hospital project is positively reflected by socio-economy development, socio-environment development and social flexibility, as well as the development of public service. The increase of employment rate (C_{11}) and satisfaction of stakeholders (C_{31}) rank at the top among all indicators. The development of NBXC hospital helps to deliver more infrastructure (i.e., hospital, transportation infrastructure, etc.) and enhance the public service (C_{22}) which would create more job opportunities and meet the demands of diverse stakeholders, for example, the patients (C_{33}) as end-users are also satisfied by the development of NBXC hospital projects. To develop this project, the military airport and nearby resident communities were relocated. Local government made a sound plan for the land use (C_{13}), e.g., business center and subway station were built nearby NBXC hospital, which helps to increase the value of the nearby land and redevelopment the landscape of the city (C_{51}), and further stimulate regional economic development (C_{12}), adjust the industrial structure (C_{14}), and improve the living standard of the local residents (C_{21}). The newly built hospital also encourages research and development (R&D) in medical technologies and education (C_{42}). Medical service provided by NBXC hospital project would enhance social security (C_{23}) and improve emergency healthcare service ability (C_{43}) of this region. However, NBXC hospital as the first tier hospital in China would attract patients of other regions of the country, which may lead to risks of exposure to diverse disease and challenge healthcare and disease prevention (C_{41}).

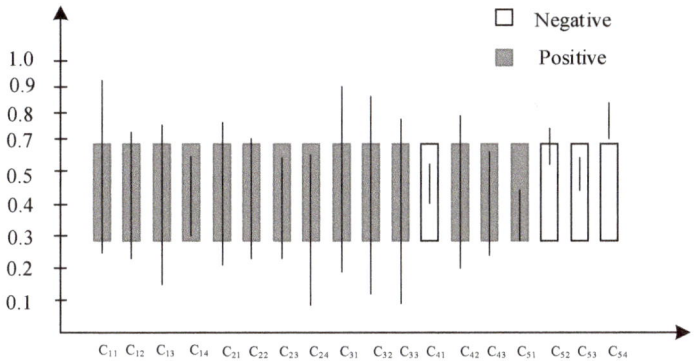

Figure 3. A modified K-chart of the research results.

The development of NBXC hospital project would negatively influence the society from the dimension of environment and resource conservation. The lifecycle of NBXC hospital project would consume enormous resource (C_{53}), and generate waste (C_{52}) (e.g., construction waste and medical waste), which would bring burden to the environment (C_{53}).

The scores of social performance evaluation of two proposed schemes show Scheme 2 would bring more social wellbeing to the society than Scheme 1 for the hospital redevelopment project. The score of social performance evaluation using Scheme 2 is 0.692, which also indicates relative high social performance of NBXC hospital project. Therefore, NBXC hospital project could be claimed to be a successful project from the social perspective and it will contribute to the sustainability of the society.

7. Conclusions

The concept of sustainable development suggests interactions of economic, environmental, and social dimensions [2]. However, social sustainability has received less appreciation than economic and environmental dimensions [28]. In the construction industry, social performance of construction projects not only contributes to social sustainability but also is critical for project success [19]. Evaluation of social performance of construction projects would help decision makers when proposing a project and project managers continuously improve social performance of construction projects.

This study developed a systematic framework for social performance evaluation of construction projects with 18 indicators developed from previous studies and a real-world hospital redevelopment project. These indicators help to evaluate social performance of the hospital redevelopment projects from five dimensions: socio-economy development, socio-environment development, social flexibility, public service development, and environment and resource conservation. In this research, public service delivered by hospital may vary from other types of infrastructure, e.g., schools and subways. However, the proposed framework could also be employed as reference to evaluate social performance of construction projects. While social impacts are intangible, and prior social performance evaluation using indicators are time-consuming and relatively subjective [13], FAHP method was introduced in this research to improve the efficiency and reduce the objectivity of the evaluation process to a certain degree.

The empirical study helps to showcase the processes to evaluate social performance of construction projects using the proposed framework and FAHP-based method. More importantly, the empirical study also helps to demonstrate how to improve social performance of construction projects based on the evaluation results. Specifically, the development of NBXC hospital project would improve the social sustainability of the society from perspectives of socio-economy development, socio-environment development, and social flexibility. Meanwhile, the development of NBXC hospital could also improve the landscape of the city due to a sound plan. However, the environment and resource conservation may be negatively impacted by NBXC hospital. As a hospital project, special considerations should be given to the associated risks of healthcare and disease prevention. By improving waste management, reducing resource depletion, and enhancing environmental protection, social performance of NBXC hospital project could be further improved.

However, this research was also subject to some limitations. For example, as stated by Vanclay [13], social impacts of construction projects are strongly correlated to the project context, while indicators for social performance evaluation are strongly dependent on the characteristics of construction projects. Therefore, future research should extend the flexibility and ensure the validity of the proposed framework.

Author Contributions: The author X.X., Y.T., and J.Y. drafted the manuscript, X.X., J.Y., and Q.L. designed the framework and research methodology. X.X. and Y.T. assisted in data collection and data analysis. T.C. and P.L. helped to polish the language. All authors have read and approved the final version of the manuscript.

Funding: The authors wish to give their sincere gratitude to the National Natural Science Foundation of China (grant Nos. 51578144, 71472037, and 71671042); the Postgraduates' Science and Innovation Foundation of Jiangsu Province(grant No. KYLX_0206); and Priority Academic Program Development of Jiangsu Higher Education Institutions. The Program for Outstanding Young Teachers of Southeast University (2242015R30009) and the Fundamental Research Funds for the Central Universities are acknowledged for financially supporting this research.

Acknowledgments: The authors also express their gratitude to experts who participated in this research and provided their expertise, as well as the anonymous reviewers, who provided constructive suggestions for the authors to improve their research and modified this paper.

Conflicts of Interest: The authors declare no conflict of interest.

References

1. Sev, A. How can the construction industry contribute to sustainable development? A conceptual framework. *Sustain. Dev.* **2009**, *17*, 161–173. [CrossRef]
2. Shen, L.Y.; Hao, J.L.; Tam, V.W.Y.; Yao, H. A checklist for assessing sustainability performance of construction projects. *J. Civ. Eng. Manag.* **2007**, *13*, 273–281.
3. Taroun, A. Towards a better modelling and assessment of construction risk: Insights from a literature review. *Int. J. Proj. Manag.* **2014**, *32*, 101–115. [CrossRef]
4. Vifell, Å.C.; Soneryd, L. Organizing matters: How 'the social dimension'gets lost in sustainability projects. *Sustain. Dev.* **2012**, *20*, 18–27. [CrossRef]
5. Wang, Y.; Han, Q.; de Vries, B.; Zuo, J. How the public reacts to social impacts in construction projects? A structural equation modeling study. *Int. J. Proj. Manag.* **2016**, *34*, 1433–1448. [CrossRef]
6. Vanclay, F. Principles for social impact assessment: A critical comparison between the international and US documents. *Environ. Impact Assess. Rev.* **2006**, *26*, 3–14. [CrossRef]
7. Tilt, B.; Braun, Y.; He, D. Social impacts of large dam projects: A comparison of international case studies and implications for best practice. *J. Environ. Manag.* **2009**, *90*, S249–S257. [CrossRef] [PubMed]
8. Li, D.; Chen, H.; Hui, E.C.M.; Yang, H.; Li, Q. A methodology for ex-post assessment of social impacts of an affordable housing project. *Habitat Int.* **2014**, *43*, 32–40. [CrossRef]
9. Shi, Q.; Liu, Y.; Zuo, J.; Pan, N.; Ma, G. On the management of social risks of hydraulic infrastructure projects in China: A case study. *Int. J. Proj. Manag.* **2015**, *33*, 483–496. [CrossRef]
10. Becker, H.A. Social impact assessment. *Eur. J. Oper. Res.* **2001**, *128*, 311–321. [CrossRef]

11. Ahmadvand, M.; Karami, E. A social impact assessment of the floodwater spreading project on the Gareh-Bygone plain in Iran: A causal comparative approach. *Environ. Impact Assess. Rev.* **2009**, *29*, 126–136. [CrossRef]
12. IAIA, International Association for Impact Assessment. Available online: http://www.iaia.org (accessed on 28 June 2018).
13. Vanclay, F. Conceptualising social impacts. *Environ. Impact Assess. Rev.* **2002**, *22*, 183–211. [CrossRef]
14. Valdes-Vasquez, R.; Klotz, L.E. Social sustainability considerations during planning and design: Framework of processes for construction projects. *J. Constr. Eng. Manag.* **2012**, *139*, 80–89. [CrossRef]
15. Strand, R. A systems paradigm of organizational adaptations to the social environment. *Acad. Manag. Rev.* **1983**, *8*, 90–96. [CrossRef]
16. Wartick, S.L.; Cochran, P.L. The evolution of the corporate social performance model. *Acad. Manag. Rev.* **1985**, *10*, 758–769. [CrossRef]
17. Salazar, J.; Husted, B.W.; Biehl, M. Thoughts on the evaluation of corporate social performance through projects. *J. Bus. Ethics* **2012**, *105*, 175–186. [CrossRef]
18. Herd-Smith, A.; Fewings, P. *The Implementation of Social Sustainability in Regeneration Projects: Myth or Reality?* Royal Institution of Chartered Surveyors (RICS): London, UK, 2008; Available online: http://www.rics.org/site/scripts/download_info.aspx (accessed on 28 June 2018).
19. Liu, B.; Li, Y.; Xue, B.; Li, Q.; Zou, P.X.; Li, L. Why do individuals engage in collective actions against major construction projects?—An empirical analysis based on Chinese data. *Int. J. Proj. Manag.* **2018**, *36*, 612–626. [CrossRef]
20. Yuan, H. A model for evaluating the social performance of construction waste management. *Waste Manag.* **2012**, *32*, 1218–1228. [CrossRef] [PubMed]
21. Shen, L.-Y.; Tam, V.W.; Tam, L.; Ji, Y.-B. Project feasibility study: The key to successful implementation of sustainable and socially responsible construction management practice. *J. Clean. Prod.* **2010**, *18*, 254–259. [CrossRef]
22. Bassioni, H.A.; Price, A.D.; Hassan, T.M. Building a conceptual framework for measuring business performance in construction: An empirical evaluation. *Constr. Manag. Econ.* **2005**, *23*, 495–507. [CrossRef]
23. Tam, V.W.; Tam, C.; Zeng, S.; Chan, K. Environmental performance measurement indicators in construction. *Build. Environ.* **2006**, *41*, 164–173. [CrossRef]
24. Kagioglou, M.; Cooper, R.; Aouad, G. Performance management in construction: A conceptual framework. *Constr. Manag. Econ.* **2001**, *19*, 85–95. [CrossRef]
25. Ullmann, A.A. Data in search of a theory: A critical examination of the relationships among social performance, social disclosure, and economic performance of US firms. *Acad. Manag. Rev.* **1985**, *10*, 540–557.
26. Trop, T. Social Impact Assessment of Rebuilding an Urban Neighborhood: A Case Study of a Demolition and Reconstruction Project in Petah Tikva, Israel. *Sustainability* **2017**, *9*, 1076. [CrossRef]
27. Zuo, J.; Jin, X.-H.; Flynn, L. Social sustainability in construction–an explorative study. *Int. J. Constr. Manag.* **2012**, *12*, 51–63. [CrossRef]
28. Almahmoud, E.; Doloi, H.K. Assessment of social sustainability in construction projects using social network analysis. *Facilities* **2015**, *33*, 152–176. [CrossRef]
29. Liu, Z.-Z.; Zhu, Z.-W.; Wang, H.-J.; Huang, J. Handling social risks in government-driven mega project: An empirical case study from West China. *Int. J. Proj. Manag.* **2016**, *34*, 202–218. [CrossRef]
30. Bititci, U.S.; Carrie, A.S.; McDevitt, L. Integrated performance measurement systems: A development guide. *Int. J. Oper. Prod. Manag.* **1997**, *17*, 522–534. [CrossRef]
31. Bassioni, H.A.; Price, A.D.; Hassan, T.M. Performance measurement in construction. *J. Manag. Eng.* **2004**, *20*, 42–50. [CrossRef]
32. Yang, H.; Yeung, J.F.; Chan, A.P.; Chiang, Y.; Chan, D.W. A critical review of performance measurement in construction. *J. Facil. Manag.* **2010**, *8*, 269–284. [CrossRef]
33. Lewis, J. *The Project Manager's Desk Reference: A Comprehensive Guide to Project Planning, Scheduling, Evaluation, and System*; McGraw-Hill: New York, NY, USA, 2000.
34. Lin, G.; Shen, Q. Measuring the performance of value management studies in construction: Critical review. *J. Manag. Eng.* **2007**, *23*, 2–9. [CrossRef]
35. Walker, A. *Project Management in Construction*; John Wiley & Sons: Chichester, UK, 2015.

36. Dixon, M. *The Association for Project Management (APM) Body of Knowledge (BoK)*; Association for Project Management: High Wycombe, UK, 2000.
37. Nitithamyong, P.; Skibniewski, M.J. Success/failure factors and performance measures of web-based construction project management systems: Professionals' viewpoint. *J. Constr. Eng. Manag.* **2006**, *132*, 80–87. [CrossRef]
38. Flyvbjerg, B.; Bruzelius, N.; Rothengatter, W. *Megaprojects and Risk: An Anatomy of Ambition*; Cambridge University Press: Cambridge, UK, 2003.
39. Atkinson, R. Project management: Cost, time and quality, two best guesses and a phenomenon, its time to accept other success criteria. *Int. J. Proj. Manag.* **1999**, *17*, 337–342. [CrossRef]
40. Hinze, J.; Thurman, S.; Wehle, A. Leading indicators of construction safety performance. *Saf. Sci.* **2013**, *51*, 23–28. [CrossRef]
41. Suárez, E.; Calvo-Mora, A.; Roldán, J.L.; Periáñez-Cristóbal, R. Quantitative research on the EFQM excellence model: A systematic literature review (1991–2015). *Eur. Res. Manag. Bus. Econ.* **2017**, *23*, 147–156. [CrossRef]
42. Chan, A.P.; Chan, A.P. Key performance indicators for measuring construction success. *Benchmarking Int. J.* **2004**, *11*, 203–221. [CrossRef]
43. Saaty, T.L. *Decision Making for Leaders: The Analytical Hierarchy Process for Decisions in a Complex Work*; Lifetime Learning Publications: Maastricht, The Netherlands, 1981.
44. Zeng, J.; An, M.; Smith, N.J. Application of a fuzzy based decision making methodology to construction project risk assessment. *Int. J. Proj. Manag.* **2007**, *25*, 589–600. [CrossRef]
45. Haider, S.; Ahmad, J.; Ahmed, M. Identifying barriers to implementation of health promoting schools in Pakistan: The use of qualitative content analysis and fuzzy analytic hierarchy process. *Int. J. Adv. Appl. Sci.* **2018**, *5*, 56–66. [CrossRef]
46. Csutora, R.; Buckley, J.J. Fuzzy hierarchical analysis: The Lambda-Max method. *Fuzzy Sets Syst.* **2001**, *120*, 181–195. [CrossRef]
47. Lutz, W.; Sanderson, W.; Scherbov, S. The coming acceleration of global population ageing. *Nature* **2008**, *451*, 716. [CrossRef] [PubMed]
48. Calvo, F.; de Oña, J.; Arán, F. Impact of the Madrid subway on population settlement and land use. *Land Use Policy* **2013**, *31*, 627–639. [CrossRef]
49. Chang, Z.; Phang, S.-Y. Urban rail transit PPPs: Lessons from East Asian cities. *Transp. Res. Part A: Policy Pract.* **2017**, *105*, 106–122. [CrossRef]
50. Yuan, J.F.; Skibniewski, M.J.; Li, Q.; Shan, J. The driving factors of china's public-private partnership projects in Metropolitian transportation systems: Public sector's viewpoint. *J. Civ. Eng. Manag.* **2010**, *16*, 5–18. [CrossRef]
51. Mantzaras, G.; Voudrias, E.A. An optimization model for collection, haul, transfer, treatment and disposal of infectious medical waste: Application to a Greek region. *Waste Manag.* **2017**, *69*, 518–534. [CrossRef] [PubMed]
52. Momtaz, S.; Kabir, S.Z. *Evaluating Environmental and Social Impact Assessment in Developing Countries*; Elsevier: Waltham, MA, USA, 2013.

© 2018 by the authors. Licensee MDPI, Basel, Switzerland. This article is an open access article distributed under the terms and conditions of the Creative Commons Attribution (CC BY) license (http://creativecommons.org/licenses/by/4.0/).

Article

Measures and Steps for More Efficient Use of Buildings

Mattias Höjer [1] and Kristina Mjörnell [2,*]

1. Division of Strategic Sustainable Studies, Department of Sustainable Development, Environmental Science and Engineering, KTH Royal Institute of Technology, 100 44 Stockholm, Sweden; hojer@kth.se
2. RISE Research Institutes of Sweden, Eklandagatan 86, 412 61 Gothenburg, Sweden
* Correspondence: kristina.mjornell@ri.se; Tel.: +46-730-88-57-45

Received: 1 May 2018; Accepted: 8 June 2018; Published: 11 June 2018

Abstract: As urbanization continues and more people move into cities and urban areas, pressure on available land for new constructions will continue to increase. This situation constitutes an incentive to review the need for interior space and uses of existing buildings. A great deal can be gained from using existing buildings more efficiently instead of constructing new ones: Reduced resource usage during construction (investments, natural resources, and energy), operation, and maintenance; more activity per square meter of buildings creates a greater basis for public transport and other services; more intensive use of buildings creates a more vibrant city without building on virgin land. The aim of this paper is to initiate a discussion regarding how digitalization can affect the demand and supply of interior space in existing buildings and elaborate on how policy can support more resource-efficient uses of space. New activity-based resource measurements intended for use in buildings are proposed, and several principles that have the potential to decrease environmental impact through more efficient usage of space are outlined. Based on these ideas for encouraging the flexible use of building spaces that are facilitated by digitalization and the new measurement approaches, a four-step principle for construction is proposed: The first step is to reduce the demand for space, the second is to intensify usage of existing space, the third is to reconstruct and adapt existing buildings to current needs, and the fourth is to construct new buildings. Urging political, municipal, construction, and real-estate decision makers to contemplate this principle, particularly in view of the new conditions that digitalization entails, will lead to more sustainable construction and, in the long term, a sustainable built environment.

Keywords: resource use; energy use; interior space utilization; buildings; sharing; digitalization

1. Introduction

In 2007, more than half of the population of the world lived in cities. In 2014, this figure was 54.4% and is predicted to reach 66% in 2050 [1]. One effect of urbanization and more people moving into cities and urban areas is an increased need for land for new buildings. One method of achieving this is to expand the urban areas (urban sprawl), another is to make the city denser (densification). A third is to use the existing building stock more efficiently. The EU Directive on the Energy Performance of Buildings (EPBD) [2] requires important measures to reduce the Union's energy dependency and greenhouse gas emissions and Energy Roadmap 2050 [3] explores the challenges posed by delivering the EU's decarbonization objective while at the same time ensuring security of energy supply and competitiveness. Buildings are responsible for 40% of energy consumption and 36% of CO_2 emissions in the EU, and it is predicted that 50% of the building stock that will exist in 2050 will have been built before 1975 [4]. Therefore, more renovation of existing buildings has the potential to lead to significant energy savings, potentially reducing the EU's total energy consumption by 5–6% and lowering CO_2 emissions by approximately 5% [4].

During 2016–2018, the Swedish Government has been focusing on five "Innovation Partnership Programs" [5], one of those being "Smart cities". Ever since the Program started, one of the authors to this paper who is a member of the advisory board to that program, has been leading a working group with participants from academia, industry and public administrations on how to use buildings more efficiently. This paper partly stems from that group's work.

While the European construction sector was hit particularly hard by the financial and economic crisis of 2008 and is still suffering from this, the Swedish construction sector faces challenges in relation to its capacity to construct new buildings. According to a forecast from the National Board of Housing, Building and Planning, Sweden needs 600,000 new homes to be constructed before 2025, along with other facilities that will provide services to inhabitants such as schools, daycare centers, shops and recreation facilities as the population grows [6]. Sweden has some of the highest construction costs in the EU after Norway, Denmark and Switzerland according to European Construction Costs (ECC) [7] and its new housing lies primarily in the upper end of the cost bracket, making it near-impossible for many groups to enter the housing market [8,9]. It is not only a matter of financing and affordability of new construction, but also of decreasing the environmental impact of construction. It is high time that options for new buildings were explored more deeply. Sweden is among the countries with the highest use of space expressed as floor area per capita, with 58 m^2/capita (16 for service and 42 for residential purposes); this follows Denmark and Cyprus, with 77 and 59 m^2/capita respectively, and is almost three times the floor area per capita in Serbia with 22 m^2/capita and Romania with 24 m^2/capita [4,10]. This is an incentive to review the need for new buildings and uses of existing buildings. A more efficient use of space means a reduced demand for building space. In practice, this could have vast implications on regional planning and ideas regarding how to develop regions in a more environmentally sustainable manner.

It is becoming increasingly important to take the environmental impact of production into consideration. In order to reduce the environmental impact of buildings, the resources used in terms of energy to extract, refine, process, transport, and fabricate a material or products must also be taken into account. In recent years, many studies have shown that so-called 'embodied energy' accounts for 45% of the total energy of a low-energy building over a lifespan of 50 years [11]. An analysis of the life-cycle energy of 60 cases from nine countries showed that operating energy represents by far the greatest energy demand for a conventional building during its life cycle, whereas low-energy buildings result in a net benefit in terms of their total life-cycle energy demand, even though there is a slight increase in the embodied energy [12]. In the future, it will likely be more a rule than an exception to consider embodied energy when designing new buildings.

Meanwhile, digitalization is changing more or less every aspect of society. Among other things, this means that the ways in which interior spaces are used are changing, and will change even more in terms of both housing, where different services make it possible to share temporary homes, and the matching of supply and demand. Digitalization is also laying the foundations for various new ways of sharing offices and office desks. Taken as a whole, much is to be gained from using spaces more efficiently and, through digitalization, new opportunities for this are appearing.

Thus, it seems there are potential environmental, social and economic benefits in using buildings more efficiently, and that digitalization can give some tools making this more feasible that previously. This paper focuses on Sweden because of the circumstances in that country described above. Even if this paper stems from Sweden, the general ideas should be relevant in all nations. It may however be more relevant for nations with a relatively high amount of building area, measured as square meter per capita. Some of the benefits of using buildings more efficiently rather than building new ones are:

(i) Resource efficiency in construction can increase.
(ii) Energy use for heating and operation can be reduced.
(iii) More people per square meter creates better support for public transport and other services.
(iv) More intensive usage of buildings creates more vibrant neighborhoods without building on virgin land.

1.1. Aim

The aim of this paper is to explore a new principle for reducing the resource use of buildings, and to combine this with a structured approach to discussing how interior space can be more efficiently used. How can these two perspectives highlight new avenues for policy development and new areas of innovation? In relation to this, several issues involving how requirements on measures for energy use in buildings can be formulated to assist in the drive for decreasing greenhouse gas emissions for the building stock, and the opportunities for change that come with digitalization, are explored.

1.2. Background

The following subsections of this paper introduce three important principles relating to reduced use of resources, the sharing of interior space, and alternative measures for energy use in buildings. These three subsections are the foundations of the 'Results' section that they precede.

1.2.1. Principles for Reducing Usage of Resources

Several sectors and areas have developed principles for increasing resource efficiency. Within transport infrastructure policy, the Swedish Transport Administration has explored the 'four-step principle' [13,14], which states that possible improvements to the transport system should be considered in a specific order. The first step is to consider measures that can affect the demand for transportation and choice of mode of transport. The second is to consider measures that could lead to more efficient uses of existing infrastructure. The third is to consider limited reconstructions. The fourth, which is only to be considered if the first three are not sufficient, is new and large-scale investment in the construction of road infrastructure [13]. Another example of guiding resource use through general policies is the 'waste management hierarchy', which is part of the EU waste directive (see Figure 1) [15]. The hierarchy states that, in order to reduce the negative impact of waste, five steps should be taken. Step 1 states that waste should be avoided through measures relating to product development and design. Steps 2–5 relate to when a product has become waste and focus on ways of reusing, recycling, and, as a last resort, disposing of products.

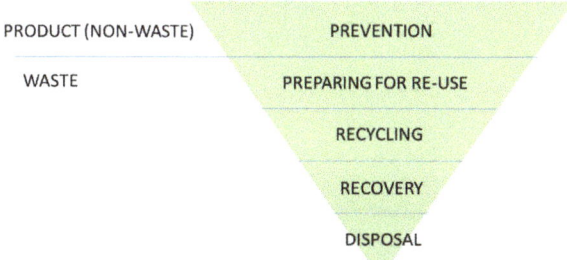

Figure 1. The 'waste management heirarchy'.

Within the area of energy use in buildings, there exist a number of strategies that resemble the 'waste management hierarchy' and the 'four-step principle' for transport. Berggren and Wall [16] identify four such strategies—'the Energy triangle', the 'Kyoto Pyramid passive energy design process ', the 'IBC Energy Design Pyramid,' and the 'passive house design principle'—to argue that reduced energy demand is the basis for such principles, positing better insulation and more efficient ventilation as possible solutions. Further measures to be taken after energy demand is reduced are use of renewable energy and clean fossil fuels insofar as possible, with smart technologies being used for both steering and heating.

1.2.2. Principles for the Sharing of Interior Spaces

An increased focus on sustainability and optimized resource usage has spurred interest in sharing things and spaces, leading to a focus on access to rather than ownership of e.g., transport modes and interior spaces.

Digitalization can assist in ensuring an increasingly efficient use of space. New sharing services have been introduced in recent years (see e.g., Reference [17]), and digital tools such as Airbnb (renting a home or property), Couchsurfing (sleeping on someone's sofa), Workaround (working in someone's office), and Coffice and Hoffice (working in a café or someone's home) allow pre-existing spaces owned by others to be utilized. These tools are part of a shift in approaches to how buildings can be used.

Different types of buildings are suitable for sharing to different extents on the part of different actors and activities. This has been studied by Brinkö et al. [18], who suggest a typology that classifies the options for sharing space and facilities within buildings for optimized use of existing buildings and minimizing the need for new ones. This was based on both literature (a large number of case studies) and interviews with stakeholders and resulted in the types of sharing being sorted into four groups. The first is sharing a specific facility—e.g., a desk or a workspace—within a semi-closed community, the second is sharing several facilities in an open or semi-closed community, the third is sharing a whole building or a physical space inside one within a closed community, and the fourth is sharing facilities between users in a network of buildings or organizations within an open, semi-closed, or closed community. These were then sorted according to a scale of sharing, from small to comprehensive. The characteristics of each type of sharing were defined by asking the following questions: What is being shared? When is it being shared? Why is it being shared? Who is sharing? How is it being shared? Digital services can be used to reveal underutilized buildings and offer space in the form of housing and premises to people and businesses who are moving into cities, thereby reducing environmental impact. CABE recommends that tenants and occupiers in office developments consider intensifying the use of space over time, taking into account increased internal and external mobility in order to achieve greater efficiency [19]. It seems like there are quite some ideas regarding how use of interior space can change, but it is not evident if the final outcome will be more or less demand for space.

1.2.3. Alternative Measures for More Efficient Energy Use in Buildings

For too long, there has been a too narrow focus on decreasing energy use in buildings by improving the energy efficiency of the existing building stock and creating new, low-energy buildings. The European building performance regulations are expressed in terms of kWh/m^2 heated area per year for buildings. This measure is advantageous for buildings that are used less frequently or by fewer users. Several recent studies, however, have explored energy use per person, rather than per square meter. Johansson [20] demonstrates that, as a performance indicator, energy use per resident is more relevant than energy performance per square meter, for all investigated building types. The average floor area per resident in multi-family buildings is lower than in 1–2 family buildings, while the energy use of many multi-family buildings is higher per square meter than that of 1–2 family buildings [15].

Sekki [21] proposes measuring energy efficiency in educational buildings by introducing indicators that take into account the number of users and the times at which a building is used. A number of new measures are suggested, taking schools as examples, including: Intensity of energy usage (energy consumption per actual number of schoolchildren using the building); specific energy consumption, adjusted for occupancy (total annual energy consumption per gross floor area times the ratio of actual daily use in relation to highest possible usage hours, i.e., 24); and specific energy consumption adjusted for usage and space efficiency (total annual energy consumption per gross area times a factor that includes the number of schoolchildren, average number of hours spent in the school per average space per schoolchild, and normal working hours—5.5 h for schools and 11.5 h for daycare centers).

2. Materials and Methods

This paper is the result of several different analyses, conducted using different research methods. First, a review of a number of approaches/tools/principles that are used as guides to reduce resource use within areas such as transport and waste was performed. Based on this, a new principle for reduced energy and resource use in buildings was developed and contextualized in relation to other relevant principles.

Secondly, several alternatives to the energy-per-square-meter metric were explored and reviewed. Based on this and a number of straight-forward examples of the forms that an intensification of space use can take, a basic categorization for the use of interior space was developed.

The combination of the principle and the categorization are proposed as something that can function as a basis for discussions regarding how space can be used more efficiently, with the potential benefits mentioned in the 'Introduction' section of this paper. Moreover, they are also proposed as potential catalysts for policy development and innovation in the 'Discussion' section, where a number of barriers to and drivers for more efficient use of interior space are summarized.

3. Results

3.1. The Four-Step Principle for Sustainable Use of Interior Space

The energy strategies in the 'Background' section are slightly less general in character than the transport and waste strategies presented. The transport and waste strategies begin with the causes of the resource use—the need for transportation and products, respectively—and then consider the aspect that is to be reduced (emissions and waste, respectively).

A corresponding approach in the energy strategies would have been to reduce the demand for climate-controlled space, and then to look at measures to reduce energy use per square meter, e.g., through insulation or more efficient heating and cooling technology.

Four steps for the sustainable use of interior space were tentatively suggested at a conference in 2017 [22], based on the four-step principle for transport infrastructure [13,14]. The four steps are further elaborated on here and contextualized in the setting of the principles for resources and energy usage in buildings, although these ignore the factor of demand for interior space. It is suggested that the four-step principle for interior spaces should be applied in any situation in which a demand for more interior space is identified. In such cases, the strategy for meeting demand should be based on a consideration of the following four steps, with the first step being the least energy- and resource-demanding:

1. Reduce the demand for space. A striking example of this is the reduction of postal and banking offices due to digitalization. Space in homes can be increased by replacing books and other physical media with digital equivalents. The storage needs of other objects could also be reduced through business-based models rather than ownership, and the highly efficient secondary market with various buy-and-sell services. Within industry, the need for remote trade is affected, and showrooms could be combined with efficient distribution via hubs to reduce the need for local space.
2. Intensify the use of existing space. If the same space requirement remains, increase space usage by using the space more intensively. This applies to office space as well as housing. With pop-up stores, it can also be applied to some extent to the retail sector. Recent decades have seen an intensification in the use of school buildings in Sweden through their being opened up to non-school activities outside of ordinary school hours. In addition, guest rooms can be shared by several households. Regardless of whether this happens on a global scale or in small groups such as within a city block, digitalization will contribute to the development of new opportunities. Households sharing space such as laundry rooms, kitchens, and social rooms allow the design and construction of small and cheap modern apartments.

3. Rebuilding can be a way of creating more useful space through the adapting of existing buildings, renovation of attics and basements into living spaces, and transformation of homes into offices or vice versa. It can also relate to creating flexible interior spaces that challenge the boundaries between housing, offices, and the retail sector. Furthermore, it is possible to create greater flexibility and, over time, adapt existing housing to changing lifestyles as the needs of households develop.
4. New construction is the last step but can be carried out in an advantageous fashion by building one or two floors onto existing buildings or parking garages, densifying the area without exploiting more virgin land at the expense of green areas. Even in new buildings, Steps 1 and 2 should be considered so that the space created can be used efficiently and is flexible in relation to future needs.

The four steps would precede the energy strategies mentioned in 1.2.1. Since they focus on the actual demand for space, whereas strategies such as the Energy triangle focus on reduced impact once the building has already been constructed.

3.2. Categorization of the Uses of Interior Space

The four-step principle above encourages greater co-use of space. This is a first step towards minimizing the environmental impact of buildings (used e.g., for housing, schools, office space). Rather than allowing large premises to be heated and ventilated to no avail, total energy use can be reduced by offering space in existing houses and premises.

In principle, the utilization of interior space can be increased by increasing the activities per space unit or per time unit. Various activities are suitable for different efficient uses of interior space. For some, it is easier to reduce the space per activity or service, while for others it may be more appropriate to use a large space for a more limited time. In the following, a categorization of different activities and services in terms of efficient use of space and their potential for sharing interior space is suggested. The uses of interior spaces for activities and services (with examples from housing, the retail sector, and offices) are divided into four categories:

1. Spacious interior spaces that are sparsely used: Retired couples ('empty nesters') with more than one house who spend considerable time abroad or at a summer house; retail buildings for the trading of bulky goods such as cars, furniture, etc.; offices with large rooms for each employee, irrespective of presence.
2. Dense interior spaces that are sparsely used: Homes for students, with minimal living space that is used mainly for sleeping; retail outlets for single goods with restricted opening hours; flexible office spaces with a low employee presence.
3. Spacious interior spaces that are frequently used: A widower remaining in a large old house or apartment, spending a lot of time at home; supermarkets with long opening hours; offices with large rooms for each employee and high presence.
4. Dense interior spaces that are frequently used: Homes for the elderly with minimal living space that are in use 24 h a day; co-operative housing with shared common rooms; service shops with long opening hours; flexible office spaces with high employee presence.

Considering these categories, using interior space more efficiently involves moving upwards and towards the right in Figure 2. Intensifying the use of interior space by densifying or extending the time of use can be achieved as follows:

Densifying use of interior space:

- Cooperative housing, sharing common areas such as a kitchen and living room, increases the number of people living in a specific space. At Tech Farm's property in Stockholm, 55 people live in 1100 m^2, which is 60% below the national average in terms of people per square meter.
- Rebuilding a house into a two-family house or generational housing (kangaroo living).

- Increasing the number of pupils in each classroom or children in each section of daycare centers.
- Open-plan offices without individual rooms.
- Several companies sharing meeting rooms, conference facilities, reception, lounge, canteen, dressing rooms, etc., within the organization or with other organizations.
- Retail outlets sharing showrooms for displaying limited collections within a small space, in an attractive location in which people can experience the products (touch, try, assess quality, colors etc.) and then order through e-shopping.

Extending the time of use:

- Gyms, shops, etc. being open 24/7.
- Shopping malls being open in the late evening or even at night during the week, which would be beneficial for those who work late or shifts and parents with small children.
- Activity-based flexible office spaces in which employees occupy a workspace only when at the office, and use different rooms depending on activities (meeting, desk work, phone calls, creative workshops).
- Activity-based schools without home classrooms.
- Sequential school use, i.e., schools used by one group of pupils for the first half of the day and by another for the second half.
- For sporting activities that require a certain space (athletics, tennis, ice hockey) and where there is no possibility to decrease the space or squeeze more people in at the same time, the time for which the space is used could be extended.
- Canteens, gymnasiums, and receptions could be shared between schools and homes for the elderly, as is practiced on the rural island of Casö, Finland.
- Churches that are used for a couple of hours a week by a small number of people could be used for other purposes during the rest of the week—as concert halls, conference centers, shelters for the homeless, and work hubs.
- Residential space is quite difficult to share, particularly over shorter time spans, although Airbnb has proven to be a successful way of sharing an apartment or house when it is not in use, as an alternative to hotels.
- Pop-up stores could be used to display goods in an attractive location for a limited time.
- Specially equipped rooms such as operating rooms, x-ray facilities, etc. could be operated 24/7, as is the case with delivery rooms (childbirth).

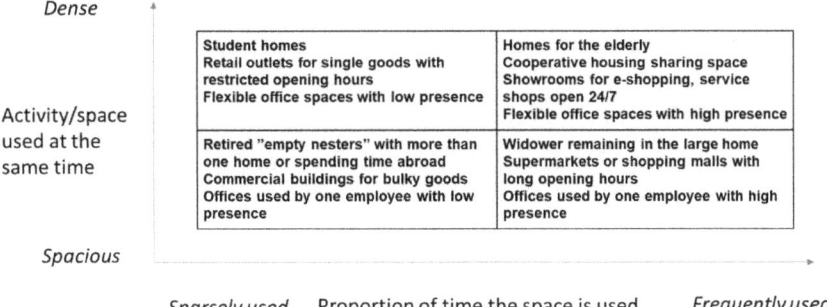

Figure 2. Categorization of use of interior space in terms of density and time of use.

It should be noted that the examples listed above function on the level of principles. The effect in practice will depend on several factors and must be thoroughly investigated. For example, the effect of

Airbnb has generally not been found to be an intensification of housing but an inflation in the housing market, resulting in loss of housing and more accommodation for visitors [23].

4. Discussion

This is a conceptual paper pointing at some new approaches to energy use in buildings. We have not provided empirical findings on effects of new measures or strategies. Rather, we are bringing thoughts from other areas into the field of energy use for buildings, with the ambition to encourage further research and empirical studies based on those thoughts.

The four-step principle developed is similar to other approaches to using resources more efficiently, such as the four-step principle for transport infrastructure. However, there are also differences, the primary one being that the four steps for buildings are more strongly interconnected and so should be regarded simultaneously. By designing with the lowest need for interior space possible in mind during the planning stage, a new building can be more efficiently used. In the case of the increasing demand for housing found in growing contemporary cities, all four steps may need to be applied at the same time. This article has presented the principle as a series of steps in order to emphasize the potential of using existing buildings more efficiently.

Giving too much consideration to the average use of space per person can be risky, as it may lead to the incorrect conclusion that everyone has surplus space in their home. Evidently this is not the case, particularly in urban areas, and factors other than the positive ones must be considered when looking at space-utilization opportunities. Table 1 summarizes several drivers of and barriers to the intensified use of interior space.

Table 1. Drivers and barriers to intensified use of space in household, company, school, and retail contexts.

	Densifying Activities (More Activities in the Same Space)	Extending the Time of Use (Use Space for a Greater Proportion of the Day, Week, Year)
Drivers (advantages)	Reduced environmental impact Cost reduction per activity/service Reduced cleaning costs for space that is not used	Reduced environmental impact Cost reduction per time of use Synergies between different users (shared equipment) Creating a vibrant atmosphere—'lively building' Increased safety and security as the building is in use more of the time
Barriers (disadvantages)	Crowded space Poorer air quality Need for more frequent cleaning Health problems Issues relating to territorial thinking, security, and privacy	Wear and tear Need for frequent refurbishment, leading to more material use and waste Complicated logistics Collisions between activities Competition for the most popular time slots Lack of demand during odd hours and off-season Need for staff 24 h a day and all year Less control over availability Issues relating to territorial thinking, security, and privacy Lack of incentives for profitability

During their lifetimes buildings are heated, ventilated, illuminated, cleaned, maintained, etc., entailing both costs and energy usage, which in turn have an impact on the environment. But do the regulations on energy use in buildings really encourage efficient use of energy in relation to the entire building stock? Although all new buildings are very energy efficient, any new buildings with heated spaces contribute to an increase in total energy usage. If buildings are used more frequently and by many users, the energy use will probably be slightly higher due to increased ventilation, lighting, consumption of hot water, etc. This has been observed in many multi-family housing areas built during the 1960s and 1970s, due to both poor energy performance and the fact that many people live in each apartment. There is undoubtedly a need to reduce energy use in these areas, but the living conditions in overcrowded apartments in socioeconomically fragile areas need to be considered. We also need to look closely at the buildings that use the most energy per person residing in each area. It may be wiser to develop policies towards resource-intensive households that have the ability to fund energy-efficiency measures to decrease their energy usage, and to encourage households that inefficiently use their living spaces to become more efficient. It is more important to reduce overall energy use than to reduce energy use per square meter.

Using indicators that represent more relevant measures of energy efficiency in buildings would be both fair and sustainable. There are, however, a number of issues that should be considered before any such measures are implemented.

Establishing the kWh-per-service or -per-activity metric, wherein a residence, workplace, school premises for a single pupil or student, daycare for a child, home for an elderly person, or place for giving birth is considered to be a service or activity, would encourage people to squeeze into smaller spaces, which means densification. The disadvantages to this are overcrowding, which can result in perceived congestion, poor air quality and health, and infections.

Establishing the kWh-per-time-used metric, wherein the built space is used for a larger proportion of the time of the day, week, or year, would encourage the use of buildings for more than one service or activity. This, however, would come with a demand on space in terms of time, which means a more intensive use of space due to the extended time of use.

The disadvantages of this are the increase in the wear and tear of the premises, as well as a lack of time for cleaning and maintenance, increased risk of collisions when changing activities, and higher incidence of double bookings and so on. The challenge would be to match different activities in terms of time, and to optimize the use of space over time. There could also be disagreements regarding interior design and furniture, limited storage possibilities, unclear division of responsibilities, and issues relating to the demands that contractual agreements place (or do not place) on co-users and co-occupants and systems of management, operation, booking, security, payment models, etc. Some premises are not attractive for use all of the time—hairdressers at night, theaters in the morning, etc.—but shopping malls, gyms, and laundry rooms could be used 24/7.

Premises for seasonal activities, such as skiing and beach resorts, could be used in other ways, such as for conferences, training courses, rehabilitation, etc. in order to extend the time of use.

As an incentive to decrease the use of interior space and hence the environmental impact of activities and utilities, this paper suggests that total GHG (greenhouse gas) emissions caused by energy use during the production of materials, transport, and operation of buildings per service provided (residence, education, work, shopping, recreation, etc.) is an accurate metric for measuring the sustainability of a building for cases in which the overall goal is to reduce GHG emissions.

Establishing a GHG-emissions-per-activity or -service metric may result in estimations of the lives of buildings that are greater than is realistic, ultimately entailing more maintenance and renovation than expected in order to continue the use of the building in the future. This would in turn increase the GHG value above the theoretical one, ultimately meaning that the initial assessment was misleading. There could also be difficulties in predicting GHG emission values for future energy production and operation of a building. Minimizing GHG emission per activity would favor the use of existing buildings as compared to new constructions.

The new dimensions and indicators that need to be developed can be based on new regulations and instruments, created for new business models for construction companies and property owners, and entail new concepts for designing housing and premises.

5. Conclusions

The aim of the research presented in this paper was to explore a new principle for reducing resource use in buildings, and to combine this with a structured approach to discussing how interior space can be used more efficiently. A four-step principle was developed further, involving (1) reducing demand for space; (2) intensifying usage of existing space; (3) rebuilding to create more useful spaces; and (4) constructing new, flexible houses. This will allow the environmental impact of the built environment and interior spaces to be decreased considerably.

A categorization of the use of interior space in terms of density and time of use has been presented. Measures for quantifying the environmental impact of buildings, expressed in terms of GHG emissions per activity rather than kWh/m^2, which would encourage decision makers to make more efficient use of interior space, have been presented.

By combining

- the four-step principle;
- the categorization of use of interior space;
- activity-based measures,

and suggesting approaches to matching the supply of spaces with demand that comes with the digitalization of society, we obtain an entirely new series of opportunities for policy makers (who wish to support reductions in energy and resource usage), innovators (who seek to find new areas to direct their attention towards), and business developers (who see the potential of new ways of providing space).

Researchers as well as decision makers in state functions, municipalities, and construction and real-estate companies are urged to contemplate the ingredients provided, particularly in view of the new conditions that digitalization entails. These can provide new recipes for lower resource usage, better premises, greater flexibility, and better meeting places. With this paper we have outlined a combination of ideas. Further steps include finding more data as well as exploring various ways of implementing the ideas both from a policy and from an innovation perspective.

Author Contributions: Concept, M.H. and K.M.; Methodology, M.H. and K.M.; Investigation, M.H. and K.M.; Analyses, M.H. and K.M.; Original draft preparation, M.H. and K.M.; Review and editing, M.H. and K.M.; Visualization, K.M.; Project Administration, M.H.

Funding: Costs for publishing the article was covered with the grant number 2013–1804 SIRen, the national research environment on sustainable integrated renovation, founded by the Swedish Research Council Formas.

Acknowledgments: We thank the reviewers and editors for providing constructive feedback on the manuscript.

Conflicts of Interest: The authors declare no conflicts of interest.

References

1. United Nations. *World Urbanization Prospects*; United Nations: New York, NY, USA, 2014; Volume 12.
2. European Parliament and Council. Directive 2010/31/EU of the European Parliament and of the Council of 19 May 2010 on the energy performance of buildings. *Off. J. Eur. Union* **2010**, *L153*, 13–35.
3. European Commission. *Energy Roadmap 2050*; COM(2011); European Commission: Brussels, Belgium, 2012; ISBN 9789279217982.
4. BPIA. *Europe's Buildings under the Microscope*; Buildings Performance Institute Europe (BPIE): Bruxelles, Belgium, 2011.
5. Swedish Government Innovation Partnership Programs. Available online: https://www.government.se/articles/2016/07/innovation-partnership-programmes--mobilising-new-ways-to-meet-societal-challenges/ (accessed on 10 June 2018).
6. Swedish Board of Housing Building and Planning. *Calculation of the Need of Housing Until 2025*; Swedish Board of Housing Building and Planning: Karlskrona, Sweden, 2017.
7. European Construction Costs Cost Index. Available online: http://constructioncosts.eu/cost-index/ (accessed on 10 June 2018).
8. Statistics Sweden Cost per sq. m. for newly Constructed Conventional Collectively Built One- or Two-Dwelling Buildings. Available online: http://www.scb.se/hitta-statistik/statistik-efter-amne/boende-byggande-och-bebyggelse/byggnadskostnader/priser-for-nyproducerade-bostader/pong/tabell-och-diagram/genomsnittlig-byggnadskostnad-per-kvm-bostadsarea-for-gruppbyggda-smahus.-ar-riket/ (accessed on 10 June 2018).

9. Swedish Board of Housing Building and Planning. *Swedish Building Costs in an International Comparison*; Swedish Board of Housing Building and Planning: Karlskrona, Sweden, 2014; ISBN 9789175630748.
10. Entranze Average Floor Area per Capita. Available online: http://www.entranze.enerdata.eu/#/average-floor-area-per-capita.html (accessed on 10 June 2018).
11. Thormark, C. A low energy building in a life cycle—Its embodied energy, energy need for operation and recycling potential. *Build. Environ.* **2002**, *37*, 429–435. [CrossRef]
12. Sartori, I.; Hestnes, A.G. Energy use in the life cycle of conventional and low-energy buildings: A review article. *Energy Build.* **2007**, *39*, 249–257. [CrossRef]
13. Swedish Government. *Infrastruktur för Framtiden—Innovativa Lösningar för Stärkt Konkurrenskraft och Hållbar Utveckling*; Swedish Government: Stockholm, Sweden, 2016.
14. Johansson, F.; Tornberg, P.; Fernström, A. A function-oriented approach to transport planning in Sweden: Limits and possibilities from a policy perspective. *Transp. Policy* **2018**, *63*, 30–38. [CrossRef]
15. European Parliament and Council. Directive 2008/98/EC of the European Parliament and of the Council of 19 November 2008 on Waste and Repealing Certain Directives. 2008. Available online: http://www.reach-compliance.eu/english/legislation/docs/launchers/waste/launch-2008-98-EC.html (accessed on 8 June 2018).
16. Berggren, B.; Wall, M. Calculation of thermal bridges in (Nordic) building envelopes—Risk of performance failure due to inconsistent use of methodology. *Energy Build.* **2013**, *65*, 331–339. [CrossRef]
17. Börjesson Rivera, M. *What Is a Sustainable Everyday Life?: Exploring and Assessing the Sustainability of Everyday Travel, Sharing and ICT, KTH, Strategic Sustainability Studies*; KTH Royal Institute of Technology: Stockholm, Sweden, 2018.
18. Brinkoe, R.; Nielsen, S.B. The characteristics to consider in municipal shared spaces. *J. Facil. Manag.* **2017**, *15*, 335–351. [CrossRef]
19. CABE. *The Impact of Office Design on Business Performance*; Commission for Architecture & the Built Environment and the British Council for Offices: London, UK, 2005.
20. Johansson, T.; Vesterlund, M.; Olofsson, T.; Dahl, J. Energy performance certificates and 3-dimensional city models as a means to reach national targets—A case study of the city of Kiruna. *Energy Convers. Manag.* **2016**, *116*, 42–57. [CrossRef]
21. Sekki, T.; Airaksinen, M.; Saari, A. Effect of energy measures on the values of energy efficiency indicators in Finnish daycare and school buildings. *Energy Build.* **2017**, *139*, 124–132. [CrossRef]
22. Höjer, M.; Ringenson, T.; Erman, M. International Conference Smart and Sustainable Planning for Cities and Regions 2015. In *Digitalisation for Environmental Sustainability in a Regional Context—A Four Step Principle*; EURAC Research: Bolzano, Italy, 2017; pp. 22–24.
23. Wachsmuth, D.; Chaney, D.; Kerrigan, D.; Shillolo, A.; Basalaev-Binder, R. *The High Cost of Short-Term Rentals in New York City*; School of Urban planning, McGill University: Montreal, QC, Canada, 2018.

© 2018 by the authors. Licensee MDPI, Basel, Switzerland. This article is an open access article distributed under the terms and conditions of the Creative Commons Attribution (CC BY) license (http://creativecommons.org/licenses/by/4.0/).

Article

Exploring Antecedents of Green Tourism Behaviors: A Case Study in Suburban Areas of Taipei, Taiwan

Judith Chen-Hsuan Cheng [1], Ai-Hsuan Chiang [2], Yulan Yuan [3] and Ming-Yuan Huang [4],*

1. Department of Applied Economics and Management, National Ilan University, Ilan City 26047, Taiwan; chengch@niu.edu.tw
2. Department of International Business, Ming Chuan University, Taipei 11103, Taiwan; eliot@mail.mcu.edu.tw
3. Department of Landscape Architecture, Tunghai University, Taichung 40704, Taiwan; yoyoyuan@thu.edu.tw
4. Department of Forestry and Natural Resources, National Chiayi University, Chiayi 60004, Taiwan
* Correspondence: myhuang@mail.ncyu.edu.tw; Tel.: +886-5-271-7518

Received: 26 April 2018; Accepted: 6 June 2018; Published: 8 June 2018

Abstract: Understanding user behaviors is the foundation to support the design and development of a sustainably built environment. This exploratory study used a mixed method to explore people's perception, motivation, intention, and behaviors of green tourism in Taiwan. The qualitative approach explored intrinsic and extrinsic factors that could influence people's intention to participate in green tourism. The quantitative approach provided evidence of influencing factors of green tourism. The findings suggested that variables, such as perception, attitudes, and self-efficacy, can indirectly influence green tourism behaviors through behavioral intention. This study suggests that government agencies should emphasize environmental education regarding the relationship between climate change and people's life; therefore, people will increase their environmental awareness regarding the urgent conditions of the environment, in addition to supporting green tourism and being more responsible for their tourism behaviors. For cities intending to accommodate tourism or Non-Governmental Organizations (NGOs) that are interested in promoting green tourism, it is critical to incorporate relevant factors, such as destination services and educational elements, into the design and development principles to built environment that supports green tourism activities.

Keywords: behavior; built environment; green tourism; intention; sustainability

1. Introduction

Over recent decades, global environmental change has been widely discussed, since many areas in the world are facing serious environmental problems. These problems are caused by industrial development, careless human behaviors, and unsustainable tourism behaviors. To help reduce these problems, every level, group, and society in the world should be involved in curbing these problems and the environmental design and planning for the tourism industry is no exception. Tourism activities directly and indirectly contribute to environmental issues because the use of airlines, tour buses, electricity, water, and many other resources requiring the consumption of natural resources and fossil fuels [1,2]. If the tourism industry continues to experience growth, it will exceed other industries and become the major source of global greenhouse gases (GHGs). To successfully achieve the reduction of GHGs, global policies, adopting environmentally friendly designs of the built environment, and a change in tourism travel behaviors are required. A number of international tourism organizations proposed their goals to reduce GHGs emissions. For example, the World Travel and Tourism Council [3] has used the CO_2 emissions of 2005 as a base line and, specifically, identified emission reduction goals of 25% to 30% by 2020 and 50% by 2035. The UK Department of Transport [4] identified sustainable low carbon travel in local areas. Some European countries and Australia have adopted methods to promote more sustainable travel programs for target populations [5]. Moreover, some studies have also suggested that people may decrease their intention to drive if the built environment can provide people with more alternatives to driving [6]. Based on the above discussion, tourism sectors and destination managers are aware of their responsibility to reduce carbon emission; therefore, the concept of green tourism has been promoted in many countries. To reduce the carbon emission resulting from the tourism industry, the Taiwan government has promoted green tourism in different regions. Various types of green tourism programs have been conducted in many counties in Taiwan [7]. The Green Tourism Association of Taiwan [8] defines green tourism as a travel mode that places an emphasis on people experiencing natural and cultural events, while leaving minimal carbon footprints and environmental impacts. Therefore, the main emphasis of green tourism is that energy consumption and CO_2 emissions, caused by the activities, products, and services of tourism, should be minimized [9,10]. However, some of these green tourism programs are not sustainable due to a lack of tourist participation. In other words, tourists might be hesitant about partaking in green tourism when they need to deal with public transportation, facing greater inconveniences, and paying more expensive travel costs [11,12].

To design better environments for green tourism promotion, it is important to understand the factors that influence tourists' intentions to participate in green tourism. Numerous studies explore tourists' willingness to participate in environmentally friendly behaviors [13,14]. However, Dolnicar, Crouch, and Long [13] suggested that most of the behavioral studies focus on ecotourism or nature-based tourism sectors instead of the general public; they argue that researchers should use samples from the general public to explore these tourists' characteristics and needs. The understanding of these needs will enable destination managers and planners to modify the built environment to encourage tourists to engage in environmental friendly behaviors and allow tourism suppliers to use those built environments to provide appropriate services and programs in the future. As such, it is imperative to explore the general public's perceptions, attitudes, and their current green tourism behaviors, as well as the factors that influence the practice of these behaviors. Moreover, most of the studies on behavioral theories focus on internal factors, such as personal attitudes and self-efficacy, that influence people's decisions to perform environmentally friendly behaviors [15,16] and rarely mention external factors, such as incentives or how the built environment can motivate people to execute certain environmental actions [17,18]. Therefore, it is imperative to investigate the external factors that could influence people's decisions concerning behaviors regarding green tourism. Based on the outlined reasons, the primary purpose of this study is to investigate the factors that influence people to participate in green tourism. The secondary purpose of this study is to explore how these factors influence people's intention to participate in green tourism. The third purpose of this research is to further explore factors that influence people's current green tourism behaviors.

First, we review the previous literature to form a basis for the investigation of possible factors that can predict tourists' intention and behaviors towards green tourism practices. We then use qualitative research methods to explore other factors that influence the general public's green tourism behaviors. Finally, we conduct quantitative measurement to explore the associations among green tourism behaviors and other factors.

2. Literature Review

In this section, we first reviewed the association between motivation and environmental behavior, followed by knowledge, perception, and environmental behaviors, and, lastly, attitudes, efficacy, intention, and environmental behaviors. The literature review explores the possible factors that could influence people's green tourism behaviors.

2.1. Motivation and Environmental Behaviors

Studies investigating motivation have long been a topic of interest in tourism and travel research. Motivation indicates people's psychological/biological needs and wants that can influence their decision-making [19,20]. Kasser and Ryan [21] described two types of motivation, which are intrinsic and extrinsic. Intrinsic motivation associates with internal feelings and instincts, such as the value or attitude toward actions, while extrinsic motivation relates to external reinforcement or social recognition, such as gaining benefits [21,22]. Previous studies showed that intrinsic motivation is a promising factor to guide environmental attitudes [23], as well as environmentally responsible behaviors [22,24,25]. This study aims to investigate the motivations that trigger people's intention to participate in green tourism behaviors. We hope to identify intrinsic and extrinsic motivations of green tourism behaviors to help providers of green tourism to understand people's needs, in addition to providing appropriate green tourism programs in the future. Therefore, we propose Hypothesis 1.

Hypothesis 1 (H1). *Motivation can positively influence people's behavior through behavioral intention.*

2.2. Knowledge, Perception, and Environmental Behaviors

Knowledge is thought to be the antecedent variable of behaviors and researchers have identified its direct and indirect influences on pro-environmental behaviors [16,26–28]. In environmental education, numerous researchers have indicated that knowledge is associated with conservation behavior [26,28]. Boubonari et al. [29] studied school teachers' knowledge, attitude, and behaviors regarding marine pollution and their findings suggested that knowledge is positively correlated with attitudes and behaviors, with the knowledge-attitude link being stronger than the knowledge-behavior link. Other researchers have identified a causal relationship between knowledge and behavior. For example, Frick, Kaiser, and Wilson [27] suggest that action-related knowledge and effectiveness knowledge (which addresses the relative effectiveness of conservation actions) have a direct effect on individuals' behavioral performance. Cheng and Monroe [16] studied children's affective attitudes toward nature and found that environmental knowledge has direct effects on children's affective attitudes, which can, in turn, influence their interests in environmentally friendly practices. Similarly, Carmi et al. [30] found that environmental knowledge can indirectly drive pro-environmental behaviors through environmental emotions. Based on the evidence of the discussed studies, environmental knowledge is a factor that is associated with the development of pro-environmental behaviors. As a result, we propose Hypothesis 2 and Hypothesis 3.

Hypothesis 2 (H2). *People's perception of sustainability can positively influence their attitudes toward green tourism*

Hypothesis 3 (H3). *People's perception of green tourism can have a positive influence on their attitudes toward green tourism.*

2.3. Attitudes, Efficacy, Intention, and Environmental Behaviors

Many researchers have explored pro-environmental behaviors to provide insights that explain, predict, and develop these behaviors [15,31–34]. Various theories have demonstrated the associations among social norms, values, attitudes, behavioral intentions, and actual behaviors [15,33,35]. These attitudes and behaviors are likely formed over time and it is possible to nurture these attitudes with environmental education programs. If programs promote environmental attitudes about behaviors, they may also succeed in increasing the intentions to perform these behaviors.

The theory of planned behavior is a classic framework that explains the relationship between predictive variables and behaviors [15]. The theory is composed of five basic concepts: Attitude toward the behavior, subjective norm, perceived behavioral control, intentions, and actual behavior, and suggests that a person's intention to perform a behavior is the direct determinant of the action. A person's intention is a function of three basic personal factors: Attitude toward the behavior, which determines a person's positive or negative evaluation of performing the behavior; subjective norm, which is a person's perception of the social pressures put on him/her to perform or not perform the behavior; and perceived behavioral control, which is a person's perception of his/her ability to perform a particular behavior. Eagly and Chaiken [36] indicated that attitude is a person's psychological inclination towards a particular entity. Numerous attitudinal studies have showed a positive relationship between environmental attitudes and behaviors. For example, studies suggest that pro-environmental attitudes were positively influential on people's willingness to pay [37,38]. Ajzen [15] indicated that perceived control is a similar form to measure one's capability to perform a certain action. Self-efficacy is considered to be an important predictor of environmentally responsible behaviors. Wood and Bandura [39] (p. 408) defined self-efficacy as the belief in one's own capacity to organize and guide the courses of action required to tackle certain situations in the immediate future. Previous studies have suggested that self-efficacy directly and indirectly influences residents' recycling behaviors in Spain [22]; self-efficacy strongly influences children's affective attitudes, as well as their intention to practice environmentally responsible behaviors [16]. The theory of plan behavior also suggests that intention can best predict a behavioral achievement. Therefore, we propose Hypothesis 4, Hypothesis 5 and Hypothesis 6.

Hypothesis 4 (H4). *People's attitudes towards green tourism can positively influence their intentions of participating in green tourism*

Hypothesis 5 (H5). *People's self-efficacy can have a positive impact on their intentions of participating in green tourism*

Hypothesis 6 (H6). *People's intention of participating in green tourism can positively influence their environmentally responsible green tourism behaviors.*

2.4. External Factors and Travel Behaviors

Some researchers have investigated the associations between the built environment and people's behaviors in residential areas. Handy, Cao, and Mokhtarian [6] studied the relationship between the built environment and travel behavior in the United States. Their finding demonstrated that, if residents were closer to their destinations and were provided with alterative travel modes, driving decreased. The results implied that the design of the built environment is influential on travel patterns in a given destination. Similarly, Titze, Stronegger, Janschitz, and Oja [18] explored the factors that influence people to use bicycles as their transportation mode in Austria. Their findings suggested that the condition of the physical environment can influence people's intention to bike. Based on the conclusions from the discussed studies, people tended to change their behaviors if there were appropriate designs. Therefore, this implied that the involvement of transportation and destination planners is critical during the planning stages of green tourism programs. Moreover, other studies suggested that nature-based recreation experiences could positively influence tourists' environmental attitudes, which further influences tourists' environmentally responsible behaviors. It is suggested that, if service providers can maintain destinations to high standards and provide positive experiences for tourists, tourists can, thus, develop positive environmental attitudes and behaviors and further contribute to the sustainable tourism industry [40].

The discussed studies present the theoretical background of factors that could influence behavioral intention and environmental behaviors. However, this study would like to examine whether there are other factors that could influence the general public's intention to participate in green tourism. Therefore, Study 1, presented hereafter, used a qualitative approach to explore motivations (both intrinsic and extrinsic factors) that could influence tourists; behavioral intentions regarding green tourism participation. Followed by Study 2, which uses a quantitative approach to explore how antecedent variables of green tourism participation influence people's green tourism behaviors.

3. Study 1

The researchers concluded that there was an insufficient understanding of factors that influence people's green tourism behaviors; therefore, it was necessary to perform an exploratory qualitative study to establish quantitative instruments that could be tested later in the quantitative stage. The specific research question for the qualitative initiative was: What are people's motivations to practice green tourism behaviors?

3.1. Study Area and Participants

The researchers selected the top five popular tourism sites that can be accessed by public transportation in New Taipei City, which are Tamsui, Jiufen, Houtung, Yingge, and Sanxia (Figure 1). The participants were 18 years or older and had travelled to these areas between March and April of 2013.

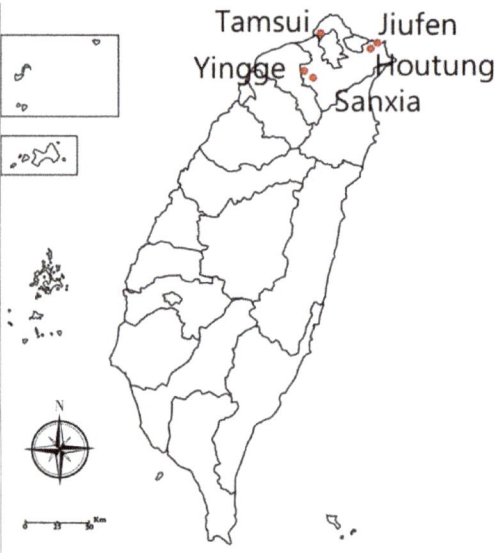

Figure 1. Interview sites.

3.2. Qualitative Initiative

During the first stage of the research, we used semi-structured interviews as the main technique to collect qualitative data. Convenience sampling was used to interview visitors. The interviewer selected the participants based on their availability. Each available participant was approached one at a time and he or she was asked of their willingness to be a research participant. After the interview with one participant was completed, the interviewer approached the next available participant. Before the interview, the researchers described the meaning of green tourism to the participants, and researchers used open-ended questions to explore the factors that could influence people's willingness of their participation in green tourism programs. The interview questions were: "Have you ever heard about green tourism?" and "What factors can influence your decision to participate in green tourism programs?" Each interview took about 8 to 10 min until the participants could not describe any new concepts.

3.3. Data Analysis

After the researchers approached 35 tourists, the data seemed saturated; therefore, we stopped the qualitative data collection. Each interview was transcribed and two researchers used Strauss and Corbin's [41] three coding steps, including open coding, axial coding, and selective coding, to explore factors that influence people's willingness to participate in green tourism.

3.4. Results

After combining items that have similar meanings and modifying the wording, two major types of factors were generated, including extrinsic factors and intrinsic factors, which are listed in Table 1.

Regarding extrinsic factors, many respondents identified an economic influence, as well as the availability of green alternatives at the destination. Some of the comments by participants were quoted below:

> It is important to have convenient public transportation. If the public transportation is not convenient or available, it is hard for people to participate in green tourism. (C6)

> If there is public transportation available between different destinations, I will take it because it is convenient. (D2)

Table 1. Factors that influence green tourism participation.

Selective Code	Axial Code	Description (Open Code)
Extrinsic factors	Economic influence	Reasonable travel costs (27)
	Destination built environment	Convenient public transportation (23) Prevalence of environmental friendly hotels (9)
	Destination experiences	The program should be interesting and attractive (16) Be close to the local environment (15) Destinations should have professional guided-services (9) To experience high quality interpretation (4)
Intrinsic factors	Social factors	To be with family and friends (10) To know people who have similar values and interests (4)
	Self-fulfilment factors	To try different types of tourism programs (11) Having environmental responsibility (8) Having environmental awareness (6) To learn new carbon reduction knowledge (5)

The participants' comments suggested that not only the availability, but also the convenience, of public transportation is critical for people to participate in green tourism. This is similar to a study presented by Larsen and Guiver [42], which reports that change towards more sustainable tourism mobility is likely to happen if tourists can value experiences at closer destinations with more available sustainable transportation.

Other respondents described the availability of environmental friendly hotels:

> If there are more green/environmental friendly hotels available, we are more likely to choose those types of hotels. (C2)

The above ideas indicate that, to promote green tourism, destinations should have sufficient facilities that could accommodate the ideas of carbon reduction. Other than destination facilities, respondents also spoke about their expectations of green tourism experiences. Their comments were as follows:

> I like traveling a lot, so if there are attractive green tourism programs or activities, I will likely participate. (A2)

> Different tourism sectors are very competitive. If you want people to participate in green tourism, the activities must be very interesting and attractive. (A4)

> I care a lot about professional guiding services, since I like to learn in-depth knowledge in the destinations. (A1)

The above quotes illustrate that, if the destinations can provide relevant services or experiences that are interesting to the visitors, people will be willing to participate in green tourism. This describes the extrinsic motivation of green tourism.

In addition to extrinsic factors, several respondents explained their motivation to participate in green tourism based on experiential factors. Some of them talked about their previous experiences. For example, one participant mentioned:

> I went to I-Lan for a green tourism program. We biked around the rice field, and had a series of in-depth environmental interpretation. The experience was memorable and interesting, so I will like to have similar experience again in the future. (C3)

From this participant, we found that experience was an important factor to attract tourists. Tourism sectors can provide memorable green tourism experiences to satisfy people. The above finding also implied that, if people have more environmental experiences, they would be more likely to participate in green tourism. Another finding was that tourists' own environmental awareness could motivate them to participate in green tourism. Some respondents pointed out:

> I am sensitive to environmental issues, so if there is a green tourism program available, I will like to participate. (A1)

> People who have interests to learn about the environment or who are having environmental responsibility will participate in these types of programs. (D2)

3.5. Discussion

We explored the factors that influence people's intentions to participate in green tourism. We found that both intrinsic and extrinsic factors can motivate people to participate in green tourism. To design and promote green tourism activities in a more effective way in the future, understanding the influential factors of green tourism behaviors is imperative. Therefore, the results of the qualitative study were used to develop a scale to reach a more general public and investigate the important factors that can influence people's willingness to participate in green tourism. Thirteen items were designed, and are listed in Table 2, based on the qualitative findings to represent intrinsic and extrinsic factors that motivate people to participate in green tourism.

Table 2. Factors that motivate people to participate in green tourism.

Items
Learning new ways of environmental conservation
Helping with environmental protection
Promotion of low carbon tourism
Awareness of environmental conservation
A skillful low carbon local tour guide
Professional low carbon guiding services
Convenient public transportation services
Attractive green tourism programs
Prevalence of green hotels
Reasonable travel cost in the destination
Diverse local cultural features
Experiencing different types of tourism
Leaving the crowed city environment

4. Study 2

The aim of Study 2 was to understand the predictors of green tourism intention and behaviors. The predictors include factors that are identified in literature review and were explored in the qualitative study. The research questions include: Can people's intention of participation in green tourism influence green tourism behaviors? Can perception, attitudes, self-efficacy, and other factors influence people's intentions to participate in green tourism?

The quantitative questionnaire included nine major sections. Section 1 included four questions that measured people's perception of sustainability. Section 2 included six questions that explored people's perceptions of green tourism. Section 3 was composed of four questions that measured people's self-efficacy, with statements such as, "I can reduce environmental impacts through participation in green tourism." Section 4 was composed of five questions that measured people's attitudes toward green tourism behaviors, with statements such as, "Green tourism can ensure the quality of the environment." Section 5 included 13 questions measuring factors that can influence people's willingness to participate in green tourism, which were developed from the qualitative study. Section 6 included one question measuring tourists' willingness to participate in green tourism in the future. The statement was, "I will follow green tourism principles when I am traveling". Section 7 included five questions that were designed following the Carbon Reduction Tourism principals provided by the Environmental Protection Agency (EPA) [43] in Taiwan. The rules included: People who participate in carbon reduction tourism should limit their footprint from transportation; they should carry light personal belongings; they should consume the local and seasonal food; they should leave no trace, etc." The example statement was, "I choose to take public transportation if available." All constructs were measured by using a 5-point Likert scale from 1, being strongly disagree, to 5, being strongly agree. The last section of the questionnaire was demographic information. Then the questionnaire was pilot tested with 60 participants to investigate the question wording and reliability. Two items from the perception of green tourism, "The carbon emission of different types of hotels is different" and "Taking public transportation can reduce carbon emission", were eliminated due to low reliability. The questionnaire items are translated in Appendix A.

4.1. Data Collection

Survey data were collected in the same five locations as visitor interviews were collected. The self-administrated questionnaire was handed to the participants. Each questionnaire took about 10 min to finish. Data were collected during June to August 2013. The study distributed 450 questionnaires and 53 were eliminated due to missing data, which left 397 surveys for data analysis.

4.2. Data Analysis

Upon completion of the data collection, data were analyzed using SPSS 20.0. Among the 397 effective samples, about 45% were men and 55% were women. The majority (78%) of tourists were from Northern Taiwan. The age distribution was 41% (20–29 years), 33% (30–39 years), 17% (40–49 years), 6% (50–59 years), and 3% (above 60 years). About 60% of visitors had a degree in college, followed by high school (19%), technical school (17%), and secondary level education (2.8%). Then data were analyzed using exploratory factor analysis to investigate factors that could predict the willingness of green tourism participation. Following the factor analysis, multiple regressions analysis was applied to discover the predictors of visitors' willingness to participate in carbon reduction tourism.

Factor analysis was used to explore the components of people's willingness to participate in green tourism behaviors. Thirteen items were factor analyzed. Researchers used the Kaiser-Meyer-Olkin (KMO) Measure of Sampling Adequacy to evaluate the appropriateness of the factor analysis. The results showed KMO = 0.72 (Barlett's test of sphericity = 1334.51, $p < 0.01$), which allowed us to further assess factor analysis. Two items with factor loading lower than 0.4 were eliminated [44], which left 11 items for the final data analysis. Three major categories were generated to represent factors that influence people to participate in green tourism behaviors: Self-environmental awareness, destination services/features, and professional interpretation (Table 3). Self-environmental awareness belongs to intrinsic motivation, while destination green service and professional interpretation belongs to extrinsic motivation.

Table 3. Motivation of green tourism participation.

Statements	Self-Environmental Awareness	Destination Green Services	Professional Interpretation
Learning new ways of environmental conservation	0.85		
Helping with environmental protection	0.78		
Promotion of low carbon tourism	0.75		
Awareness of environmental conservation	0.66		
Convenient public transportation services		0.76	
Attractive green tourism program		0.75	
Prevalence of green hotels		0.60	
Diverse local culture and features		0.55	
Reasonable travel costs in the destination		0.41	
Skillful low carbon local tour guide			0.91
Professional low carbon guiding services			0.90
Eigen values	3.28	1.86	1.40
Percentage of variance explained	24.26	18.29	16.86
Cumulative variance explained	24.26	42.55	59.41
Cronbach's alpha	0.78	0.87	0.63

4.3. Testing of Hypotheses

After factors analysis, motivation of green tourism participation was categorized into three factors; therefore, the previous Hypothesis 1, was modified to Hypothesis 1a, Hypothesis 1b and H1c. Professional interpretation can influence people's intention of participating in green tourism. The hypothetical model of this study is explained in Figure 2. The study used the statistical package, Statistical Product and Service Solutions (SPSS), as well as Analysis of MOment Structure (AMOS), as our data analysis tool and a 95% of confidence level was implemented for the statistical analysis.

Hypothesis 1a (H1a). *Self-environmental awareness can positively influence people's intention of participating in green tourism.*

Hypothesis 1b (H1b). *Destination green service can positively influence people's intention of participating in green tourism.*

Hypothesis 1c (H1c). *Professional interpretation can influence people's intention of participating in green tourism.*

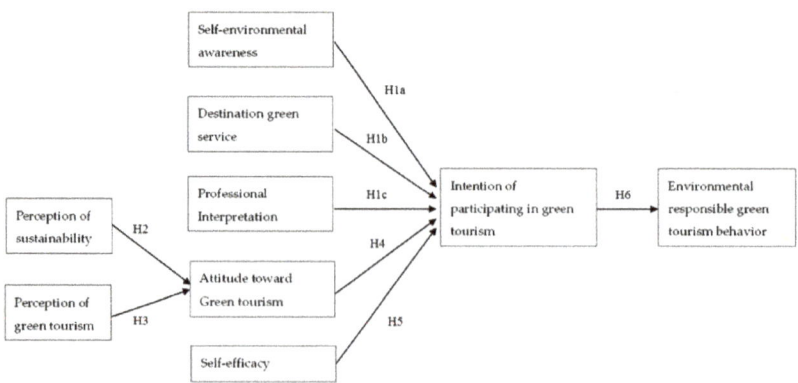

Figure 2. Proposed model of green tourism behaviors.

4.4. Results

The results were divided into two parts: The analyses of reliability and validity, and hypotheses testing. The following section shows the reliability and validity, as well as whether the results support our research hypotheses.

4.4.1. Reliability and Validity

A series of analyses were performed to test the reliability and validity of the constructs. Table 4 summarizes the descriptive statistics of the study variables. All correlations are significant at the 0.05 significance level. Besides, composite reliabilities are in italics in the diagonal. For composite reliabilities, a minimum value of 0.7 is considered acceptable.

The study also used confirmatory factor analysis (CFA) to test our measurement model. The fit indices of CFA were: $\chi^2/df = 2.685$; NNFI = 0.801; CFI = 0.826; AGFI = 0.810; SRMR = 0.066; RMSEA = 0.065. Therefore, our research model is acceptable, indicating convergent validity [45]. The confirmatory factor analysis confirmed if the items belong to the constructs. The results demonstrate construct unidimensionality.

The study examined discriminant validity among constructs using a procedure provided by Fornell and Larcker [46], whose criterion for discriminant validity is that the variance shared by a construct with its indicators should be greater than the variance shared with other constructs in the model. As shown in Table 4, the square root of the average variance extracted (ranging from 0.574 to 0.893) was greater than almost all of the corresponding correlations, indicating adequate discriminant validity.

The diagonal values in italics to the right of the slash are the square root of the average variance extracted (AVE) for each construct; the values to the left are the composite reliabilities.

Table 4. Descriptive statistics and correlations.

Variables	Mean (SD)	1	2	3	4	5	6	7	8	9
1. Perception of sustainability	3.095 (0.790)	0.691/0.631								
2. Perception of green tourism	3.761 (0.699)	0.314 **	0.801/0.681							
3. Attitudes toward green tourism	3.695 (0.549)	0.133 **	0.388 **	0.778/0.574						
4. Self-efficacy	3.386 (0.750)	0.126 *	0.404 **	0.528 **	0.693/0.644					
5. Self-environmental awareness	3.426 (0.691)	0.180 **	0.299 **	0.463 **	0.381 **	0.834/0.711				
6. Destination green service	3.864 (0.619)	0.124 *	0.126 *	0.218 **	0.302 **	0.327 **	0.703/0.589			
7. Professional interpretation	3.731 (0.855)	0.048	0.025	0.071	0.056	0.283 **	0.061	0.896/0.893		
8. Intention of participating in green tourism	3.496 (0.855)	0.131 **	0.291 **	0325 **	0.271 **	0.378 **	0.303 **	0.175 **	-/-	
9. Environmental responsible green tourism behavior	2.904 (0.699)	0.339 **	0.365 **	0.379 **	0.449 **	0.482 **	0.241 **	0.121 *	0.268 **	0.748/0.611

Note: $N = 398$. * Correlations at the 0.05 significant level, ** Correlations at the 0.01 significant level.

4.4.2. Hypotheses Testing

Table 5 presents the AMOS estimates for our propose model. The fit indices for the proposed model were: $\chi^2 = 1257.926$ (df = 441; $p < 0.001$); NNFI = 0.774, CFI = 0.799; AGFI = 0.797, SRMR = 0.081; RMSEA = 0.068. The fit indices suggest that the data fit our model well. The results reveal that environmental self-awareness significantly impacted on intentions of participating in green tourism ($\beta = 0.254$, $p < 0.001$). Meanwhile, destination green services are also significantly related to intentions of participating in green tourism ($\beta = 0.155$, $p < 0.001$). However, the results do not support the hypothesis that professional interpretation is positively related to intentions of participating in green tourism ($\beta = 0.035$, $p > 0.05$). Furthermore, the results show that perceptions of sustainability are not significantly related to attitudes toward green tourism ($\beta = -0.021$, $p > 0.05$). The results confirm that perceptions of green tourism are significantly related to attitudes toward green tourism ($\beta = 0.443$, $p < 0.001$). The results reveal that attitudes toward green tourism are significantly related to intentions of participating in green tourism ($\beta = 0.169$, $p < 0.05$). The results support the conclusion that self-efficacy is significantly related to intentions of participating in green tourism ($\beta = 0.307$, $p < 0.001$). Moreover, the results confirm that intentions of participating in green tourism significantly impacts on environmentally responsible green tourism behavior ($\beta = 0.966$, $p < 0.001$).

Table 5. Path estimates for the hypothesized model.

Path	Path Estimates
H1a: Environmental self-awareness→Intentions of participating in green tourism	0.254 ***
H1b: Destination green service→Intentions of participating in green tourism	0.155 ***
H1c: Professional interpretation→Intentions of participating in green tourism	0.035
H2: Perception of sustainability→Attitudes toward green tourism	−0.021
H3: Perception of green tourism→Attitudes toward green tourism	0.443 ***
H4: Attitudes toward green tourism→Intentions of participating in green tourism	0.169 *
H5: Self-efficacy→Intentions of participating in green tourism	0.307 ***
H6: Intentions of participating in green tourism→Environmentally responsible tourism behavior	0.966 ***

Note: * $p < 0.05$, *** $p < 0.001$. The fit indices: $\chi^2/df = 2.852$; NNFI = 0.774, CFI = 0.799; AGFI = 0.797, SRMR = 0.081; RMSEA = 0.068.

After hypotheses testing, we have several major findings. First, the perception of green tourism can significantly predict attitudes toward green tourism. Second, attitudes toward green tourism, self-efficacy, self-environmental awareness, and destination services can significantly predict intentions of participating in green tourism. Third, intentions of participating in green tourism can significantly predict environmentally responsible tourism behaviors.

4.5. Discussion

This study used both factors that were identified from the literature and the factors generated from qualitative results to predict people's willingness to participate in green tourism. Our study explored people's actual tourism behaviors and found that people who have stronger intentions to participate in green tourism are more likely to perform environmentally responsible travel behaviors. This finding is consistent with previous studies that intention is one of the strongest predictors of behaviors [15,47]. Also consistent with previous literature, perception, attitudes, and self-efficacy are also predictors of behavioral intentions [15,16,39].

One interesting finding of our study is that destination services, which is an external factor, can significantly influence people's intention to participate in green tourism. While previous literature has focused on people's internal motives of performing certain behaviors, our study explained that external factors can play an important role for people to perform certain behaviors. One possible explanation is that, if the destinations are attractive and have services that can be convenient or easy to use, people will be more willing to participate in green tourism, which is similar to previous suggestions in the literature, which indicate that external factors, such as economic, social, and cultural, and design of the built environment might influence people's behavioral decisions [6,17,18,28]. However, we further investigated the predictability of other independent variables of people's willingness to participate in green tourism programs and found that professional interpretation is not a significant predictor. The reason may be because people who want to participate in green tourism programs are more interested in local resources rather than professional guides and interpretation services.

5. Conclusions and Implications

This exploratory study used both qualitative and quantitative methods to find out factors that can influence people's willingness to participate in green tourism, as well as the factors that influence people's responsible green tourism behaviors. The findings show that people's own environmental awareness and destination services are influential on people's willingness to participate in green tourism. These findings provide suggestions for government agencies, tourism sectors, NGOs, and city planners and developers that want to promote green tourism programs in the future. We suggest that government agencies should emphasize environmental education regarding the relationships between climate change and people's lives; in doing so, people will increase their environmental awareness regarding the urgent conditions of the environment and will support green tourism efforts, as well as being more responsible in their tourism behaviors. For the tourism sectors or NGOs that are interested in promoting green tourism, it is critical to consider factors, such as destination services and educational elements, during the program developmental process, since our findings suggest that, if the destination can provide more green tourism services, people are more likely to participate in green tourism in the future. We also suggest that city planners and developers provide opportunities for public involvement and consider environmentally friendly designs during their planning processes so that they will know the actual demands for the users. For example, different stakeholders are able to participate in the planning stage and make it possible to have more convenient public transportation for green tourism promotion.

The quantitative results suggested that self-environmental awareness, self-efficacy, and destination services, as well as their attitudes toward green tourism, are critical factors for people to make decisions about participating in green tourism. The results are consistent with previous studies, which indicate that both intrinsic and extrinsic motivations were associated with environmental behaviors [48,49]. We suggest future research should use more in-depth qualitative research methods to explore people's motivations and thoughts regarding green tourism participation; therefore, countries that want to promote green tourism may benefit from these findings. This study was conducted in sub-urban areas in Taiwan, and, thus, we also suggest researchers should explore how these factors influence people to participate in green tourism programs in different cultures and countries. More diverse data could be beneficial for green tourism promotion all over the world.

6. Limitations

This study has several contextual and design limitations that may affect the results. The qualitative study used convenient sampling on site. Although, after interviewing 35 people, the data had been saturated, on-site interviews may have several limitations. First, people do not have a large period of time to explain their thoughts. Second, people do not have detailed ideas regarding green tourism so their responses may be limited. Therefore, we suggest further in-depth interviews can be conducted to explore the factors that influence people's decision to participate in green tourism programs. Third, this study investigated the general public's green tourism behaviors; some participants might not have in depth knowledge regarding green tourism, which might result in scoring low in the actual green tourism behaviors. A more precise questionnaire can be designed to investigate green tourism participants in the future.

Author Contributions: J.C.-H.C. led coordinated the research; M.-Y. Huang and Y.Y. reviewed relevant literatures and developed implications; J.C.-H.C. and A.-H.C. analyzed the data; All Authors contributed to the manuscript preparation.

Acknowledgments: The research is funded by the Ministry of Science and Technology, Taiwan, R.O.C. under Grant No. NSC 100-2511-S-032-005-MY2. The authors would like to thank for the former research assistant Ming-Hsun Yeh for helping to conduct the research and the funding support from National Science Council in Taiwan.

Conflicts of Interest: The authors declare no conflict of interest.

Appendix A

Section 1: Perception of sustainability	Mean	SD
I understand carbon footprint	2.86	1.17
I understand carbon cycle	2.77	1.15
I understand green consumption	3.14	1.01
I understand sustainable development	3.61	0.96
Section 2: Perception of green tourism		
Consuming local food helps reduce carbon emission	3.61	1.01
Reducing use of disposable products helps reduce carbon emission	3.93	0.83
Wearing lightweight clothing helps reduce carbon emission	3.60	0.91
Carrying your own toothbrush and towels helps reduce carbon emission	3.89	0.89
Taking public transportation can reduce carbon emission	4.01	0.82
The carbon emission of different types of hotels is different	3.90	0.73
Section 3: Self-efficacy		
I can reduce environmental impacts through participation in green tourism	3.74	0.89
I have a responsibility to reduce carbon emission during my travel journey	3.35	0.94
I have a responsibility to persuade those who are damaging natural resources	3.08	1.07
Carbon emission reduction from green tourism cannot be neglected	3.55	1.05
Section 4: Attitudes toward green tourism		
Green tourism can ensure the quality of the environment	3.60	0.83
Green tourism can promote the local tourism industry	3.51	0.79
Green tourism has educational meaning	3.87	0.91
Green tourism can enhance my travel experiences	3.72	0.75
Green tourism can enhance my knowledge of resource conservation	3.77	0.83
Section 5: Influencing factors of willingness of participation in green tourism		
Self-environmental awareness		
Learning new ways of environmental conservation	3.54	0.93
Helping with environmental protection	3.43	0.87
Promotion of low carbon tourism	3.40	0.80
Awareness of environmental conservation	3.33	0.95
Destination green service		
Convenient public transportation services	3.99	0.83
Attractive carbon reduction tourism program	3.91	0.89
Prevalence of green hotels	3.73	0.90
Diverse local cultural and features	3.82	0.90
Reasonable travel cost in the destination	3.85	0.96
Experience different types of programs (eliminate due to low factor loading)	3.65	0.83
Escaping from urban areas (eliminate due to low factor loading)	3.69	0.93

Professional interpretation		
Skillful low carbon local tour guide	3.66	0.94
Professional low carbon guiding services	3.80	0.88
Section 6: Intention of participation in green tourism		
I will follow green tourism principles when I am traveling	3.50	0.85
Section 7: Environmental responsible tourism behaviors		
I choose to take public transportation if available	3.10	1.06
I carry my own water bottles when traveling	2.85	1.08
I carry my own shopping bags when traveling	2.55	0.98
I carry my own toothbrush and towels when traveling	2.69	1.02
I choose to purchase souvenirs with less packaging	3.33	0.85

References

1. Becken, S.; Patterson, M. Measuring national carbon dioxide emissions from tourism as a key step towards achieving sustainable tourism. *J. Sustain. Tour.* **2006**, *14*, 323–338. [CrossRef]
2. Scott, D.; Peeters, P.; Gössling, S. Can tourism deliver its "aspirational" greenhouse gas emission reduction targets? *J. Sustain. Tour.* **2010**, *18*, 393–408. [CrossRef]
3. World Travel and Tourism Council (WTTC). *Leading the Challenge on Climate Change*; WTTC: London, UK, 2009.
4. Transportation for Transport. *Delivering Sustainable Low Carbon Travel: An Essential Guide for Local Authorities*; Crown Copyright: London, UK, 2009; ISBN 978-1-84864-040-5.
5. Brög, W.; Erl, E.; Mense, N. Individualised marketing changing travel behaviour for a better environment. Presented at the OECD Workshop: Environmentally Sustainable Transport, Berlin, Germany, 5 June 2002; pp. 3–12.
6. Handy, S.; Cao, X.; Mokhtarian, P. Correlation or causality between the built environment and travel behavior? Evidence from Northern California. *Transp. Res. Part D Transp. Environ.* **2005**, *10*, 427–444. [CrossRef]
7. Pingling County Government. Low Carbon Tourism. Available online: http://lowcarbon.epd.ntpc.gov.tw/dispPageBox/Tpclc/TpcCp.aspx?ddsPageID=TPCLCH11& (accessed on 14 July 2017).
8. The Green Tourism Association of Taiwan. Available online: https://www.facebook.com/groups/117446668294810/about/ (accessed on 2 June 2018).
9. Chiesa, T.; Gautam, A. *Towards a Low Carbon Travel & Tourism Sector*. World Economic Forum Report Prepared with the Support of Booz & Company. 2009. Available online: http://www.greeningtheblue.org/sites/default/files/Towards%20a%20low%20carbon%20travel%20&%20tourism%20sector.pdf (accessed on 14 July 2017).
10. Filimonau, V.; Dickinson, J.; Robbins, D.; Huijbregts, M.A. Reviewing the carbon footprint analysis of hotels: Life Cycle Energy Analysis (LCEA) as a holistic method for carbon impact appraisal of tourist accommodation. *J. Clean. Prod.* **2011**, *19*, 1917–1930. [CrossRef]
11. Nawijn, J.; Peeters, P.M. Travelling 'green': Is tourists' happiness at stake? *Curr. Issues Tour.* **2010**, *13*, 381–392. [CrossRef]
12. Horng, J.-S.; Liaw, Y.-J. Can we enhance low-carbon tour intentions through climate science or responsibility sharing information? *Curr. Issues Tour.* **2017**, 1–25. [CrossRef]
13. Dolnicar, S.; Crouch, G.I.; Long, P. Environment-friendly Tourists: What Do We Really Know about Them? *J. Sustain. Tour.* **2008**, *16*, 197–210. [CrossRef]
14. Han, H. Travelers' pro-environmental behavior in a green lodging context: Converging value-belief-norm theory and the theory of planned behavior. *Tour. Manag.* **2015**, *47*, 164–177. [CrossRef]
15. Ajzen, I. From intentions to actions: A theory of planned behavior. In *Action Control*; Springer: Berlin, Germany, 1985; pp. 11–39.
16. Cheng, J.C.-H.; Monroe, M.C. Connection to nature: Children's affective attitudes toward nature. *Environ. Behav.* **2012**, *44*, 31–49. [CrossRef]

17. Hunecke, M.; Blöbaum, A.; Matthies, E.; Höger, R. Responsibility and environment: Ecological norm orientation and external factors in the domain of travel mode choice behavior. *Environ. Behav.* **2001**, *33*, 830–852. [CrossRef]
18. Titze, S.; Stronegger, W.J.; Janschitz, S.; Oja, P. Association of built-environment, social-environment and personal factors with bicycling as a mode of transportation among Austrian city dwellers. *Prev. Med.* **2008**, *47*, 252–259. [CrossRef] [PubMed]
19. Dann, G.M. Tourist motivation an appraisal. *Ann. Tour. Res.* **1981**, *8*, 187–219. [CrossRef]
20. Uysal, M.; Hagan, L.A.R. Motivation of pleasure travel and tourism. *Encycl. Hosp. Tour.* **1993**, *21*, 798–810.
21. Kasser, T.; Ryan, R.M. Further examining the American dream: Differential correlates of intrinsic and extrinsic goals. *Personal. Soc. Psychol. Bull.* **1996**, *22*, 280–287. [CrossRef]
22. Tabernero, C.; Hernández, B. Self-efficacy and intrinsic motivation guiding environmental behavior. *Environ. Behav.* **2011**, *43*, 658–675. [CrossRef]
23. Chow, A.S.Y.; Cheng, I.N.Y.; Cheung, L.T.O. Self-determined travel motivations and ecologically responsible attitudes of nature-based visitors to the Ramsar wetland in South China. *Ann. Leisure Res.* **2017**, *25*, 1–20. [CrossRef]
24. De Young, R. New Ways to Promote Proenvironmental Behavior: Expanding and Evaluating Motives for Environmentally Responsible Behavior. *J. Soc. Issues* **2000**, *56*, 509–526. [CrossRef]
25. Deci, E.L.; Ryan, R.M. The "What" and "Why" of Goal Pursuits: Human Needs and the Self-Determination of Behavior. *Psychol. Inq.* **2000**, *11*, 227–268. [CrossRef]
26. Boerschig, S.; De Young, R. Evaluation of Selected Recycling Curricula: Educating the Green citizen. *J. Environ. Educ.* **1993**, *24*, 17–22. [CrossRef]
27. Frick, J.; Kaiser, F.G.; Wilson, M. Environmental knowledge and conservation behavior: Exploring prevalence and structure in a representative sample. *Personal. Individ. Differ.* **2004**, *37*, 1597–1613. [CrossRef]
28. Kollmuss, A.; Agyeman, J. Mind the Gap: Why do people act environmentally and what are the barriers to pro-environmental behavior? *Environ. Educ. Res.* **2002**, *8*, 239–260. [CrossRef]
29. Boubonari, T.; Markos, A.; Kevrekidis, T. Greek Pre-Service Teachers' Knowledge, Attitudes, and Environmental Behavior toward Marine Pollution. *J. Environ. Educ.* **2013**, *44*, 232–251. [CrossRef]
30. Carmi, N.; Arnon, S.; Orion, N. Transforming Environmental Knowledge into Behavior: The Mediating Role of Environmental Emotions. *J. Environ. Educ.* **2015**, *46*, 183–201. [CrossRef]
31. Kaplan, S. New ways to promote proenvironmental behavior: Human nature and environmentally responsible behavior. *J. Soc. Issues* **2000**, *56*, 491–508. [CrossRef]
32. Schultz, P. New environmental theories: Empathizing with nature: The effects of Perspective taking on concern for environmental issues. *J. Soc. Issues* **2000**, *56*, 391–406. [CrossRef]
33. Stern, P.C. New environmental theories: Toward a coherent theory of environmentally significant behavior. *J. Soc. Issues* **2000**, *56*, 407–424. [CrossRef]
34. Stern, P.C.; Dietz, T. The value basis of environmental concern. *J. Soc. Issues* **1994**, *50*, 65–84. [CrossRef]
35. Schwartz, S.H. Normative influences on altruism1. In *Advances in Experimental Social Psychology*; Elsevier: New York, NY, USA, 1977; Volume 10, pp. 221–279.
36. Eagly, A.H.; Chaiken, S. *The Psychology of Attitudes*; Harcourt Brace Jovanovich College Publishers: San Diego, CA, USA, 1993.
37. Hultman, M.; Kazeminia, A.; Ghasemi, V. Intention to visit and willingness to pay premium for ecotourism: The impact of attitude, materialism, and motivation. *J. Bus. Res.* **2015**, *68*, 1854–1861. [CrossRef]
38. Shin, Y.H.; Moon, H.; Jung, S.E.; Severt, K. The effect of environmental values and attitudes on consumer willingness to pay more for organic menus: A value-attitude-behavior approach. *J. Hosp. Tour. Manag.* **2017**, *33*, 113–121. [CrossRef]
39. Wood, R.; Bandura, A. Impact of conceptions of ability on self-regulatory mechanisms and complex decision making. *J. Personal. Soc. Psychol.* **1989**, *56*, 407–415. [CrossRef]
40. Lee, T.H.; Jan, F.-H. The effects of recreation experience, environmental attitude, and biospheric value on the environmentally responsible behavior of nature-based tourists. *Environ. Manag.* **2015**, *56*, 193–208. [CrossRef] [PubMed]
41. Strauss, A.; Corbin, J. *Basics of Qualitative Research: Procedures and Techniques for Developing Grounded Theory*; Sage: Thousand Oaks, CA, USA, 1998.

42. Larsen, G.R.; Guiver, J.W. Understanding tourists' perceptions of distance: A key to reducing the environmental impacts of tourism mobility. *J. Sustain. Tour.* **2013**, *21*, 968–981. [CrossRef]
43. Environmental Protection Agency. Low Carbon Sustainable Home Information Network. Available online: https://lcss.epa.gov.tw/LcssOldData/Default.aspx?Action=EBB53A55B1BBAF77 (accessed on 14 July 2015).
44. Chen, J.S.; Hsu, C.H. Developing and validating a riverboat gaming impact scale. *Ann. Tour. Res.* **2001**, *28*, 459–476. [CrossRef]
45. O'Leary-Kelly, S.W.; Vokurka, R.J. The empirical assessment of construct validity. *J. Oper. Manag.* **1998**, *16*, 387–405. [CrossRef]
46. Fornell, C.; Larcker, D.F. Evaluating structural equation models with unobservable variables and measurement error. *J. Mark. Res.* **1981**, 39–50. [CrossRef]
47. Lee, T.H.; Jan, F.-H. Ecotourism Behavior of Nature-Based Tourists: An Integrative Framework. *J. Travel Res.* **2017**, 0047287517717350. [CrossRef]
48. Cheung, L.T.; Chow, A.S.; Fok, L.; Yu, K.-M.; Chou, K.-L. The effect of self-determined motivation on household energy consumption behaviour in a metropolitan area in southern China. *Energy Effic.* **2017**, *10*, 549–561. [CrossRef]
49. Webb, D.; Soutar, G.N.; Mazzarol, T.; Saldaris, P. Self-determination theory and consumer behavioural change: Evidence from a household energy-saving behaviour study. *J. Environ. Psychol.* **2013**, *35*, 59–66. [CrossRef]

© 2018 by the authors. Licensee MDPI, Basel, Switzerland. This article is an open access article distributed under the terms and conditions of the Creative Commons Attribution (CC BY) license (http://creativecommons.org/licenses/by/4.0/).

Article

Analysis of the Land Use and Cover Changes in the Metropolitan Area of Tepic-Xalisco (1973–2015) through Landsat Images

Armando Avalos Jiménez [1,*]**, Fernando Flores Vilchez** [2]**, Oyolsi Nájera González** [2] **and Susana M. L. Marceleño Flores** [2]

1. Posgrado en Ciencias Biológico Agropecuarias, Unidad Académica de Agricultura, Universidad Autónoma de Nayarit (Autonomous University of Nayarit), Carretera Tepic-Compostela Km 9, Xalisco C.P. 63780, Nayarit, México
2. Secretaría de Investigación y Posgrado, Universidad Autónoma de Nayarit, Ciudad de la Cultura s/n, Col. Centro, Tepic C.P. 63000, Nayarit, Mexico; vilchez@hotmail.com (F.F.V.); oyolsi92@gmail.com (O.N.G.); smlmarcel@hotmail.com (S.M.L.M.F.)
* Correspondence: armand18_a@hotmail.com; Tel.: +52-311-122-88-48

Received: 21 April 2018; Accepted: 31 May 2018; Published: 4 June 2018

Abstract: Land use and cover changes (LUCC) have been identified as one of the main causes of biodiversity loss and deforestation in the world. Fundamentally, the urban land use has replaced agricultural and forest cover causing loss of environmental services. Monitoring and quantifying LUCC are essential to achieve a proper land management. The objective of this study was to analyze the LUCC in the metropolitan area of Tepic-Xalisco during the period 1973–2015. To find the best fit and obtain the different land use classes, supervised classification techniques were applied using Maximum Likelihood Classification (MLC), Support Vector Machines (SVMs) and Artificial Neural Networks (ANNs). The results were validated with control points (ground truth) through cross tabulation. The best results were obtained from the SVMs method with kappa indices above 85%. The transition analysis infers that urban land has grown significantly during 42 years, increasing 62 km^2 and replacing agricultural areas at a rate of 1.48 km^2/year. Forest loss of 5.78 km^2 annually was also identified. The results show the different land uses distribution and the dynamics developed in the past. This information may be used to simulate future LUCC and modeling different scenarios.

Keywords: Maximum Likelihood Classification; Support Vector Machines; Artificial Neural Networks; significant transitions; urban growth; Nayarit (Mexico)

1. Introduction

Terrestrial ecosystems are important components of nature since they have biological and functional effects on climate regulation, the hydrologic cycle and as a source of natural resources to satisfy human needs. However, during the last 300 years, the planet has suffered big transformations [1]. The ecosystems have been subject to accelerated processes of land use and cover changes (LUCC) [2], which have been identified as one of the main factors contributing to global environmental change [3–5], as a result of major current environmental problems [6] such as land loss and degradation, climate change, biodiversity loss, deforestation [7] and ecosystems fragmentation [4,8,9], which in turn cause loss of associated environmental services [10] to such a degree that more than half of the world's forest cover has been lost, and around 30% of these ecosystems face degradation processes. Anthropogenic activities are one of the main elements that contribute to land use changes [11].

Mexico is known as a megadiverse country, consisting of large diversity of organisms, landscapes and terrestrial ecosystems [12]. Forests, jungles and other natural vegetation are distributed all over the country [13] covering 74% of the national territory [14], about 146 million hectares [12]. The distribution of natural vegetation has been studied for monitoring LUCC, but most of the research conducted at a regional scale has focused on the analysis of losses in natural vegetation and deforestation [15,16]. The deforestation rates recorded show a difference from 260,000 ha/year to 775,000 ha/year (i.e., 2600 km^2/year to 7760 km^2/year) [15]. Since local scale studies on LUCC have been scarce [14], the present work was centered on monitoring LUCC locally, paying particular attention to the urban land use in the metropolitan area of Tepic-Xalisco as a starting point for further research on urban growth simulation and future scenarios design.

The advance in using Geographic Information Systems (GIS) and remote sensing techniques has proven to be very useful to get accurate and coherent information according to the spatial reality [16]; these tools are widely used to analyze the distribution, patterns and trends of the LUCC processes via different methods to obtain several land use classes in the territory, as well as diverse approaches to detect temporary differences, such as the traditional method of cross tabulation [17,18].

In this context, the objective of this study was to analyze the dynamics on urban LUCC at local scale. The methodology was developed through the analysis and processing of four Landsat satellite images corresponding to the years 1973, 1985, 2000 and 2015. To find the best fit and obtain the different land use classes, three supervised classification methods were applied: Maximum Likelihood Classification (MLC), Support Vector Machines (SVMs) and Artificial Neural Networks (ANNs). The results were validated with control points (ground truth). Then, to identify the significant transitions between different land uses—especially in the urban land use changes—losses, gains, changes and interchanges were obtained through the cross-tabulation matrix and according to the methodology of Pontius [19].

The remainder of the paper is organized into four sections. Section 2 describes the study area and the satellite images that were used to obtain different classes of cover and land use. Section 3 outlines the procedure that was followed to classify the satellite images using two supervised classification methods and LUCC analysis. Section 4 describes and discusses the results obtained with respect to other similar works. Finally, the conclusions are presented in the Section 5.

2. Materials and Methods

2.1. Study Area

The metropolitan area of Tepic-Xalisco is located in the central part of the state of Nayarit (Mexico), as presented in Figure 1. The study area comprises two of the main localities of the state that are linked by commercial and administrative activities through the Tepic-Xalisco highway, which in turn provoked a conurbation process that was formalized as metropolitan area in 2006 by the National Institute of Statistics and Geography (INEGI, for its abbreviation in Spanish), the National Population Council (CONAPO, for its abbreviation in Spanish) and the Ministry of Social Development (SEDESOL, for its abbreviation in Spanish).

The study area was delimited through a 900 km^2 quadrant (30 km × 30 km polygon) including the metropolitan area, a polygon wide enough to locally observe and analyze the processes of LUCC during a 42 year-period.

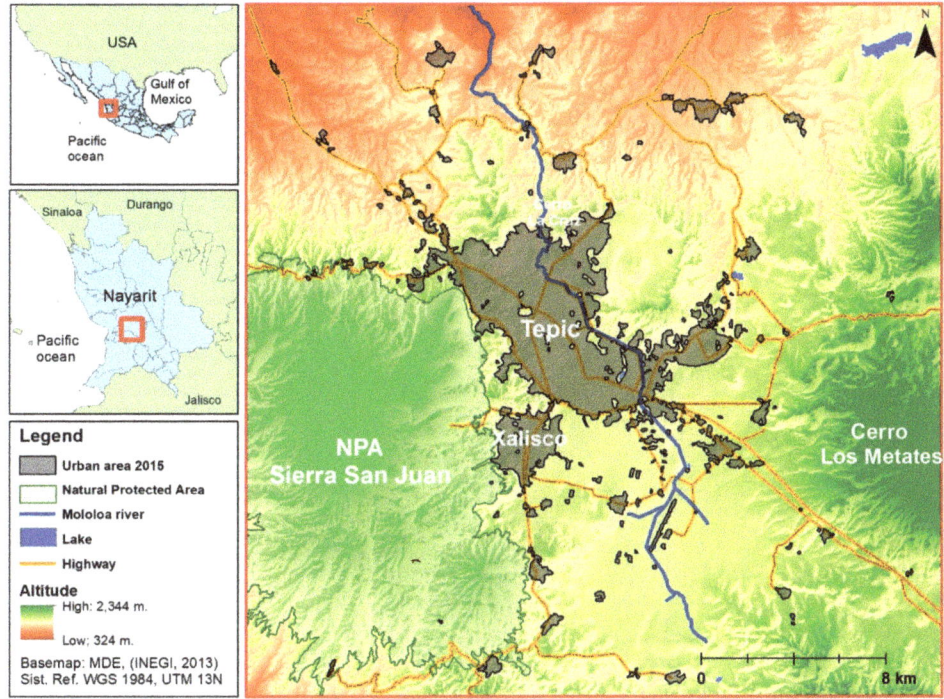

Figure 1. Localization and delimitation of the study area. Source: Own elaboration based on INEGI's data.

The surrounding zone of the study area contains a diversity of land use where are predominant intensive agricultural activities—mainly devoted to sugarcane, maize, mango and jicama crops—as well as farming activities. To the West of the study area is located the Natural Protected Area (NPA) Sierra de San Juan with over 50% of cover consisting of pine trees and live oaks forest, a great variety of natural resources that provide diverse environmental services susceptible of exploitation.

The metropolitan area has suffered important changes in land use as a result from urban growth. The forest is endangered by the indiscriminate clearcutting activities and mining performed in the east of the mountain, at the boundaries of the locality of Xalisco. Therefore, a historical analysis is necessary to show the distribution of land use and the change processes during a period of 42 years. Such information may be used to simulate future land use changes, such as the urban growth and modeling several scenarios.

2.2. Data

To set a temporary standardized thematic nomenclature for the analysis of the different land use, the four Landsat satellite images described in Table 1 were used. The images were taken from the United States Geological Survey (USGS) official website (http://glovis.usgs.gov).

Table 1. Landsat images used for mapping land uses in the study area.

Description	Image	Description	Image
Landsat 1 (1973) Multispectral Scanner System (MSS) Sensor LM10320451973043GDS03 Scene Spatial resolution 60 m Acquisition date 12 February 1973 Composition V-A-R		Landsat 5 (1985) Thematic Mapper (TM) Sensor LT50300451985139AAA03 Scene Spatial resolution 60 m Acquisition date 5 May 1985 Composition NIR-SWIR-R	
Landsat 7 (2000) Enhanced Thematic Mapper (ETM) Sensor LE70300452000045EDC00 Scene Spatial resolution 30 m Acquisition date 14 February 2000 Composition NIR-SWIR-R		Landsat 8 (2015) Operational Land Imager (OLI) Sensor LO80300452015062LGN01 Scene Spatial resolution 30 m Acquisition date 3 April 2015 Composition NIR-SWIR-R	

Source: United States Geological Survey (USGS) official website.

2.3. Methodology

Figure 2 shows the methodological process that was followed to analyze urban land use changes in the metropolitan area during a period of 42 years. First, supervised classification techniques were applied through Maximum Likelihood Classification (MLC), Support Vector Machines (SVMs) and Artificial Neural Networks (ANNs); from the preparation of the images (pre-processing) to the application of three supervised classification methods (processing), and validation of classifications (post-processing). Then, to detect changes between different land uses, the periods 1973–1985, 1985–2000, 2000–2015 and 1973–2015 were analyzed using cross tabulation. Finally, to identify significant changes, transitions analysis was conducted using the method of Pontius [19]. All processes were performed using GIS with ENVI 5.3, Arcgis 10.3 and Focus (PCI Geomatics, 2015) applications.

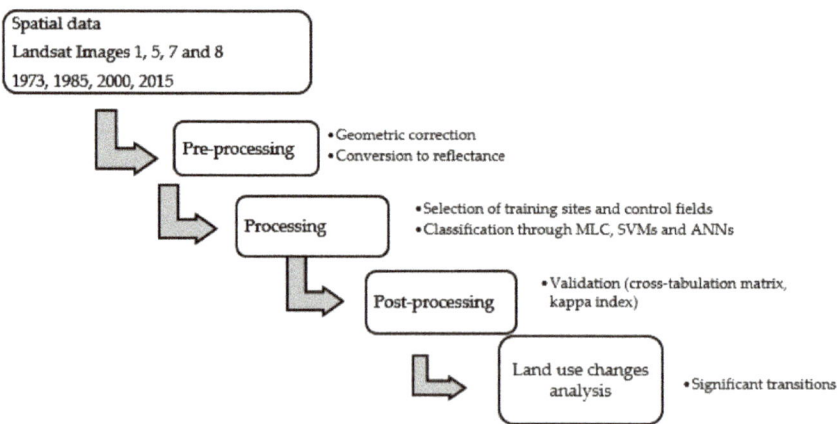

Figure 2. Landsat images classification methodology for LUCC analysis. Source: Own elaboration.

3. Satellite Images Classification

3.1. Pre-Processing

The scenes of Landsat images 1, 5, 7 and 8 were pre-processed to classify the different types of land use during the identification of changes in the urban areas for each analyzed period. Prior to images processing, the study area was cropped for each satellite image by means of a layer (30 km × 30 km polygon) wide enough to visualize urban land use changes in the metropolitan area during a 42-year period; the polygon was measured from the center of the urban land of the metropolitan area.

In the region of interest, it was necessary to verify and standardize the pixel size and the dimensions of each image by means of geometric correction in the WGS-1984 reference system, UTM projection in the zone 13 N. The validation of the geometric correction was obtained with the Mean Squared Error (MSE) for the control points, using one of the images as reference and comparing it (in pairs) to get the best geometric adjustment. To standardize a spatial resolution to 30 m, the four Landsat images were standardized by resampling the pixel size, especially the 1 and 2 Landsat images.

Spectral bands were selected for each image, in particular the ones with optical spectrum, Near Infrared (NIR) and Short-Wave Infrared (SWIR); the panchromatic bands were omitted due to high atmospheric influence, as well as the thermal bands, especially in the 7 and 8 Landsat images. To easily visualize the several classes of covers and land uses, a RGB composition in false color was used, highlighting the strong green areas with the NIR, SWIR and R combination; in addition, 2% of linear contrast highlighting was applied to enhance the visualization and identification of the training sites.

The conversion to reflectance was performed to obtain the terrestrial area spectral reflectance values for the different covers and land uses, giving the spectral value to each pixel. The conversion to reflectance was conducted considering the method of Chavez [20] through the following equation (Equation (1)).

$$\rho_k = \frac{d^2 * \pi * \partial_{1,k} * (ND_k - ND_{min,\,k})}{E_{0,k} * sin\,\theta_e * \tau_{k,i}} \qquad (1)$$

where ρ_k is the reflectance for the k band; d is the factor that considers the solar variation from the Earth–Sun distance, calculated from the Julian Day; $\partial_{1,k}$ is the conversion to radiance multiplicative coefficient; θ_e is the solar elevation angle; and $E_{0,k}$ is the solar irradiance in the top of atmosphere for the k band. The data to make the conversion to reflectance were obtained from the header files of each satellite image.

3.2. Processing

For processing the images, spectral signatures were created from selecting training sites based on the identification of similar areas in different covers and land uses, combining the knowledge of the area for a proper selection of the regions of interest (ROI). To identify the different land use classes, some visual patterns such as tone, texture and the influence areas were used.

During the identification of the training sites, the separability of the spectral signatures was verified for the five land use classes described in Table 2.

Table 2. Description of identified covers and land uses.

Class No.	Class	Description
1	Urban	Includes urban and industrial areas.
2	Agricultural	Periodic and temporary irrigation agriculture.
3	Water bodies	Water bodies, lakes and rivers.
4	Secondary vegetation	Includes arbustive (scrub and grassland) and arboreal vegetation of low or scarce density.
5	Forest	High density arboreal vegetation.

Source: Own elaboration.

While selecting the training sites, control fields verified in situ were also set for validation of each classified image (post-processing).

To obtain the different land use classes, three supervised classification methods were used. First, the Maximum Likelihood Classification (MLC) method, as the most widely used in the scientific literature, is fast, easy to apply and enables a clear interpretation of the results [21]. This algorithm can obtain a spectral image of each land use class through variance and covariance statistics of the set of training sites identified in the image and calculates the probability of belonging to each class according to the spectral signature; this method has been proven in works such as those of [22–25], with satisfactory results.

The Support Vector Machines (SVMs) algorithm was the second method applied. This automatic learning algorithm trains linear and non-linear learning functions by transforming the original data into a different space with a function (kernel) to obtain the hyperplane which maximizes the margin of separation between two or more classes to be classified [26]. Currently, the SVMs algorithm is among the most reliable methods; therefore, it is used in many works [27–29] with satisfactory results. For the classification of images the Radial Basis Function (RBF) for not-linearly separable data was used.

Finally, the third method applied was the Artificial Neural Networks (ANNs), an automatic learning method that predicts a complex behavior from a sample of observed inputs and outputs. The network structure is based on a simplified model of the human brain consisting of three layers: input, hidden and output. This structure is trained to recognize the result from input values and classify the rest according to the given rules [30,31]. Neural networks have been applied to classify satellite images with good results [32,33].

The ANNs classification was applied with a hidden layer of standard backpropagation for supervised learning by means of the logistic activation function for non-linear classification.

3.3. Post-Processing

To obtain a better representation of the land use mapping, each classified image was subject to a series of auxiliary processes: a process of majority filtering (3 × 3 pixels) and a method of generalization of polygons less than one hectare—as they are few representatives with respect to the minimum mapping unit—were applied, which reduced the image noise and eliminated the isolated polygons, resulting in the land use mapping for each year of analysis. Finally, to standardize and confirm the location of urban areas, a visual inspection of the mapped urban localities and the population census with historical data from INEGI for the same analyzed periods was carried out.

To validate the obtained results, the classified images were compared against the control fields through a cross-tabulation matrix for the different dates. At the same time the following were obtained: the kappa index, which shows the degree of similarity between a set of control fields and the classified image; the general accuracy, which indicates the percentage of pixels properly classified; the percentage of producer's accuracy, which sets the percentage of a kind of particular land use change correctly classified in the image; and the percentage of user's accuracy, which provides the percentage of a land use class in the image that matches with the class that corresponds in the land.

The model validated as the one with higher accuracy was used to represent the cover and land use mapping for the years 1973, 1985, 2000 and 2015.

3.4. Analysis of Land Use and Cover Changes

The analysis of LUCC was conducted through the cross-tabulation matrix to obtain losses, gains and interchanges between the different covers and land uses and after that, the significant transitions analysis was conducted according to the methodology proposed by [19]. For each cover and land use the cross-tabulation matrix enables to obtain, through the diagonal values, the stable areas between two dates, as well as the losses (below the diagonal) and the gains (above the diagonal). Such significant transitions on each cover and land use were obtained by comparing the gains and/or real losses against the gains and/or expected losses randomly, divided by the gains and/or expected losses. This comparison gave as a result the transition index where the values less than one indicated non-significant changes among covers and land uses, while the positive values more than one, indicated significant transitions.

4. Results and Discussion

4.1. Satellite Images Classification

Figure 3 shows the mapping of covers and land uses obtained for each classified Landsat image. The urban land use has mainly replaced agricultural land use where the productive activities has been focused on sugarcane, maize, mango and jicama crops at rate of 1.48 km^2/year.

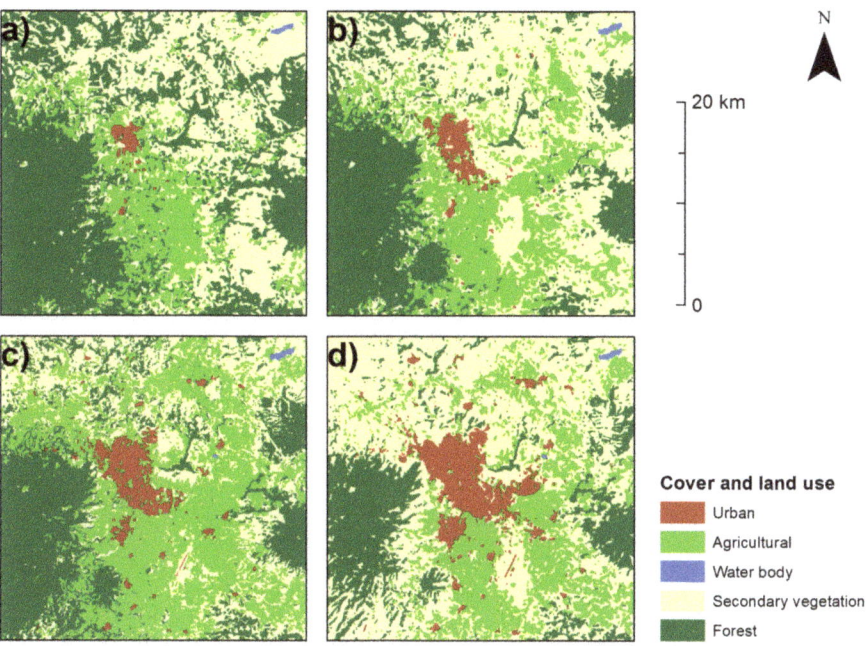

Figure 3. Image classified from Landsat images 1, 5, 7 and 8 with SVMs; Cover and land use: (**a**) 1973; (**b**) 1985; (**c**) 2000; and (**d**) 2015. Source: Own elaboration from Landsat images 1, 5, 7 and 8.

4.2. Classification Validation

Table 3 shows the results from the process of classified images validation through the cross-tabulation matrix and the parameters obtained: kappa index, general accuracy, producer's accuracy and user's accuracy. Validation statistics show better results when using the SVMs classification method, from which general accuracy above 85% is recorded for the four classified Landsat images 1, 5, 7 and 8, unlike the maximum likelihood classification and the artificial neural networks methods.

Table 3. Classified images validation.

Year Evaluated	SVMs		MLC		ANNs	
	General Accuracy	Kappa Index	General Accuracy	Kappa Index	General Accuracy	Kappa Index
1973	98.7%	0.98	97.7%	0.96	97.7%	0.96
1985	89.0%	0.85	92.5%	0.90	96.5%	0.95
2000	89.3%	0.85	82.1%	0.76	92.7%	0.90
2015	90.4%	0.87	86.1%	0.81	86.1%	0.81

Source: Own elaboration.

When selecting training fields, separability problems were identified between the secondary vegetation and the agricultural land use classes, which is reflected in the producer's accuracy percentage with values below 90%. The best fit was recorded when using the SVMs classification method, where the accuracy percentages average 96% for the user, and 95% for the producer; as shown in the Table 4.

Table 4. Percentages of producer's accuracy and user's accuracy.

Classified Image	Class	SVMs		MLC		ANNs	
		Producer's Accuracy (%)	User's Accuracy (%)	Producer's Accuracy (%)	User's Accuracy (%)	Producer's Accuracy (%)	User's Accuracy (%)
Landsat 1 MSS (1973)	Urban	100	100	100	100	100	100
	Agricultural	97	100	96	99	95	100
	Water body	100	100	100	100	100	100
	Secondary vegetation	100	95	100	93	100	98
	Forest	99	100	96	100	98	94
Landsat 5 TM (1985)	Urban	100	100	100	100	83	199
	Agricultural	100	100	100	100	100	98
	Water body	100	100	100	100	100	90
	Secondary vegetation	68	100	82	100	99	94
	Forest	100	74	100	83	93	100
Landsat 7 ETM (2000)	Urban	100	100	100	56	100	39
	Agricultural	100	100	93	100	89	98
	Water body	100	100	100	100	100	100
	Secondary vegetation	56	100	63	100	86	100
	Forest	100	67	100	71	100	89
Landsat 8 OLI (2015)	Urban	100	100	100	100	94	89
	Agricultural	100	80	100	94	95	84
	Water body	100	100	100	100	100	95
	Secondary vegetation	72	100	44	100	71	69
	Forest	100	96	100	76	83	93
Mean		95	96	94	94	94	96

Source: Own elaboration.

4.3. Analysis of Land Use Changes

Table 5 shows the changes in area for each class of cover and land use occurred during 1973–2015. The urban land use has increased 7% from the overall analyzed area with an annual rate of 1.48 km^2/year. At the same time, the forest cover has lost 28% of area in 42 years, with an annual rate of 5.78 km^2/year.

Sustainability **2018**, *10*, 1860

Table 5. LUCC during 1973–2015 according to the SVMs method.

Classification Method	Class	Description	1973 Area (km^2)	1973 Area (%)	1985 Area (km^2)	1985 Area (%)	2000 Area (km^2)	2000 Area (%)	2015 Area (km^2)	2015 Area (%)	Annual Rate (km^2)
SVMs	1	Urban	6.8	1	19.2	2	40.0	4	68.8	8	1.48
	2	Agricultural	151.1	17	215.1	24	342.8	38	211.3	23	1.43
	3	Water body	1.1	0	1.4	0	1.5	0	1.4	0	0.01
	4	Secondary vegetation	322.7	36	383.7	43	243.7	27	442.8	49	2.86
	5	Forest	418.3	46	280.6	31	272.1	30	175.7	20	5.78 *
MLC	1	Urban	6.8	1	19.2	2	39.9	4	68.6	8	1.47
	2	Agricultural	179.5	20	258.1	29	234.7	26	218.5	24	0.93
	3	Water body	1.2	0	1.7	0	1.4	0	1.5	0	0.01
	4	Secondary vegetation	396.0	44	319.6	36	372.7	41	401.7	45	0.13
	5	Forest	316.4	35	301.3	33	251.3	28	209.7	23	2.54 *
ANNs	1	Urban	6.8	1	19.2	2	39.9	4	68.8	8	1.48
	2	Agricultural	140.1	16	221.3	25	314.2	35	331.1	37	4.55
	3	Water body	1.1	0	2.6	0	2.1	0	1.8	0	0.02
	4	Secondary vegetation	281.2	31	413.4	46	283.6	32	284.4	32	0.08
	5	Forest	470.8	52	243.6	27	260.2	29	213.9	24	6.12 *

* Loss. Source: Own elaboration.

The annual rates of change for each class of cover and land use are shown in Figure 4. Differences were identified depending on the classification methods applied. When applying the SVMs and ANNs methods, the urban land use presented a similar trend of 1.48 km^2/year of change. The three methods recorded differences regarding agricultural land use, where the ANNs method registered a rate of 4.55 km^2/year, while the SVMs method obtained rates of 1.43 km^2/year. The secondary vegetation cover also presented important differences; while the SVMs method registered a rate of 2.86 km^2/year, the results of the MLC and ANNs methods presented a rate below 0.13 and 0.08 km^2/year, respectively. With regard to the forest, the methods SVMs and ANNs obtained similar annual rates of 5.78 and 6.12 km^2/year; the negative rate of change is due to the loss of surface.

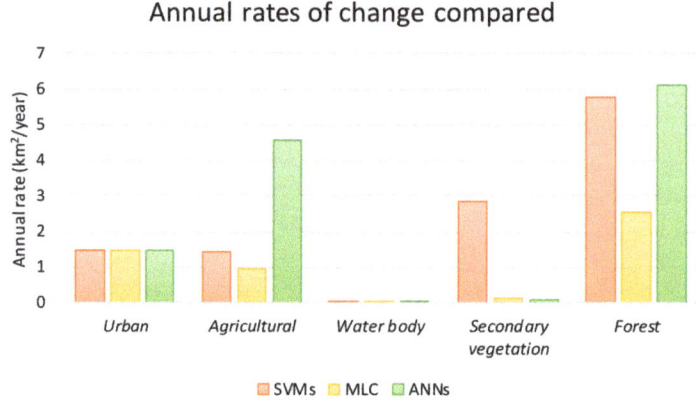

Figure 4. Annual rates of change according to the classification methods applied. Source. Own elaboration.

Since the SVMs method proved the best fit, the results from applying this method were used to analyze significant transitions and land use changes in the metropolitan area.

The changes occurred in this same period due to losses and gains are shown in Table 6. The main results show that the urban land use in the metropolitan area has increased 62 km^2 since 1973. The agricultural area and the secondary vegetation present most of the interchanges due to crop rotation, which is confirmed when observing the losses and gains of these same land use classes.

Table 6. Land use losses, gains and interchanges between periods with SVMs.

Period	Class	Area (km²)							
		Total (t1)	Total (t2)	Steady (E)	Gains (G)	Losses (L)	Interchanges (I)	Net Change (NT)	Total Change (CT)
1973–1985	Urban	6.8	19.2	6.8	12.4	0.0	0.0	12.4	12.4
	Agricultural	151.1	215.1	86.6	128.5	64.5	129.0	64.0	193.0
	Water body	1.1	1.4	1.0	0.4	0.0	0.1	0.3	0.4
	Secondary vegetation	322.7	383.7	229.0	154.7	93.7	187.4	61.0	248.4
	Forest	418.3	280.6	263.0	17.6	155.3	35.2	137.7	172.9
1985–2000	Urban	19.2	40.0	19.2	20.8	0.0	0.0	20.8	20.8
	Agriculture	215.1	342.8	171.3	171.5	43.8	87.6	127.7	215.3
	Water body	1.4	1.5	1.3	0.2	0.1	0.2	0.1	0.3
	Secondary vegetation	383.7	243.7	185.2	58.5	198.5	116.9	140.1	257.0
	Forest	280.6	272.1	227.1	45.0	53.6	90.1	8.5	98.6
2000–2015	Urban	40.0	68.8	40.0	28.9	0.0	0.0	28.9	28.9
	Agricultural	342.8	211.3	182.3	29.0	160.5	58.1	131.5	189.5
	Water body	1.5	1.4	1.3	0.1	0.3	0.2	0.2	0.4
	Secondary vegetation	243.7	442.8	203.7	239.1	40.0	79.9	199.1	279.1
	Forest	272.1	175.7	161.8	13.9	110.3	27.8	96.4	124.2
1973–2015	Urban	6.8	68.8	6.8	62.0	0.0	0.0	62.0	62.0
	Agriculture	151.1	211.3	64.4	146.9	86.7	173.4	60.3	233.6
	Water body	1.1	1.4	1.0	0.3	0.0	0.0	0.3	0.3
	Secondary vegetation	322.7	442.8	194.2	248.6	128.5	257.0	120.1	377.1
	Forest	418.3	175.7	166.4	9.3	251.9	18.5	242.7	261.1

Source: Own elaboration.

The transitions analysis for each cover and land use is summarized in Table 7. Significant transitions for the analyzed periods are registered particularly in the agricultural land use change to urban land use and secondary vegetation. The increase of urban area and secondary vegetation is because the agricultural area is being replaced. During a 42-year period, 33.1 km² of agricultural use transformed into urban areas and 53.3 km² into secondary vegetation. The forest also has been affected with transitions, although non-significant. They recorded ecological changes of 49.1 km² to agricultural land and 195.3 km² to secondary vegetation.

Table 7. Significant transitions analysis.

From	Area (km²)				To
	1973–1985	1985–2000	2000–2015	1973–2015	
Agricultural	8.9 *	7.6 *	24.7 *	33.1 *	Urban
	0.0	0.1	0.1	0.0	Water body
	51.3 *	29.6	133.2	53.3	Secondary vegetation
	4.2	6.5	2.5	0.2	Forest
Water body	0.0	0.0	0.0	0.0	Urban
	0.0	0.0	0.1	0.0	Agricultural
	0.0	0.0	0.2	0.0	Secondary vegetation
	0.0	0.0	0.0	0.0	Forest
Secondary vegetation	2.7	12.8	3.9	21.4	Urban
	77.3	147.2	24.6	97.8	Agricultural
	0.4	0.1	0.0	0.2	Water body
	13.3	38.5	11.4	9.0	Forest
Forest	0.7	0.4	0.2	7.5	Urban
	51.2	24.3	4.4	49.1	Agricultural
	0.0	0.1	0.0	0.1	Water body
	103.4	28.8	11.4	195.3	Secondary vegetation

* Significant transition. Source: Own elaboration.

The methodology applied in this study is similar to the one used by Aguayo et al. [1], Lopez and Plata [25] and Antillón et al. [34], who applied the maximum likelihood algorithm to get the different land use classes of the study area and analyzed LUCC through cross-tabulation method or confusion matrix. The results of said investigations have had the same trend as obtained in the present study as regards to urban areas replacing agricultural lands, as well as a decrease in the forest area.

According to the scientific literature, the SVMs method has been used in several studies such as those developed by Mountrakis et al. [27]; Lu et al. [28]; and Xie et al. [29] with good results on image classification, similar to the results obtained in this work where the SVMs method registered the best fit. Some works have compared different supervised classification methods; for example, Pal and Mather [35] used the same three classification methods that were applied in this study (MLC, ANNs and SVMs), obtaining the best fit results through SVMs. On the other hand, Otukei and Blaschke [36] compared three methods of classification, MLC, SVMs and Decision Trees (DT), and affirmed that the best results were obtained when applying SVMs. In addition, Mondal et al. [37] compared the SVMs and MLC methods, and concluded that when preparing land use maps, the SVMs method is more appropriate than the MLC. Finally, in a more recent study, Wu et al. [38] compared the SVMs, ANNs and DT methods; although radial base and polynomial functions were used for SVMs, this method obtained better results with kappa indices (0.72 and 0.79 for each function, respectively). Both the results in the above mentioned research and in this work obtained the best fit applying the SVMs method; therefore, the results from applying this method were used for the LUCC analysis in the study area.

In Mexico, several studies on LUCC have been conducted to achieve a better understanding on the dynamics and processes of land use change (e.g., [2,8,14,24,25,34,39–43]). These studies have been oriented to identify forest areas loss rates, unlike this research that was focused on analyzing the urban land use during three periods with the intention of identifying the historical dynamics of urban growth to be able to build urban growth simulation models and development of future scenarios.

Particularly, Cano et al. [44] analyzed urban land use in Hidalgo State (Mexico). From satellite images, they identified urban growth of 72.3 km^2 in a 14-year period, equivalent to an annual growth rate of 1.8%. On the other hand, by means of digital and visual techniques classification, Lopez and Plata [25] analyzed LUCC in the metropolitan area of Mexico City regarding urban expansion detecting an urban growth of 202 km^2 equivalent to 16% in a 10-year period. In comparison with the present work, in the metropolitan area of Tepic-Xalisco it was possible to quantify an urban growth of 62 km^2 during a 42-year period, with an annual rate of 1.48 km^2, which means a relatively low growth regarding to the metropolitan area of Mexico City.

For the study area, the research on cover and land use analysis conducted by Nájera et al. [45] was identified for the Mololoa River watershed, which determined natural vegetation losses of 41.67 ha/year with deforestation rates of 0.1% and urban growth of 74.86 ha/year, which is nearly half that obtained by this study that registered an increase of urban land of 148 ha/year (1.48 km^2/year). These results may be attributed to the difference in the boundaries set for the study areas, and to the methodology used to obtain the different land use classes.

5. Conclusions

The validation results from the classifications developed suggest that the SVMs method gives the best fit and offers greater certainty on the distribution and quantification of the different classes of cover and land use obtained.

The urban land use in the metropolitan area of Tepic-Xalisco has experienced an important increase within a period of 42 years, exceeding ten times the urban area recorded in 1973, with a rate of 1.48 km^2/year. This growth has produced significant changes in land use with transitions towards agricultural and secondary vegetation land use. The forest cover also has been affected, since it has experienced considerable losses of area with transformation trends towards secondary vegetation. In addition, the agricultural land use has been replaced as a result of urban growth. This situation has caused functional implications on ecosystems and to date losses of agricultural productive area are present, as well as deforestation processes.

The applied methodology enabled learning about the historical dynamics and quantifying the LUCC during a 42-year period, identifying the transitions between each land use. This information will help to establish land planning strategies, promote management and develop land use conservation policies.

Author Contributions: The first author, A.Á.J., designed the research and wrote this paper. F.F.V., O.N.G. and S.M.L.M.F. validated the information that was used. All authors read and approved the final version of this manuscript.

Acknowledgments: The authors express their gratitude to the Autonomous University of Nayarit and the National Council on Science and Technology (CONACYT, for its acronym in Spanish) of Mexico by financing the basic scientific research project 2015 No. 258991: Simulation models of urban growth through cellular automata in the city of Tepic, Nayarit, from which this study is part.

Conflicts of Interest: The authors declare no conflict of interest.

References

1. Aguayo, M.; Pauchard, A.; Azócar, G.; Parra, O. Cambio del uso del suelo en el centro sur de Chile a fines del siglo XX. Entendiendo la dinámica espacial y temporal del paisaje. *Rev. Chil. Hist. Nat.* **2009**, *82*, 361–374. [CrossRef]
2. Mas, J.F.; Velázquez, A.; Reyes, D.G.J.; Mayorga, S.R.; Alcántara, C.; Bocco, G.; Castro, R.; Fernandez, T.; Perez, V.A. Assessing land use/cover changes: A nationwide multidate spatial database for Mexico. *Int. J. Appl. Earth Obs. Geoinf.* **2004**, *5*, 249–261. [CrossRef]
3. Mendoza, E.; Dirzo, R. Deforestation in Lacandonia (southeast Mexico): Evidence for the declaration of the northernmost tropical hot-spot. *Biodivers. Conserv.* **1999**, *8*, 1621–1641. [CrossRef]

4. Lambin, E.F.; Turner, B.L.; Helmut, J.; Geist, S.B.; Agbola, S.B.; Arild, A.; Bruce, J.W.; Coomes, O.T.; Dirzo, R.; Fischer, G.; et al. The causes of land-use and land-cover change: Moving beyond the myths. *Glob. Environ. Chang.* **2001**, *11*, 261–269. [CrossRef]
5. Turner, B.L.; Lambin, E.F.; Reenberg, A. The emergence of land change science for global environmental change and sustainability. *Proc. Natl. Acad. Sci. USA* **2007**, *104*, 20666–20671. [CrossRef] [PubMed]
6. Lambin, E.F.; Geist, H.J. (Eds.) *Land-Use and Land-Cover Change: Local Processes and Global Impacts*; Springer: Berlin, Germany, 2008.
7. FAO. *Evaluación de los Recursos Forestales Mundiales: Informe Nacional México*; FAO: Roma, Italy, 2010; Available online: http://www.fao.org/docrep/013/al567S/al567S.pdf (accessed on 31 May 2018).
8. Bocco, G.; Mendoza, M.; Masera, O.R. La Dinámica del Cambio del Uso del Suelo en Michoacán: Una Propuesta Metodológica para el Estudio de los Procesos de Deforestación. *Investig. Geogr.* **2001**, *44*, 18–36. Available online: http://www.redalyc.org/articulo.oa?id=56904403 (accessed on 31 May 2018). [CrossRef]
9. Velázquez, A.; Durán, E.; Ramírez, I.; Mas, J.F.; Bocco, G.; Ramírez, G.; Palacio, J.L. Land use-cover change processes in highly biodiverse areas: The case of Oaxaca, México. *Glob. Environ. Chang.* **2003**, *13*, 175–184. [CrossRef]
10. Velázquez, A.; Mas, J.F.; Díaz, G.J.R.; Mayorga, S.R.; Alcántara, P.C.; Castro, R.; Fernández, T.; Bocco, G.; Ezcurra, E.; Palacio, J.L. Patrones y tasas de cambio de uso del suelo en México. *Gaceta Ecol.* **2002**, *62*, 21–37.
11. Millennium Ecosystem Assessment. Ecosystems. 2003. Available online: https://www.millenniumassessment.org/en/index.html (accessed on 31 May 2018).
12. SEMARNAT. *Informe de la Situación del Medio Ambiente en México*; SEMARNAT: México City, México, 2012. Available online: http://apps1.semarnat.gob.mx/dgeia/informe_12/pdf/Informe_2012.pdf (accessed on 31 May 2018).
13. Challenger, A.; Soberón, J. *Los Ecosistemas Terrestres, en Capital Natural de México, vol. I: Conocimiento Actual de la Biodiversidad*; CONABIO: México City, México, 2008; pp. 87–108. Available online: http://www.biodiversidad.gob.mx/pais/pdf/CapNatMex/Vol%20I/I03_Losecosistemast.pdf (accessed on 31 May 2018).
14. Galicia, L.; García, R.A.; Gómez-Mendoza, L.; Ramírez, M.I. Cambio de uso del suelo y degradación ambiental. *Ciencia* **2007**, *584*, 50–59.
15. Mas, J.F.; Velázquez, A.; Couturier, S. La evaluación de los cambios de cobertura/uso del suelo en la República Mexicana. *Investigación Ambiental Ciencia y Política Pública* **2009**, *1*, 23–39.
16. Chuvieco, E. Teledetección Ambiental: La Observación de la Tierra Desde el Espacio (Editorial Ariel). 2008. Available online: https://drive.google.com/file/d/0B0KUmy_fthbuX09sUE9RejJJX1U/view (accessed on 31 May 2018).
17. Civco, D.L.; Hurd, J.D.; Wilson, E.H.; Song, M.; Zhang, Z. A Comparison of Land Use and Land Cover Change Detection Methods. In Proceedings of the ASPRS-ACSM Annual Conference, Washington, DC, USA, 22–26 April 2002; Available online: https://www.researchgate.net/profile/Daniel_Civco/publication/228543190_A_comparison_of_land_use_and_land_cover_change_detection_methods/links/5570b54608aedcd33b292ec1.pdf (accessed on 31 May 2018).
18. Monjardín, A.S.A.; Pacheco, A.C.E.; Plata, R.W.; Corrales, B.G. La deforestación y sus factores causales en el estado de Sinaloa, México. *Madera y Bosques* **2017**, *231*, 7–22. [CrossRef]
19. Pontius, R.G.; Shusas, E.; McEachern, M. Detecting important categorical land changes while accounting for persistence. *Agric. Ecosyst. Environ.* **2004**, *101*, 251–268. [CrossRef]
20. Chavez, P.S. Image-based atmospheric corrections. Revisited and improved. *Photo-Gramm. Eng. Remote Sens.* **1996**, *62*, 1025–1036.
21. Bolstad, P.; Lillesand, T.M. Rapid maximum likelihood classification. *Photogramm. Eng. Remote Sens.* **1991**, *571*, 67–74.
22. Hassan, Z.; Shabbir, R.; Ahmad, S.S.; Malik, A.H.; Aziz, N.; Butt, A.; Erum, S. Dynamics of land use and land cover change (LULCC) using geospatial techniques: A case study of Islamabad Pakistan. *SpringerPlus* **2016**, *5*, 1–11. [CrossRef] [PubMed]
23. Tahir, M.; Iman, E.; Hussain, T. Evaluation of land use/land cover changes in Mekelle City. Ethiopia using Remote Sensing and GIS. *Comput. Ecol. Softw.* **2013**, *3*, 9–16.
24. García Mora, T.J.; Mas, J.F. Comparación de metodologías para el mapeo de la cobertura y uso del suelo en el sureste de México. *Investig. Geogr.* **2008**, *67*, 7–19. [CrossRef]

25. López, V.V.H.; Plata, R.W. Análisis de los cambios de cobertura de suelo derivados de la expansión urbana de la Zona Metropolitana de la Ciudad de México, 1990–2000. *Investig. Geogr.* **2009**, *68*, 85–101. [CrossRef]
26. Vapnik, V.N. *The Nature of Statistical Learning Theory*; Springer: New York, NY, USA, 1995.
27. Mountrakis, G.; Im, J.; Ogole, C. Support vector machines in remote sensing: A review. *ISPRS J. Photogramm. Remote Sens.* **2011**, *66*, 247–259. [CrossRef]
28. Lu, D.; Weng, Q.; Moran, E.; Li, G.; Hetrick, S. *Remote Sensing Image Classification*; CRC Press; Taylor and Francis: Boca Raton, FL, USA, 2011; pp. 219–240.
29. Xie, L.; Li, G.; Xiao, M.; Peng, L.; Chen, Q. Hyperspectral image classification using discrete space model and support vector machines. *IEEE Geosci. Remote Sens. Lett.* **2017**, *14*, 374–378. [CrossRef]
30. Richards, J.A. *Remote Sensing Digital Image Analysis*; Springer: Berlin, Germany, 1999; p. 240.
31. Atkinson, P.M.; Tatnall, A.R.L. Introduction Neural Networks in Remote Sensing. *Int. J. Remote Sens.* **1997**, *184*, 699–709. [CrossRef]
32. Kavzoglu, T.; Mather, P.M. The use of backpropagating artificial neural networks in land cover classification. *Int. J. Remote Sens.* **2003**, *24*, 4907–4938. [CrossRef]
33. Verbeke, L.P.C.; Vabcoillie, F.M.B.; Dewulf, R.R. Reusing back-propagating artificial neural network for land cover classification in tropical savannahs. *Int. J. Remote Sens.* **2004**, *25*, 2747–2771. [CrossRef]
34. Antillón, V.M.Y.; Corral, G.G.M.; Alatorre, C.L.C. Análisis de los Cambios de Cobertura y uso de Suelo en los Márgenes de la Laguna de Bustillos, Chihuahua: Efectos de la Expansión Agrícola. Memorias de Resúmenes en Extenso SELPER-XXI-México-UACJ-2015. 2015. Available online: http://selper.org.mx/images/Memorias2015/assets/m005.pdf (accessed on 31 May 2018).
35. Pal, M.; Mather, P.M. Support vector machines for classification in remote sensing. *Int. J. Remote Sens.* **2005**, *26*, 1007–1011. [CrossRef]
36. Otukei, J.R.; Blaschke, T. Land cover change assessment using decision trees, support vector machines and maximum likehood classification algorithms. *Int. J. Appl. Earth Obs. Geoinf.* **2010**, *12*, S27–S31. [CrossRef]
37. Mondal, A.; Kundu, S.; Chandniha, S.K.; Shukla, R.; Mishra, P.K. Comparison of support vector machine and maximum likelihood classification technique using satellite imagery. *Int. J. Remote Sens. GIS* **2012**, *1*, 116–123.
38. Wu, W.; Li, A.D.; He, X.H.; Ma, R.; Liu, H.B.; Lv, J.K. A comparison of support vector machines, artificial neural network and classification tree for identifying soil texture classes in southwest China. *Comput. Electron. Agric.* **2018**, *144*, 86–93. [CrossRef]
39. Camacho, S.J.M.; Pérez, J.; Isabel, J.; Pineda, J.N.B.; Cadena, V.E.G.; Bravo, P.L.C.; Sánchez, L.M. Cambios de cobertura/uso del suelo en una porción de la Zona de Transición Mexicana de Montaña. *Madera y Bosques* **2015**, *21*, 93–112.
40. Evangelista, O.V.; López, B.J.; Caballero, N.J.; Martínez, A.M.Á. Patrones espaciales de cambio de cobertura y uso del suelo en el área cafetalera de la sierra norte de Puebla. *Investig. Geogr.* **2010**, *72*, 23–38. [CrossRef]
41. Ramírez, R.I. Cambios en las Cubiertas del Suelo en la Sierra de Angangueo, Michoacán y Estado de México, 1971-1994-2000. *Investig. Geogr.* **2001**, *45*, 39–55. Available online: http://www.redalyc.org/pdf/569/56904504.pdf (accessed on 31 May 2018). [CrossRef]
42. Guerrero, G.; Masera, O.; Mas, J.F. Land use/land cover change dynamics in the Mexican highlands: Current situation and long-term scenarios. In *Modelling Environmental Dynamics*; Springer: Berling, Germany, 2008; pp. 57–76.
43. Pineda, J.N.B.; Bosque, S.J.; Gómez, D.M.; Plata, R.W. Análisis de cambio del uso del suelo en el Estado de México mediante sistemas de información geográfica y técnicas de regresión multivariantes: Una aproximación a los procesos de deforestación. *Investig. Geogr.* **2009**, *69*, 33–52.
44. Cano, S.L.; Rodríguez, L.R.; Valdez, L.J.R.; Acevedo, S.O.A.; Beltrán, H.R.I. Detección del crecimiento urbano en el estado de Hidalgo mediante imágenes Landsat. *Investig. Geogr.* **2017**, *92*, 1–10. [CrossRef]
45. Nájera, G.O.; Bojórquez, S.J.I.; Cifuentes, L.J.L.; Marceleño, F.S. Cambio de cobertura y uso del suelo en la cuenca del río Mololoa, Nayarit. *Revista Bio Ciencias* **2010**, *1*, 19–29. [CrossRef]

© 2018 by the authors. Licensee MDPI, Basel, Switzerland. This article is an open access article distributed under the terms and conditions of the Creative Commons Attribution (CC BY) license (http://creativecommons.org/licenses/by/4.0/).

Article

Morphology Dependent Assessment of Resilience for Urban Areas

Kai Fischer [1,*], Stefan Hiermaier [1,2], Werner Riedel [1] and Ivo Häring [1]

1. Fraunhofer Institute for High-Speed Dynamics, Ernst-Mach-Institut, 79104 Freiburg, Germany; stefan.hiermaier@inatech.uni-freiburg.de (S.H.); werner.riedel@emi.fraunhofer.de (W.R.); ivo.haering@emi.fraunhofer.de (I.H.)
2. Institute for Sustainable Technical Systems, University of Freiburg, 79110 Freiburg, Germany
* Correspondence: kai.fischer@emi.fraunhofer.de; Tel.: +49-7628-9050-628

Received: 17 April 2018; Accepted: 28 May 2018; Published: 30 May 2018

Abstract: The formation of new threats and the increasing complexity of urban built infrastructures underline the need for more robust and sustainable systems, which are able to cope with adverse events. Achieving sustainability requires the strengthening of resilience. Currently, a comprehensive approach for the quantification of resilience of urban infrastructure is missing. Within this paper, a new generalized mathematical framework is presented. A clear definition of terms and their interaction builds the basis of this resilience assessment scheme. Classical risk-based as well as additional components are aligned along the timeline before, during and after disruptive events, to quantify the susceptibility, the vulnerability and the response and recovery behavior of complex systems for multiple threat scenarios. The approach allows the evaluation of complete urban surroundings and enables a quantitative comparison with other development plans or cities. A comprehensive resilience framework should cover at least preparation, prevention, protection, response and recovery. The presented approach determines respective indicators and provides decision support, which enhancement measures are more effective. Hence, the framework quantifies for instance, if it is better to avoid a hazardous event or to tolerate an event with an increased robustness. An application example is given to assess different urban forms, i.e., morphologies, with consideration of multiple adverse events, like terrorist attacks or earthquakes, and multiple buildings. Each urban object includes a certain number of attributes, like the object use, the construction type, the time-dependent number of persons and the value, to derive different performance targets. The assessment results in the identification of weak spots with respect to single resilience indicators. Based on the generalized mathematical formulation and suitable combination of indicators, this approach can quantify the resilience of urban morphologies, independent of possible single threat types and threat locations.

Keywords: resilience quantification; resilience engineering; multiple threat assessment; urban form

1. Introduction

Cities are key drivers for technological, organizational, social and economic innovation and well-being for individuals and the society. Sustainable progress in these domains depends on the availability of infrastructures and buildings. This also holds true for the scale of interaction and connectivity.

Agglomerated areas comprise a high degree of critical infrastructure. At the same time, critical infrastructures specify significantly the resilience and the robustness [1]. It is clearly observable that systems, cities and infrastructures will become more complex and interconnected [2]. Due to this change, the failure of a single element increases the probability to produce cascading effects with unexpected consequences [3] as well as emergent threats.

Industrialization and population growth are the main reasons for an increasing urban population. This fact is clearly observable in different studies, as stated in the report of the United Nations [4]. It results in a changing density of population, a higher degree of urbanization and an increasing focus on hazard vulnerability reduction and resilience [5].

A further challenge is the formation of new threats. According to Branscomb [1], cities are increasingly vulnerable to three kinds of disasters:

- natural, like hurricanes, floods or earthquakes,
- technogenic, resulting from human error and failing infrastructure, like a power failure, and
- terrorism, "a growing and important asymmetric threat which can pick targets anywhere" [6].

In summary, the rising urbanization, growing complexity of critical infrastructure and formation and increase of new and old threats lead to the need to manage possible hazardous events and their corresponding consequences in an ever-increasing number, especially in urban areas. These aspects motivate the need of sustainable cities, which are able to cope with adverse events.

Based on the new challenges for urban areas, this paper focuses on how the built urban environment, urban spaces, buildings and infrastructure can better cope with potentially adverse and disruptive events. The overarching aim is to contribute from an engineering–technical science driven perspective to the sustainability of urban areas and infrastructures. Achieving sustainability requires the strengthening of resilience [7].

The concept of resilience is used in a great variety of interdisciplinary work concerned with the interaction of people and nature [8]. Examples can be found in social sciences [9,10] or engineering disciplines [11–13]. There are approaches which give a holistic overview but results in qualitative measures [14]. Other methods have no detailed information concerning the recovery behavior after a disruptive event [15,16] or focus on single scenarios and objects [17], which require a manual reapplication, if there are deviations or uncertainties. In summary, there is a need for the development of a clear analysis scheme for the quantification of resilience of urban areas. The assessment focusses on the identification on weak spots to circumvent a pure scenario-driven approach.

An urban area reflects a complex and dynamic composition of different zonings and functions, which defines the urban form and have a lasting effect on the sustainability, the resilience [18,19] as well as the coping capacity with disruptive events [10]. Within this paper, different urban footprints are evaluated to investigate the resilience depending on the morphology. Building density, building dimensions, construction types and object use are main parameter, which will be varied within the investigations and the introduced framework can give contributions to new development plans to reflect or incorporate resilience indicators and to shape a sustainable environment.

Section 2 introduces a generalized framework for the evaluation of resilience and Section 3 follows with a mathematical definition of single components of that framework. Certain analysis examples for different urban forms and the discussion of the assessment scheme are shown in Section 4. A summary and conclusion is given in Section 5.

2. Generalized Framework to Quantify Expected Losses and Recovery Processes

Findings from different approaches to evaluate resilience are sighted, compared and consolidated to propose a novel framework with the aim to quantify resilience, which requires a clear definition of terms. Based on the interdisciplinary research in the field of resilience, there are different interpretations concerning the definition and of that term [20]. Within the present work, the resilience cycle (Figure 1 left) according to Thoma [21] is used as definition. Therefore, resilience is defined as:

"The ability to repel, prepare for, take into account, absorb, recover from and adapt ever more successfully to actual or potential adverse events. Those events are either catastrophes or processes of change with catastrophic outcome, which can have human, technical or natural causes."

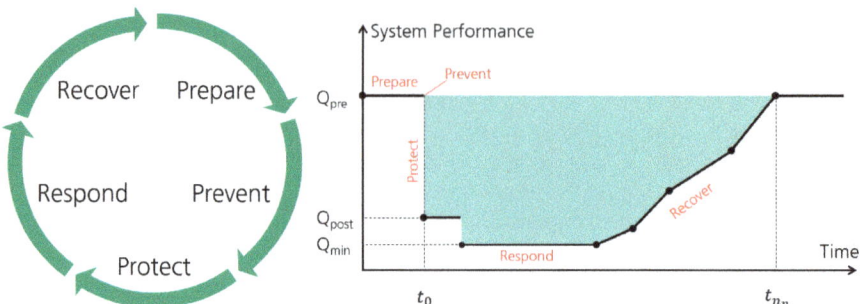

Figure 1. Five phases of the resilience cycle (**left**) according to [21] and their interpretation within a performance-time relation (**right**) for the quantification of resilience.

As shown by Bruneau [11] or Kröger [3], a performance-time relation can be used to describe the resilience of a system. A generalized and simplified shape of one such relation shows the right picture of Figure 1. A catastrophic event causes a disruption and a sudden performance loss at time t_0, which is followed by a response stabilization and recovery process. Resilience can be measured by the size of the expected degradation of performance over time, as indicated with the green area in the right diagram of Figure 1. Single phases of the resilience cycle can be assigned using performance-time diagrams for suitable system parameter and determine the effectiveness of single indicators and, if approved, to measure the resilience of the system. As indicated in Figure 1, measures of preparation and prevention will extend the time before disruptive events or avoid them completely. The drop of the system performance indicators is a measure of the level of protection and vulnerability. Efficient response decreases the degree of disruption and helps to start to bounce back quickly after the shock event. Finally, the resilience phase recovery describes all the aspects of relaxation, recovery and possible learning and the preparation for future events.

Based on the definitions in [22] and Figure 1, the aim of the proposed framework is the characterization of a performance target over time as basis for resilience quantification. Several components are integrated to achieve this objective and a generalized overview is given in Figure 2. The assessment scheme can be separated into two main parts. Under the assumption of a threat occurrence, the deterministic part uses physical models to quantify the intensity of a hazard source and the corresponding damage effects (vulnerability). A certain degree of recovery is required based on the resulting damage effects. The deterministic realm is applicable to derive a performance-time relation for a single threat, but requires the definition of a decisive scenario. Based on uncertainties that a certain threat event occurs, the deterministic part is coupled with a probabilistic realm. Stochastic methodologies are applied to evaluate the frequency and the exposition of a threat within the susceptibility approach. The combination of susceptibility and potential damage effects results in a risk-based vulnerability. Averaged results for multiple threat scenarios moves the approach from a scenario driven to a consequence based analysis for the identification of weak spots. The combination of weighted (risk-based) vulnerabilities and corresponding recovery processes consider a multitude of random scenarios and results in an averaged performance-time relation to characterize the resilience of a system, e.g., an urban surrounding. Single components of this framework can operate single phases of the resilience cycle (Figure 1).

Bruneau [11] states that resilience can be conceptualized "as encompassing four interrelated dimensions: technical, organizational, social and economic". With regard to Bruneau, the introduced framework including the quantification of susceptibility and vulnerability cannot cover all aspects concerning the resilience of urban areas but can give essential contributions.

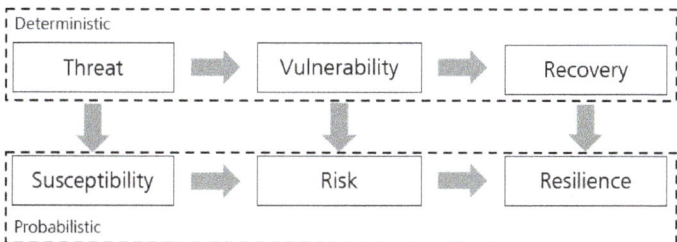

Figure 2. Proposed framework to assess the occurrence of adverse events and the expected losses as basis for resilience quantification according to [22].

First results within the susceptibility and vulnerability approach of Figure 2 are published in [23] to evaluate terroristic explosive events in urban areas. Pre-defined construction types are applied to assess the physical damage effects of buildings with the use of engineering models [24,25], if an adverse event occurs. The susceptibility, i.e., the frequency of a hazardous event and the exposition of single urban objects, is derived with historical statistical data from the Terror Event Database [26,27], depending on the object use, the threat type and the region. Based on these essential findings, the approach is in this paper subsequently enlarged to compare the risk-based results between different morphologies in a quantitative way. Furthermore, the consideration of time scales allows the assessment of a recovery process to result in a single quantity for the resilience of urban objects. Subsequent, a mathematical formulation will introduce the interaction of single components of the presented framework.

3. Mathematical Formulation

In alignment to the introduced framework in Figure 2, an abstract model of an urban area U is defined as a superset including a finite number of subsets, like free spaces a_m or buildings b_k

$$U = (a_m, b_k), \ m = 1, \ldots, n_{area}; \ k = 1, \ldots, n_{building}. \tag{1}$$

A single building b_k, $k = 1, \ldots, n_{building}$ is characterized by a position $\vec{r}_{b_k}$, a spatial extension dimension $L(b_k)$ and a type of object use $u_l(b_k)$, $l = 1, \ldots, n_{object\ type}$, like residential or office, for example.

A security relevant event, such as an explosion source or an earthquake within or close to an urban environment is defined as threat T_i. A threat can have different forms and the various threat types are expressed with the running index $i = 1, \ldots, n_{threat}$. A threat can occur at a number $j = 1, \ldots, n_{position}$ of possible locations $\vec{r}_j$. The physical hazard potential of a threat is described within a hazard model $H\left(T_i, \vec{r}_j; P\right)$ [22], as indicated in Equation (2). This model relates the threat type T_i and the event location $\vec{r}_j$ to the urban environment U. The physical properties are defined within the attribute P to characterize the (time dependent) hazard potential, like the magnitude of an earthquake, for example.

Depending on the intensity and the exposition, the occurrence of a threat can cause a certain type of consequences D_g, $g = 1, \ldots, n_{consequence\ type}$ at different locations in the urban surrounding $\vec{r}_o$, $o = 1, \ldots, n_{consequence\ position}$. Possible consequences of type D_g, like direct structural or non-structural damage at a building, at location $\vec{r}_k$ are characterized within the local what-if vulnerability $V\left(\vec{r}_k, D_g\right)$. An exemplary assessment of structural building damage can be realized with the use of single degree of freedom models [24] as basis for the collapse behavior of buildings [25]. Further details of the vulnerability assessment are described in [22].

Based on the degree of damage or loss of functionality, a certain degree of recovery is required to reach normal community activities and the initial performance of the investigated system, like an urban environment. The rebuild and recovery function $Q_{n_p}(t)$ characterize the time-dependent behavior as a stepwise linear function considering n_p, $p = 1, \ldots, n_{phases}$ recovery phases. A generalized sketch of this function is shown in the right picture of Figure 1.

This causal chain of threat occurrence, resulting vulnerability and required time-dependent recovery is summarized as

$$H\left(T_i, \vec{r}_j; P\right) \to V\left(\vec{r}_k, D_g\right) \to Q_{n_p}(t) \quad (2)$$

and expresses the deterministic part of the introduced methodology in Figure 2. This mathematical expression is valid to describe arbitrary threat types and investigated systems. The application of physical or engineering models results in quantitative measures as basis for decision makers. In particular, for each building type and damage level, a recovery function with respective recovery phases is defined, e.g., by resorting to typical planning and construction times and respective subsystem availabilities. Exemplary construction type dependent recovery times are shown in the Appendix A in Table A1.

The prediction of a single threat type scenario can be fraught with inaccuracies because it is difficult to estimate the threat position and the threat intensity can vary. Based on this fact and in alignment to the generalized framework in Figure 2, the frequency that a certain threat T_i occurs at a certain position $\vec{r}_j \in A_j$ is summarized within the susceptibility $S(T_i, A_j)$ and hence the causal chain in Equation (2) can be weighted with a probability that such an event occurs on A_j in the urban surrounding. This step incorporates the probabilistic realm of the assessment scheme.

The introduction of an averaged time-dependent recovery process (Equation (3)) considers multiple threat types and intensities (index i), threat positions (index j) and urban objects (index k). Each combination is weighted with the corresponding susceptibility $S(T_i, A_j)$. Equation (3) quantifies the averaged loss and recovery with respect to all possible threat events and urban objects, if a single event occurs:

$$Q(t; n_p, D_g) = \sum_i \sum_j \sum_k Q_{n_p}\left(t \middle| V\left(H\left(T_i, \vec{r}_j; P\right), \vec{r}_{b_k}, D_g\right)\right) \cdot S(T_i, A_j) . \quad (3)$$

The summation of the performance-time relations in Equation (3) results in a single quantity to describe the resilience of urban environments. The recovery function Q for a single scenario is characterized with the deterministic part of the framework in Figure 2. The consideration of multiple scenarios and the corresponding probabilistic susceptibility weighting transfers the approach from a scenario driven to a consequence based approach.

The introduced framework combines statistical data and physical approaches to evaluate urban environments with respect to the region and the geo-spatial information of the urban surrounding as well as properties of single urban objects, like the object use, constructional details, person densities or the asset value.

Single elements of the introduced approach are validated in [22] and enable a postulation of a resilience quantity for an arbitrary city. Furthermore, single resilience phases, like preparation, prevention, protection or recovery can be evaluated with this structured methodology. In particular, the susceptibility quantity, a generalized frequency of event an exposure measure, is an indicator for preparation and prevention, the vulnerability quantity, a generalized damage expression characterizes robustness and the recovery quantity characterizes response and recovery.

The presented framework intends to provide a quantitative methodology to achieve more robust and sustainable cities. Subsequently, different urban forms are investigated with the introduced approach. Based on the fact of a growing urbanization, the results should give insights for a sustainable growth of agglomerated areas.

4. Analysis Examples of Different Urban Forms

An urban environment reflects a complex interaction of different zonings and functions. Physical footprints are categorized in buildings, open spaces, traffic routes and landscapes and characterize the morphology [28]. Zonings distinguish between residential, retail, commercial, financial, industrial or educational objects. The variables of a city describe the physical and social characteristics and are dynamic in nature. Examples are physical constraints, growth, population (e.g., age, education or health), economic activities, environmental characteristics or community facilities.

The dynamic growth influences the footprint of a city and can result in a change of the zonings and variables. Urban planning processes encompass a variety of technical and political challenges to characterize and design the environment with respect to the well-being of the population and the natural habitat. To manage the growth and the shaping for tomorrow, new development plans are multi-dimensional in nature and match different variables.

The economic size of a city depends on the available infrastructure, for example. The mixture of different social groups results in a stable community of an urban area (social cohesion). A pedestrian area can give space for human interaction and creativity. The presented resilience framework can be integrated to consider safety and security aspects. The challenge is to achieve a secure and sustainable environment, which still allow for convenient living conditions. New development plans should include policies and objectives to reflect or incorporate the needs of the five phases of the resilience management cycle according to Figure 1 left. These aims have to be adopted depending in particular on the physical layout and the dynamic urban variables, i.e., demographic, social, economic or environmental.

4.1. Characterization and Modelling of the Urban Surroundings

The introduced framework is applied to three different examples representing main urban forms which are oriented to existing cities. The comparison of a compact and a linear city model investigates the resilience depending on the urban footprint. A further assessment describes a central business district to investigate variations of object and construction types. The generated information is intended to provide decision support for urban planning activities to integrate security and sustainability aspects aligned with a dynamic and sustainable growth of agglomerated areas.

As introduced in Section 3, single buildings of the urban surrounding can be abstracted characterized with a certain number of attributes. The consideration of each individual building within a city would exceed the effort of investigation. The introduced framework uses, a set of 10 pre-defined and fully designed buildings [29]. Possible designs are oriented to the categories in the left bar diagram of Figure 3. For each construction type, the physical properties are available to characterize the robustness and hence certain structural damage effects in case of a disruptive event occurrence [25]. Furthermore, the periods for planning and construction are available depending on the building type and result in quantities to estimate the required recovery for the introduced formulation in Equations (2) and (3). Table A1 in the Appendix A gives a detailed overview of the used building types and time scales to estimate the recovery process.

Figure 3 (left) compares the three application cases concerning their construction types in accordance to the list of pre-defined buildings. Based on the high connectivity and the mixture of residential with other uses, the compact city includes a high degree of multi-family houses with commercial use in the ground floor, indicated with "multi-family house +" in the left diagram of Figure 3. Due to the clear separation of zonings, the linear city includes areas with a higher number of industrial buildings and residential areas with single- or multi-family houses. Characteristic for a district with specific task assignment, the central business district includes an increased number of office buildings and office towers.

Next to the constructional characterization, a further description of the three city models includes the description of building use types. The properties of the compact city become apparent by comparison of the object use types, as shown in the right diagram of Figure 3. A high degree of residential objects is mixed with a wide range of different other types within all sectors. Dual use objects (residential and commercial) are considered as commercial use and result in the higher degree of objects for retail and service. The linear city is dominated by approximately two thirds of residential objects. Besides this object type, buildings dedicated to finance, trading, retail and service dominate the central business district.

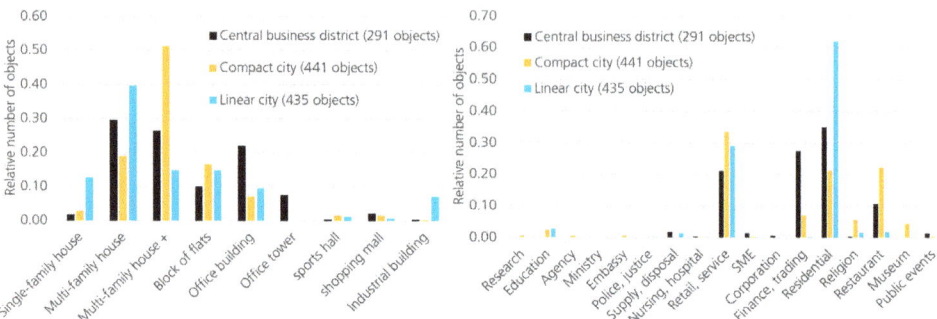

Figure 3. Distribution of construction types (**left**) and object use types (**right**) depending on the investigated city model.

4.2. Empirical and Spatial Distributed Evaluation of Disruptive Events

In alignment to the already published parts of the approach [22,23], the introduced framework is applied to evaluate possible terroristic events in the considered urban forms. As shown in Figure 2, a first step includes the quantitative susceptibility analysis. Statistical information from the Terror Event Database [26,27] are combined with the geospatial characteristics of a city to evaluate possible threat positions. Figure 4 visualizes the results with historical data of Western Europe and give the information of possible locations with higher susceptibilities. The color code indicates locations with the highest probability, if a single event occurs.

The results underline the characteristics of the three urban forms. A compact city includes a high mixture of object types, no clear zonings and a higher building density, which is apparent in the upper picture of Figure 4. There are several hotspots with a relative high susceptibility. The clear separation of different zonings within the linear city results in area-covering susceptibilities within an elongated area as shown in the middle picture of Figure 4. Broader areas with residential use generate low criticalities. Finally, the lower picture shows the empirical area distributed results for the central business district. Based on the clear assignment of object types, there are many objects with a similar criticality. In opposite to the linear model, the derived susceptibility is slightly higher and there is no local maximum.

The susceptibility approach in Figure 4 allows an efficient evaluation of possible threat types at certain locations within a single quantity $S(T_i, A_j)$. The frequency of a disruptive event depends on the empirical data of the threat type and object use. This information is distributed on possible event positions in alignment to the investigated city models [22]. This probability quantity is combined with the vulnerability approach to evaluate expected damage effects at certain positions $\vec{r}_o$ in the city model of a certain damage type D_g, as introduced in the general overview of Figure 2.

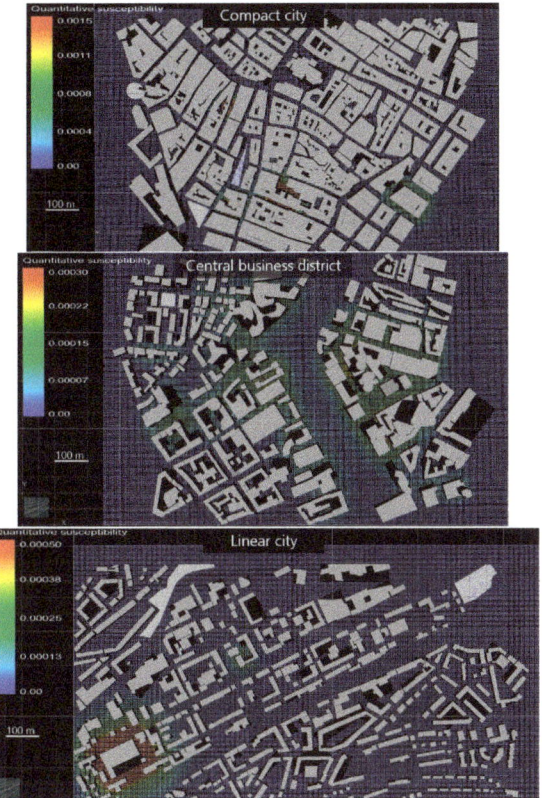

Figure 4. The quantitative susceptibility analysis combines statistical data per object use with the geospatial information of an urban surrounding to evaluate possible threat positions. Comparison of the three investigated urban forms. The susceptibility is given as probability of any dangerous event per area of size, in case of an occurring disruptive event.

4.3. Quantitiative Risk Assessment in F-N Diagrams

The introduced methodology [22,23] uses physical and engineering models to assess expected damage effects. Each combination of threat position A_j, as shown in Figure 4, threat type and intensity T_i causes a certain degree of damage and is weighted with the derived susceptibility $S(T_i, A_j)$ that this event occurs. The results can be counted to events with N or more damaged buildings and a corresponding cumulated frequency of occurrence. This information is collected for each investigated urban form of the application examples within a frequency-number (F-N) diagram, see Figure 5. Based on the investigated urban areas, combinations of buildings, threat types and threat positions results in tens of millions possible combinations with respect to the formulations in Equations (2) and (3) which exceeds the capacity of spreadsheet applications. Therefore, the investigated city models are separated into single areas, which results in certain dispersion for each city in the diagram of Figure 5.

The dotted lines separate the diagram into regions for acceptable (green line) or not acceptable (red line) risk quantities. The area between these two criteria marks the "ALARP" region, meaning as low as practicable possible and optional mitigation measures should be considered in relation to their efficiency [30]. Different criteria define the level of acceptance [31]. In this diagram, the "Groningen criterion" according to [30] is chosen.

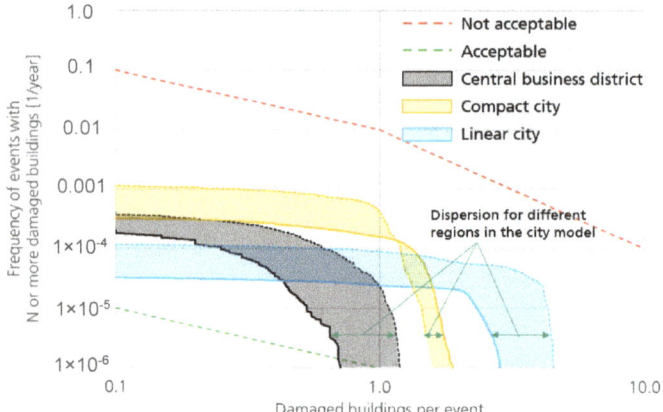

Figure 5. Comparison of the investigated city models within a frequency-number diagram concerning the expected damage and the comparison to criteria of acceptance. Based on the composition of construction types and object uses, each city model results in varying probability that an event with N or more damaged buildings occurs.

The central business district is characterized with many equal object types, which result in a relatively uniform distribution of susceptibility, as indicated in the empirical approach in Figure 4. This city model includes many robust construction types, like the high-rise buildings and results in low vulnerabilities with rather restricted local building damage effects and these characteristics result in the lowest quantitative risks of the three considered city models.

A larger proportion of construction types with a high vulnerability, e.g., single-family houses, results for the linear city in the highest criticality of all three examples. The large areal dispersion and the higher percentage of uncritical object types result in the lowest frequencies.

The compact city model includes a high mixture of different object types, a high building density and results in several local hotspots. Different construction types lead to a varying severity concerning the vulnerability. The lowest criticality of the compact city is higher rated than the maximum values of the central business district, which is observable with the yellow and black curves on the ordinate in Figure 5, for example.

Beside the derived risk quantities, the introduced framework enables the consideration of recovery processes, which is quantitatively described as performance over time [22]. The performance is oriented to the usability of an object and depends on the derived degree of damage.

4.4. Analysis of Recovery and Overall Resilience

The results for the response and recovery processes depending on the urban configuration are compared in Figure 6. The application of Equation (3) is visualized in this diagram, meaning the averaged loss of performance for a certain city model, if an adverse event occurs with respect to all possible threat types, threat positions and buildings. The dispersion of risk, as shown in the risk diagram of Figure 5 is eliminated in the performance-time relation by building the average.

In Figure 6, the high vulnerability for the linear city model is present with the strongest drop of performance at the time of the impact. The smallest discontinuity at $t = 0$ underlines the robust behavior of the central business district, which is congruent with the results of the risk analysis in Figure 5. Full recovery time of the building usability of the central business district is twice as long as in case of the linear model, which shows a relative short recovery behavior. This circumstance is due to the fact that high-rise buildings have a longer construction time than multi-functional and single-family houses, as shown in the overview in Table A1. Hence, the mixture of different object and construction types of the compact city is also reflected with the performance-time relations.

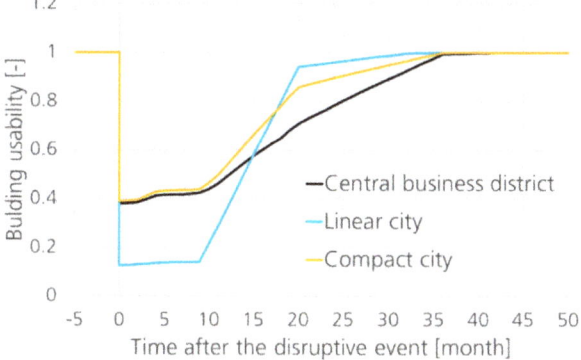

Figure 6. Comparison of the performance-time relation of the three investigated city models based on the average of the derived risk quantities in combination with expected recovery times.

Per the definition, as shown in Figure 1 (right), an often used single quantity to measure the resilience of a system is realized by integration of the performance loss over time or the loss per time is often used [11,32]. The recovery capabilities are determined in the following with respect to a performance loss function per duration time of disruption (second division by $t_{n_p} - t_1$), a kind of performance function loss gradient,

$$R_Q = \frac{1}{Q_{max}\left(t_{n_p} - t_1\right)^2} \sum_{p=0}^{n_p-1} \int_{t_p}^{t_{p+1}} (Q_{max} - Q(t))dt \qquad (4)$$

Additionally, to the pure risk assessment, the recovery behavior influences the target quantity of acceptable resilience. The elongated form of the linear city results in areas, which will be not affected by critical objects and hence small susceptibilities. Many of the residential objects are constructed as a single-family house with masonry wall constructions and result in strong vulnerabilities and hence in critical risk values. The simple construction types include short recovery to reach the initial usability, which has a lasting effect on the defined resilience quantity. The application of Equation (4) shows that the linear city ($R_Q = 0.27$ [loss per time]) has almost the same averaged performance loss per time than the central business district ($R_Q = 0.26$ [loss per time]). Because of the larger proportion of similar object use types, the central business district includes no single hotspots. The averaged risk depends mainly on the vulnerability. The considered construction types to the larger extent of office buildings or towers include longer recovery phases, which is the reason for similar resilience quantities compared to the linear city.

Small free spaces and a high mixture of object types are the main reasons for an aerial susceptibility with several hot spots within the compact city model. A great variety of construction types result accordingly in a broad spread concerning expected damage effects but also in parts with robust behavior and short recovery phases. Within the three applied models and in alignment to the introduced methodology, the outcomes point out that the compact city ($R_Q = 0.22$ [loss per time]) results in the smallest quantities concerning the averaged performance loss over time. This is a very interesting result, since the compact city is also favored from many other perspectives including sustainability and quality of living.

The three application examples underline the benefit of a susceptibility, vulnerability driven and risk-informed resilience assessment. The extension on the further dimension of recovery allows a more precise and deeper evaluation compared to classical risk assessment schemes. Low vulnerability or a high susceptibility results in critical risk values. However, in combination with short recovery phases, such systems can still be comparatively resilient despite critical risk quantities. From a risk perspective, the costs of the overall recovery phase have to be quantified adequately. From a resilience management perspective, classical susceptibility, vulnerability and risk cover only parts of the resilience management cycle. Hence, a correlation between risk and resilience is not mandatory.

The new framework allows a quantification which resilience phase is more effective for the considered urban area. Based on the multi-dimensional and complex characteristics of a certain city type, generalized statements about a most effective resilience improvement measure are not available and requires an individual investigation per city and the examination of different resilience phases. If the assessment results in relatively high susceptibilities, preparation or prevention measures will be more powerful. Protection measures are adequate, if the considered system exhibits high vulnerabilities. Decreasing damage effects result in smaller recovery efforts and require lower efforts concerning the response.

The response and recovery perspective, with focus on reconstruction offers the additional quantification of resilience in terms of recovery times, recovery slopes and expected loss. A steeper slope of the performance function results in a faster recovery and is considered in the applied expression to give an idea of rapidity within the recovery phase. The introduced formulation in Equation (4) results in a single quantity and gives the option of comparability between different cities or resilience improvements.

5. Summary and Conclusions

Within this paper, a risk-based method, as introduced in [22], is applied to three different urban forms and expanded to the aspect of recovery to get insights concerning the resilience of urban areas. Based on a decisive definition of the terminology, the present paper introduces a mathematical concept for the quantification of resilience. Different quantities are identified to be most relevant for the five resilience management phases. Urban modelling quantities have a lasting effect on the preparation phase for resilience. The susceptibility analysis is able to evaluate the prevention phase. Protection measures can be matched with the vulnerability and risk quantities. Recovery processes can be matched with the estimation of time spans for different urban objects. Response measures are only indirectly matched with consideration of other resilience management phases.

The combination of risk quantities and recovery processes deliver a time dependent estimation of performance to quantify resilience. The application of different enhancement measures allows then the evaluation of the effectiveness for single resilience management phases. Preparation, prevention and protection measures can be directly addressed. The management phases of response and recovery are indirectly supported. An increased robustness results in smaller damage effects and hence in smaller efforts concerning response and recovery, for example.

Complex and mostly qualitative social aspects are not considered, but the derived approach delivers a precise estimation of expected losses in terms of loss and degree of recovery of built functionalities of urban objects. However, the performance of buildings is not yet quantified and prized.

The risk-based resilience approach is applied to three typical urban forms concerning the damage type building collapse. The building density, the mixture of object types and the applied construction types determine the various resilience management quantities. City quarters with a clear and homogeneous allocation of use types result in an approximately uniform distribution of susceptibility and the risk depends mainly on the vulnerability effects, as shown with the results of the central business district. Therefore, the application of protection measures would be most effective to result in a resilient surrounding.

The extension on recovery as a further resilience dimension shows that an increasing robustness or low risk values alone are not sufficient to qualify resilient systems. The example of the linear city results in stronger damage effects but a similar resilience quantity compared to the central business district, based on shorter recovery times. The approach allows a quantitative comparison, how effective the investigation of further resilience phases, like preparation of prevention, which could be an option if there are several hotspots at risk.

The high mixture of object and construction types within a compact city results in several hot spots but the composition results in a robust behavior an short recovery and hence in the most resilient morphology in terms of recovery behavior without application of enhancement measures. The inhomogeneous distribution repeals scenarios with strong damage effects or long recovery phases. This fits nicely with the often-attributed sustainability and societal acceptance of compact city forms.

Building density or the distribution of objects, free spaces, construction type or the use of a building are main attributes, which will influence the resilience of an urban surrounding. The results deliver information on how growing agglomerations can be sustainably designed also with regards to new threats. The overall framework and calculation methods builds a possible basis for urban planners, decision makers or insurance companies to analyze and optimize designs of city areas.

Within this paper, terroristic threats are exemplary evaluated. Based on the clear definition, this framework allows also an evaluation of other main kind of disasters. This requires the availability of statistical data and appropriate models to assess expected damage effects. Examples could be models in the range of earthquake events [33] or flood risks [34].

The introduced framework uses validated engineering models and the comparison to real events underline the accuracy of the statistical data. The estimation of recovery phases based on expert knowledge and results in capable quantities to postulate resilience. A possible deviation of recovery times is currently not considered and will be a point of reference for future research. Similar to social aspects, which are currently only indirect matched.

Author Contributions: Besides the work of K.F., the following contributions are provided by the coauthors: S.H. supervised the research that leads to the content of this paper. W.R. gave contributions within the application examples and the definition of the city models. I.H. helped to define the mathematical formulations, presented in this paper.

Acknowledgments: The research leading to these results has received funding form the European Commission's 7th Framework Programme within the EU project EDEN under grant agreement no. 313077 and VITRUV under grant agreement no. 261741. The contribution of Andreas Bach and Ingo Müllers (Schüßler-Plan Engineering GmbH) for the provision of the list with pre-defined construction types including the estimated time scales is gratefully acknowledged.

Conflicts of Interest: The authors declare no conflict of interest.

Appendix A

Table A1. Overview of pre-defined building types, their construction and time-scales for planning and construction to estimate the recovery process [29].

Construction Type	Construction	Number of Floors	Recovery Time [month]	
			Planning, Approval	Construction
Single-family house	Masonry (walls) Reinforced concrete (slabs, beams)	3	9	11
Multi-family house	Reinforced concrete (walls, slabs, beams)	6	15	13
Block of flats	Reinforced concrete (walls, slabs, columns)	17	18	18

Table A1. *Cont.*

Construction Type	Construction	Number of Floors	Recovery Time [month]	
			Planning, Approval	Construction
Industrial building	Steel (beams) trapezoidal steel profiles (roof, walls)	1	18	14
Multi-family house, mixed use	Reinforced concrete (walls, columns, slabs) Steel (columns) Timber (girders)	4	15	13
Sports hall	Reinforced concrete (walls, columns, slabs) Steel (roof construction)	1	15	18
Shopping mall	Reinforced concrete (walls, slabs, columns)	5	20	22
Office building		7	18	18
Office tower		13	20	22
Public transport terminal	Reinforced concrete (walls, slabs, columns) Steel (roof)	1–2	24	24

References

1. Branscomb, L. Sustainable cities: Safety and Security. *Technol. Soc.* **2006**, *28*, 225–234. [CrossRef]
2. The Minerals, Metals and Materials Society (TMS). *Engineering Solutions for Sustainability, Materials and Rescources—Workshop Report and Recommendations*; John Wiley & Sons: Hoboken, NJ, USA, 2012.
3. Kröger, W.; Zio, E. *Vulnerable Systems*; Springer: London, UK, 2011.
4. Department of Economic and Social Affairs. *World Urbanization Prospects*; The 2014 Revision; United Nations: New York, NY, USA, 2014.
5. Cross, J. Megacities and small towns: Different perspectives on hazard vulnerability. *Environ. Hazards* **2001**, *3*, 63–80.
6. Lin, C.; Liou, D.; Wu, K. Opportunities and challenges created by terrorism. *Technol. Forecast. Soc. Chang.* **2007**, *74*, 148–164. [CrossRef]
7. Tamvakis, P.; Xenidis, Y. Comparative Evaluation of Resilience Quantification Methods for Infrastructure Systems. *Procedia Soc. Behav. Sci.* **2013**, *74*, 339–348. [CrossRef]
8. Carpenter, S.; Walker, B.; Anderies, J.; Abel, N. From Metaphor to Measurement: Resilience of What to What? *Ecosystems* **2001**, *4*, 765–781. [CrossRef]
9. Cutter, S.; Bernes, L.; Berry, M.; Burton, C.; Evans, E.; Tate, E.; Webb, J. A place-based model for understanding community resilience to natural disasters. *Glob. Environ. Chang.* **2008**, *18*, 598–605. [CrossRef]
10. Gallopín, G. Linkages between vulnerability, resilience, and adaptive capacity. *Glob. Environ. Chang.* **2006**, *16*, 293–303. [CrossRef]
11. Bruneau, M.; Chang, S.; Eguchi, R.T.; Lee, G.C.; O'Rourke, T.D.; Reinhorn, A.M.; Shinozuka, M.; Tierney, K.; Wallace, W.A.; von Winterfeldt, D. A Framework to Quantitatively Assess and Enhance the Seismic Resilience of Communities. *Earthq. Spectra* **2003**, *19*, 733–752. [CrossRef]
12. Chang, S.; Shinozuka, M. Measuring Improvements in the Disaster Resilience of Communities. *Earthq. Spectra* **2004**, *20*, 739–755. [CrossRef]
13. Cimellaro, G.; Reinhorn, A.; Bruneau, M. Framework for analytical quantification of disaster resilience. *Eng. Struct.* **2010**, *32*, 3639–3649. [CrossRef]
14. Jabareen, Y. Planning the resilient city: Concepts and strategies for coping with climate change and environmental risk. *Cities* **2013**, *31*, 220–229. [CrossRef]
15. Federal Emergency Management Agency. HAZUS—Methodology for Estimating Potential Losses from Disasters. Available online: http://www.fema.gov/hazus (accessed on 3 August 2015).
16. Quiel, S.; Marjanishvili, S.; Katz, B. Performance-Based Framework for Quantifying Structural Resilience to Blast-Induced Damage. *J. Struct. Eng.* **2015**, *142*, C4015004. [CrossRef]

17. Federal Emergency Management Agency. *FEMA-426: Reference Manual to Mitigate Potential Terrorist Attacks Against Buildings*, 2nd ed.; U.S. Department of Homeland Security: Washington, DC, USA, 2011.
18. Curdes, G. Stadtmorphologie und Klimawandel—Welche Stadtstrukturen können den Klimawandel überleben? In Proceedings of the 17th International Seminar on Urban Form, Hamburg, Germany, 20–23 August 2010.
19. Jabareen, Y. Sustainable Urban Forms—Their Typologies, Models, and Concepts. *J. Plan. Educ. Res.* **2006**, *26*, 38–52. [CrossRef]
20. Adger, W. Social and ecological resilience. Are they related? *Prog. Hum. Geogr.* **2000**, *24*, 347–364. [CrossRef]
21. Thoma, K. *Resilien-Tech-Resilience-by-Design: Strategie für die Technologischen Zukunftsthemen*; Acatech, Deutsche Akademie der Wissenschaften: Berlin, Germany, 2014.
22. Fischer, K. Resilience Quantification of Urban Areas, an Integrated Statisitcal-Empirical-Physical Approach for Man-Made and Natural Disruptive Events. Ph.D. Thesis, University of Freiburg, Freiburg, Germany, 2018.
23. Fischer, K.; Häring, I.; Riedel, W.; Vogelbacher, G.; Hiermaier, S. Susceptibility, vulnerability and averaged risk for resilience enhancement of urban areas. *Int. J. Protect. Struct.* **2016**, *7*, 45–76. [CrossRef]
24. Fischer, K.; Häring, I. SDOF response model parameters from dynamic blast loading experiments. *Eng. Struct.* **2009**, *31*, 1677–1686. [CrossRef]
25. Müllers, I.; Fischer, K.; Nawabi, A.; Riedel, W. Design against Explosions and Subsequent Progressive Collapse. *Struct. Eng. Int.* **2015**, *25*, 319–325. [CrossRef]
26. Siebold, U.; Ziehm, J.; Häring, I. Terror Event Database and Analysis Software. In Proceedings of the 4th Security Research Conference, Karlsruhe, Germany, 29th September–1st October 2009.
27. Fischer, K.; Siebold, U.; Vogelbacher, G. Empirical analysis of security critical events in urban areas. *Bautechnik* **2014**, *91*, 262–273. [CrossRef]
28. Valente-Pereira, L. *Urban Form Definition in Urban Planning*; Simplissimo: Lisbon, Portugal, 2014.
29. Schüßler Plan Ingenieurgesellschaft mbH. *D4.4: Vulnerability Analysis of Generic Structures and Functional Units*; EU project VITRUV: Düsseldorf, Germany, 2013. Available online: www.vitruv-project.eu (accessed on 28 May 2018).
30. Proske, D. *Catalogue of Risks; Natural, Technical, Social and Health Risks*; Springer: Berlin, Germany, 2008.
31. Hunter, P.; Fewtrell, L. Acceptable Risk. In *Water Quality: Guidelines, Standards and Health*; Fewtrell, L., Bartram, J., Eds.; WHO by IWA Publishing: London, UK, 2001.
32. Cimellaro, G. *Urban Resilience for Emergency Response and Recovery—Fundamental Concepts and Applications*; Springer: Berlin, Germany, 2016.
33. Krawinkler, H.; Miranda, E. Performance-based earthquake engineering. In *Earthquake Engineering, from Engineering Seismology to Performance-Based Engineering*; Bozorgniam, Y., Bertero, V.V., Eds.; CRC-Press: Boca-Raton, FL, USA, 2004; pp. 1–59.
34. Büchele, B.; Kreibich, H.; Kron, A.; Thieken, A.; Ihringer, J.; Oberle, P.; Merz, B.; Nestmann, F. Flood-risk mapping: Contributions towards an enhanced assessment of extreme events and associated risks. *Nat. Hazards Earth Syst. Sci.* **2006**, *6*, 485–503. [CrossRef]

© 2018 by the authors. Licensee MDPI, Basel, Switzerland. This article is an open access article distributed under the terms and conditions of the Creative Commons Attribution (CC BY) license (http://creativecommons.org/licenses/by/4.0/).

Article

Effects of Commercial Activities by Type on Social Bonding and Place Attachment in Neighborhoods

Byungsuk Kim [1] and Jina Park [2,*]

1 Department of Urban Planning, Hanyang University, Seoul 04763, Korea; bsk0728@nate.com
2 Department of Urban Planning and Engineering, Hanyang University, Seoul 04763, Korea
* Correspondence: paran42@hanyang.ac.kr; Tel.: +82-2-2220-0332

Received: 20 March 2018; Accepted: 24 May 2018; Published: 29 May 2018

Abstract: Place attachment is an emotion that people experience in connection to a specific place and it is needed to maintain a sustainable neighborhood community. The emotion is affected by various factors, such as experience, function, environment, and satisfaction. This study focuses on commercial structures, which are one feature that characterizes the physical environments of neighborhoods. The aim of this study is to determine the effects of commercial activities in different commercial environments on social bonding and place attachment in residents. Two sites were selected for analysis due to their different commercial environments, and path analysis was used to examine the relationships among factors. The results indicate that commercial activities, which can vary according to commercial type, had both direct effects and indirect effects through social bonding between residents on place attachment. These results suggest that the commercial environment is an important element affecting the community and place attachment of residents in neighborhoods.

Keywords: place attachment; commercial types; commercial activities; social bonding; physical activities

1. Introduction

In light of the emotions, intimacy, and happiness that people feel in connection to their neighborhoods, there is more than functional and physical meaning to neighborhoods. The emotional ties of residents to the area in which they live can be defined as place attachment to a neighborhood. Analysis of such has been extensively addressed in looking at the emotional experiences of people and people's ties to places in terms of various factors of function and satisfaction [1–4]. Place attachment is a concept similar to residential satisfaction in terms of cognition about a physical environment, with one important distinction. If residential satisfaction is a functional evaluation of a place of residence from the viewpoint of the people living there, then place attachment is an emotional evaluation of the place of residence. In order to foster a salubrious neighborhood, it is necessary to consider not only the functional aspects of a place, but also the psychological and emotional demands of human beings. In this context, place attachment can be an important criterion of residential environment evaluation [5]. Since place attachment is formed when a person is psychologically connected to a specific place, it can be a source of relief for residents amid the rapid changes of modern urban environments.

Place attachment can be formed by various factors. In the work of Lewicka [6], which analyzed the studies of place attachment within the past 40 years, various variables including socio-demographic, social, and physical predictors are identified in place attachment. Many of the previous studies related to place attachment in neighborhoods have been conducted in terms of the influence of physical environment characteristics on place attachment in local residents [7–11]. This is because place attachment is basically an emotional bond that a person has with a place. On the other hand, some studies [4,7,9] have analyzed the influence of social bonds on place attachment, focusing on the social ties of residents of a neighborhood. Additional other studies [7,12,13] have investigated the influence of people's personal characteristics on place attachment.

Overall, research results show that physical environment characteristics such as place, personal characteristics of individual residents, and characteristics of social relationships between residents influence the formation of place attachment among locals. Based on these results, the focus of this study is the relationships among these various physical environment characteristics and place attachment. Beyond the influence of place attachment on individual variables, it is necessary to clarify the mechanism of the relationship of structural influence among the features of physical environments, the personal characteristics of residents, and social bonding. In order to establish the relationship between place attachment and its influencing variables, it is necessary to understand the interaction between people and place—as well as between people and people—in the physical features of neighborhoods. In this study, we focus on "activities" as a key parameter in our understanding of the relationships between the variables affecting place attachment.

From the perspective of environmental psychology, human activities arise in the context of an environment. Many studies have demonstrated the relationships between physical environment and people's activities [14–17]. If the neighborhood environment affects people's activities, and if the activities affect the level of place attachment, then it is possible to discuss which characteristics of a neighborhood environment ultimately promote place attachment. If the unique physical environment of a neighborhood increases the amount and types of activities pursued by its residents—thereby positively impacting place attachment—then physical environment characteristics stand to provide urban design implications for a salubrious neighborhood.

Among various activities that occur in a neighborhood, this study focuses on commercial activities. Commercial activities are basic, essential activities in people's daily lives. Depending on the type of commercial facility, however, the style of activities can be very different. Various types and characteristics of commercial facilities are identified based on their location, usage, and surroundings, but commercial facilities can be compared in two basic forms: street shops, which are located along the streets of residential environments, and mall-type shops, describing a configuration in which shops are concentrated in specific buildings (in contrast to street shops). In terms of the commercial environment of a neighborhood, this difference in commercial form can impact the commercial activities of residents. As a result, the distinction can manifest as a difference in neighborhood activities. In a neighborhood with small shops, small quantities of goods are frequently purchased. In contrast, in a neighborhood with large marts, large amounts of goods are purchased less frequently. Additionally, if an individual uses a car to purchase a large quantity of goods, then the individual may have fewer opportunities to come into contact with her or his neighbors and have less interaction with the neighborhood environment than when walking on foot. Thus, these differences in commercial activities may lead to differences in face-to-face opportunities among human beings. The activities of the residents in a neighborhood can be a driving force for local initiatives such as local revitalization and community spirit.

In this respect, commercial forms affect the physical activities of people and are, therefore, an important subject of study in terms of social bonding among residents in a neighborhood and the promotion of place attachment. Various physical elements of neighborhoods have been studied, including walkability [17,18], street connectivity [19], land-use mix [15,16], pedestrian and traffic safety [20], and recreation facilities and parks [16]. In contrast, however, few studies have focused on the physical elements of commercial activities and commercial types—one of the basic activities of people's daily lives. Indeed, there are no studies to analyze relationships among the influencing factors of place attachment, such as the effects of commercial form on people's commercial activities and the effects of commercial activities on social bonding and place attachment. Therefore, this study investigates the effects of "activities" as a function of commercial form in relation to the physical environment of a neighborhood, social bonding between residents, and place attachment.

2. Previous Studies

2.1. Place Attachment

The concept of attachment can be applied to explain the relationship between human beings and the environment (that is, our interest in a specific place or community), although the concept is mainly described in terms of people's connection to other people, such as infants and parents or family and friends. Recent research has focused on attachments and so-called place attachments in terms of "friendliness to physical places" in contrast to the relationship networks between people that are emphasized in sociological research. Place attachment can be understood in two dimensions: place identity and place dependence [21]. Proshansky [22] defines place identity as "a complex pattern of beliefs, values, feelings, expectations, and preferences relevant to the nature of the physical world," which is a complex cognitive structure of a person at a specific place. Place dependence is a functional relationship with a person's residence (or other specific area) and may be explained by comparing one place with another according to individual needs. In this sense, place dependence is defined in terms of whether a particular area or facility functions in accordance with a user's activities [23].

Place attachment is a complex concept that describes the interaction of emotional or symbolic relationships, emotions that are formed in a single physical environment, and human interrelationships and feelings occurring in that particular place. In other words, place attachment shows the ways in which a place is more than just a physical environment. Place attachment arises not only from the place itself, but also from emotions due to consciousness, experience, psychological reaction, symbolism, and other complex functions of cognition that people associate with the place [24]. Thus, we can define place attachment as being caused by empirical experiences that occur when people consistently interact with a specific place. From this point of view, if the residents in a neighborhood engage in continuous visits and activities in a specific place, a positive relationship may be found between people and place in the form of place attachment.

2.2. Place Attachment in Neighborhoods

Place attachment refers to an emotional factor between people and physical spaces. In this context, research trends related to place attachment have examined personal, social, and regional differences in place attachment, together with the factors affecting place attachment. Existing research mainly deals with personal and social variables (such as race, age, and economic power), time variables (such as duration of residence and satisfaction with the local environment), and spatial characteristics. This multifactorial lens suggests the complexity of the feeling of place attachment.

It is generally accepted that levels of satisfaction with people's residential environments are highly related to place attachment [7–11]. In order to maintain place attachment, certain neighborhood environment standards must be achieved. In declining residential areas, the level of attachment of people to their place of residence will decline, and relocation will be considered due to deterioration of the quality of life [25]. Place attachment occurs mainly through neighborhood environments and in bonding with neighbors. Neighborhood satisfaction, which is a passive, direct experience—as opposed to social bonding with neighbors (which needs to be actively pursued during settlement in a new local environment or following neighborhood redevelopment)—plays an important role in the formation of place attachment [4]. In addition, place attachment has different characteristics depending on the scale of a place, such as a city versus a smaller neighborhood [2,26].

In the physical environment of a neighborhood, place attachment changes according to the personal and social characteristics of the people directly experiencing place attachment. There are differences depending on whether people reside in a home of their own, on race [7], and on whether people are indigenous or immigrants [12]. If there is a link between residents and a neighborhood wherein place attachment is formed, positive effects are noted in the neighborhood environment. Home ownership, race parity, and indigenous people inspire a sense of belonging to an area. In this respect, bonding with neighbors is an important factor in place attachment. When this bond is strong, place attachment is positively affected [4,7,9]. On the contrary, if there is a low level of solidarity among residents and people move frequently, the formation of place attachment can be difficult [3]. To increase social bonding between neighbors, contact with neighbors and time spent together must increase. In other words, the accumulation of experience in the neighborhood is important, and experience is generally proportional to time. A variable that represents this relationship in the context of neighborhoods is duration of residence. Many studies have shown that residence period has an impact on place attachment [7,9,27,28].

In order to foster strong social bonding and place attachment among residents, residents should be active in their neighborhood, and the neighborhood environment should reflect their desires. The physical environment of various neighborhoods is related to the activities of people. Especially, focusing on commercial environments, it has been found that mixed-land use [15], retail floor area ratio [29], and commercial facilities [30,31] affect people's walking activities. Walking also positively affects resident bonding [32], and walking-friendly neighborhood conditions can improve people's sense of community [33,34]. This correlation explains the importance of a neighborhood's physical environment relative to the amount and types of activities that occur therein in the context of place attachment. Accordingly, this study aims to investigate the effects of commercial activities on social bonding by neighborhood commercial types, together with the impact on place attachment.

3. Methods

3.1. Study Areas

Neighborhoods characterized by different types of commercial structures were selected in order to analyze the relationship among commercial types, commercial activities, social bonding, and place attachment. The distinction is between commercial types of street shops or a shopping mall. First, Lancry in Paris, France, was selected to represent the street shops type (see. Figure 1). The whole area is characterized by medium-rise buildings with an average of six to seven floors. On the first floors of buildings, various small shops including commercial, service, and manufacturing shops are located along the street. The upper parts of the buildings are composed of residential living spaces. Lancry is included in a grouping of 11 regions that comprise the business district of Vital'Quartier, initiated in 2004 by the Société d'Economie Mixé d'Aménagement de l'Est de Paris (SEMAEST) as part of the commercial revitalization project in Paris. It can be seen as a commercial type of street shop.

Figure 1. Lancry.

The opening of large shopping malls in Paris is regulated by law as "La loi Royer" (1973) and "La loi Raffarin" (1996) and permission is required to open stores over 300 m^2 by Raffarin law. For this reason, mall-type commercial areas should be chosen outside of Paris. La Défense was chosen as a representative shopping mall type among commercial areas (as compared to street shop type). La Défense (see Figure 2) is a representative new town in France and has two large shopping malls (CNIT and Quaten Temps). Thus, unlike the street shops seen in Paris, large, mall-type commercial facilities are used by residents of La Défense for various shopping activities.

Figure 2. La Défense.

The average age of residents in Lancry is 37 years old and the population density is 38,160/km^2. There are 6.1 stores per 100 m among bars, cafes, and restaurants, and 26.8 stores per 100 m among all kinds of shops. The housing density in Lancry is 244 log./ha [35]. La Défense is distributed across three zones—Courbevoie, Puteaux, and Nanterre. In this study, the average values for Fb de l'Arche (Courbevoie), Gambetta (Courbevoie), and La Défense (Puteaux) were used. In the case of La Défense, the average age is 35.33 years with a population density of 17,836.67/km^2. There are, on average, 1.57 shops for bars, cafes and restaurants, with an average of 9.53 stores per 100 m. The housing density in La Défense is 94.67 log./ha [36]. This information indicates that Lancry has a higher density of shops than La Défense, which means that the residents in Lancry have a higher accessibility to shops than La Défense.

3.2. Variable Settings

First, the two commercial types were changed to dummy variables (street: 1, mall: 0) to determine the effects of the commercial types. The weekly shopping frequency of residents depending on the commercial type was measured to represent levels of commercial activities (1 to 5 points). Aspects of social bonding were established as variables to determine whether bonding among residents affects place attachment, and the items were related to levels of closeness among neighbors. Place attachment variables were measured using place identity (measured by a fundamental question about the extent to which residents feel attached to their neighborhood) and place dependence, which is a functional necessity.

By definition, place attachment is affected by levels of attachment in residents to a specific place. Accordingly, neighborhood environments and place attachment have been the main subjects of previous research. In this study, the following variables were set as environmental factors in order to scrutinize environmental factors through an analytical lens. First, parameters of "satisfaction with commercial infrastructure" and "satisfaction with commercial quality" were set in terms of commercial environment satisfaction. Two variables were set to determine whether the physical features of commercial environments influence place attachment or whether qualitative factors of shops have effects on place attachment. In addition, the variable of "satisfaction with neighborhood facilities" was set in order to confirm the effects of neighborhood environments in the findings of previous research on place attachment.

3.3. Questionnaires and Data Collection

Table 1 shows that questionnaires consisted of items about place attachment, social bonding, commercial satisfaction, and resident satisfaction with neighborhood facilities. The items related to place attachment and social bonding were reconstructed based on previous studies [9,11,37]. The items on commercial satisfaction and neighborhood satisfaction were constructed based on the activities and direct experiences of the residents in the neighborhoods. Complete questionnaires included five items on place identity, five items on place dependency, five items on social bonding, eight items on commercial satisfaction, and six items on neighborhood satisfaction. All of the above items were measured on a five-point scale.

Table 1. Questionnaire items.

Factors		Items
Place attachment	Place identity	This neighborhood is important in my life. I say that I live in this neighborhood when I introduce myself. If someone asks me about this neighborhood, I can answer the questions. I am proud to live in this neighborhood. This neighborhood is special to me.
	Place dependence	This neighborhood is suitable to my line of work. This neighborhood is better to live in than other neighborhoods I do many activities around this neighborhood. Leaving this neighborhood causes me to feel sad. I would live in this neighborhood even if I had the chance to move to other areas.
Social bonding		I know the residents of my neighborhood well. I have friendly neighbors to talk to. I have many friends in the neighborhood. I attend neighborhood gatherings often. I attend the event of neighborhood often.
Satisfaction with commerce		Number of shops, types of shops, necessary shops, price of products, service of shops, type and quality of products, distance to shops.
Commercial activities		Shopping frequency.
Satisfaction with neighborhood facilities		Green space, public and cultural facilities, public transportation, education services, safety, pedestrian environment.
Commercial type		Street shops-type neighborhood: Lancry. Mall type-neighborhood: La Défense.

Note. Place identity and place dependence items, as well as social bonding items, are in a designated order (e.g., identity 1, identity 2); Commercial type: Street shops = 1, Mall = 0.

Six surveyors performed surveys from 10 am to 7 pm in Lancry (26–27 June 2015). In La Défense, the same six surveyors conducted surveys from 10 am to 7 pm (30 June 2015). Sampling was conducted using a convenience sampling method, and data were collected evenly across age group and gender. Before starting the surveys, the surveyors ensured there was "agreement to participate in the questionnaire" and that the participants were "were residents of Lancry or La Défense." Here, 60% of respondents participated in the survey and a total of 164 questionnaires were collected. Of the returned questionnaires, 94 were collected from residents of Lancry and 70 from residents of La Défense. In Lancry, the questionnaire was distributed among local residents passing through the streets where the shops are located, whereas in La Défense, the survey was distributed in residential areas, squares, or parks in the greater La Défense area rather than the shopping mall itself. Since the large shopping mall in La Défense is a commercial center with mass appeal, local residents, in addition to people coming from distant areas to shop, were potentially involved in the survey. For accuracy of communication, the questionnaire was a non-English version conducted in French. The collected questionnaires were coded using SPSS 21 statistical software (IBM, NEWYORK, USA).

3.4. Research Design

This study focuses on the influence of commercial activities on social bonding and place attachment among residents of a neighborhood. To accomplish this, we performed the following two-step model setting process. First, note that commercial activities will affect social bonding between residents as well as place attachment; this is because a larger amount of activity in a neighborhood allows for more opportunities to meet neighbors and experience the neighborhood. Second, note that commercial activities are influenced by satisfaction with the commercial and physical environments according to different commercial types (e.g., street shops or mall types), which are likely to affect commercial activities.

This study uses two models. Model 1 analyzes each site individually to determine the effects of commercial activities on social bonding, as well as the direct and indirect effects of commercial activity on place attachment. A larger amount of commercial activities leads to higher activity in the neighborhood. The relationships between shop owners and neighbors are formed through the spaces in the neighborhood, such as shops and streets, and social bonding is an important factor that leads to attachment to the neighborhood. In addition, if people continue to use a space, they will develop feelings about that space, which can lead to place attachment. Model 2 analyzes the two sites combined to determine the different commercial types that affect commercial activities. The type of commercial environment, consisting of street shops or malls, can affect the commercial activities depending on accessibility. In the case of street shops, the shops are located near the houses and a pattern of frequent purchases of small quantities of goods will appear. However, accessibility in a mall-type commercial environment is lower than that for street shops, and instead exhibits a pattern of buying many goods at once.

3.5. Data Analysis Using Path Analysis

Based on the collected survey data, the following analysis process is conducted to confirm the purpose of this study. First, factor analysis is conducted based on questionnaire items to derive analysis factors. Second, correlation analysis is performed to analyze the correlation between factors and to remove factors with high relevance. Finally, analysis of the research model shown in Figure 3 is conducted through path analysis using the Amos program.

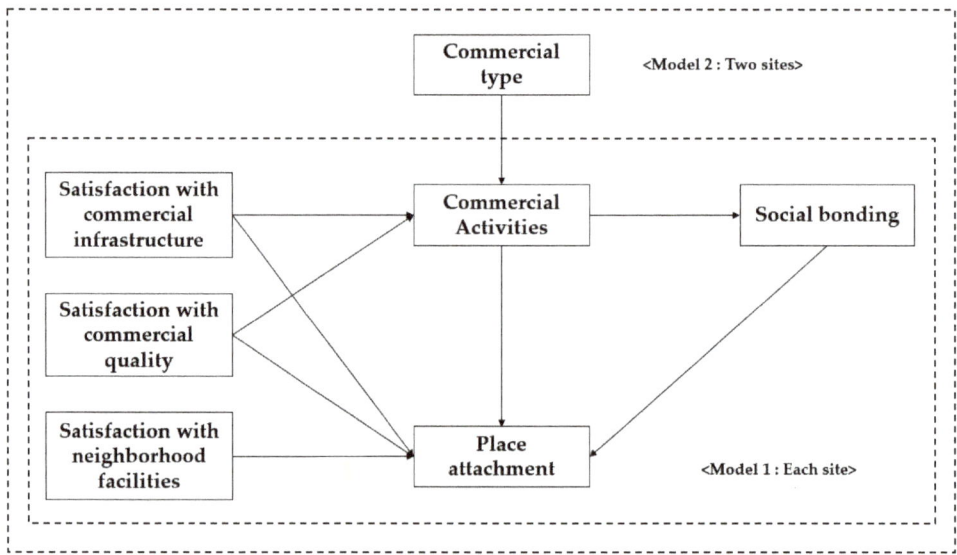

Figure 3. Research models.

Path analysis is a technique for explaining the causal relationships between variables in a non-experimental situation. The validity of the causal relationships between variables is examined using the collected data. The effects of any one variable on another variable are called direct effects, while the effects of one variable on one (or more) variable(s) by way of other variables are called indirect effect. The purpose of this study is to analyze direct and indirect effects of commercial activities on social bonding and place attachment, and effects of commercial types on commercial activities using path analysis based on collected data.

4. Results

4.1. Demographics and Factor Analysis

The demographic characteristics of the subjects were as follows (see Table 2). Of 164 total subjects, 87 were males and 77 were females. In terms of age distribution, the highest proportion of the sample was people in their 30s (24.4%), followed by those in their 20s (23.8%), and those in their 40s (14.6%). The average age of participants is 39.89 years old in Lancry and 34.22 years old in La Défense. The highest proportion of the sample claimed a residence period of 4–10 years in a neighborhood (31.7%), followed by 1–3 years (21.3%). In addition, 47% of the residents had lived for longer than 10 years in the neighborhood. Distribution according to site was 57.3% in Lancry and 42.7% in La Défense.

Table 2. Demographics.

Factors		Lancry N (%)	La Défense N (%)	Total N (%)
Sex	Males	51 (54.3)	36 (51.4)	87 (53)
	Females	43 (45.7)	34 (48.6)	77 (47)
	Total	94 (100)	70 (100)	164 (100)
Age	10s	7 (7.4)	15 (21.4)	22 (13.4)
	20s	23 (24.5)	16 (22.9)	39 (23.8)
	30s	20 (21.3)	20 (28.6)	40 (24.4)
	40s	17 (18.1)	7 (10.0)	24 (14.6)
	50s	16 (17.0)	4 (5.7)	20 (12.2)
	60s and over	11 (11.7)	8 (11.4)	19 (11.6)
	Total	94 (100)	70 (100)	164 (100)
Residence period	1–3 years	23 (24.5)	12 (17.1)	35 (21.3)
	4–10 years	26 (27.7)	26 (37.1)	52 (31.7)
	11–15 years	13 (13.8)	16 (22.9)	29 (17.7)
	16–20 years	16 (17.0)	10 (14.3)	26 (15.9)
	21 years	16 (17.0)	6 (8.6)	22 (13.4)
	Total	94 (100)	70 (100)	164 (100)

Factor analysis was performed twice. The first group included items related to commerce and neighborhood facilities, and the second group included items related to place attachment. The final factors are "satisfaction with commercial infrastructure" (SCI), "satisfaction with commercial quality" (SCQ), "satisfaction with 'neighborhood facilities" (SNF), "place attachment" (PA), "social bonding" (SB) (see Table 3), and "commercial activities" (CA).

Table 3. Factor analysis.

Factors		Factor Loading	Cronbach's α
Satisfaction with commercial infrastructure	Necessary shops	0.840	0.758
	Number of shops	0.691	
	Distance to go to shops	0.683	
Satisfaction with commercial quality	Types of shops	0.676	0.618
	Type and quality of products	0.814	
	Price of products	0.689	
	Service of shops	0.654	
Satisfaction with neighborhood facilities	Green space	0.853	0.658
	Public and cultural facilities	0.760	
	Pedestrian environment	0.654	
Duration of residence	Place identity 5	0.797	0.891
	Place dependence 2	0.765	
	Place identity 4	0.764	
	Place identity 1	0.758	
	Place dependence 3	0.745	
	Place dependence 4	0.713	
	Place identity 3	0.663	
	Place identity 2	0.660	
	Place dependence 5	0.594	
Social bonding	Community 3	0.845	0.778
	Community 2	0.797	
	Community 1	0.757	
	Community 4	0.626	

Table 4 shows that the averages for the factors are different according to the site. For SNF, the value for La Défense (3.55) was higher than the value for Lancry (3.14). However, for the other factors of PA (Lancry: 3.52; La Défense: 3.04), SB (Lancry: 2.98; La Défense: 2.75), SCI (Lancry: 3.65; La Défense: 3.42), SCQ (Lancry: 3.44; La Défense: 3.23), CA (Lancry: 2.80; La Défense: 2.31), and residence period (Lancry: 13.38; La Défense: 10.54), the values for Lancry were higher than the values for La Défense. Especially, CA in Lancry (2.80) was higher than in La Défense (2.31), which means that CA can vary depending on the commercial presence.

Table 4. Average value of factors (t-test).

Factors	Site	N	Avg.	S.D.	S.E.	t
Satisfaction with commercial infrastructure	Lancry	94	3.65	0.63087	0.06507	2.140 **
	La Défense	70	3.42	0.76084	0.09094	
Satisfaction with commercial quality	Lancry	94	3.44	0.58294	0.06013	2.248 **
	La Défense	70	3.23	0.57972	0.06929	
Satisfaction with neighborhood facilities	Lancry	94	3.14	0.76911	0.07933	−3.487 **
	La Défense	70	3.55	0.69710	0.08332	
Social bonding	Lancry	94	2.98	0.75956	0.07834	1.846 *
	La Défense	70	2.75	0.83839	0.10021	
Place attachment	Lancry	94	3.52	0.67531	0.06965	4.495 ***
	La Défense	70	3.04	0.69590	0.08318	
Commercial activities	Lancry	94	2.80	1.08039	0.11143	2.956 ***
	La Défense	70	2.31	1.02918	0.12301	
Duration of residence	Lancry	94	13.38	11.51913	1.18811	1.931 *
	La Défense	70	10.54	7.25356	0.86709	

* $p < 0.1$, ** $p < 0.05$, *** $p < 0.01$.

4.2. Path Analysis and Indirect Effects

Path analysis was used to examine the effects of individual factors. Model 1 (Lancry model: $p = 0.303$, CMIN/df = 4.851, RMR = 0.032, RMSEA = 0.048, GFI = 0.983, AGFI = 0.912, NFI = 0.964,

IFI = 0.994, CFI = 0.993; La Défense model: $p = 0.882$, CMIN/df = 0.295, RMR = 0.020, RMSEA = 0.000, GFI = 0.994, AGFI = 0.970, NFI = 0.974, IFI = 1.068, CFI = 1.000) is presented in Figure 4. CA (Lancry: $\beta = 0.179$, $p < 0.1$, La Défense: $\beta = 0.340$, $p < 0.01$) had a positive effect on SB, and SB (Lancry: $\beta = 0.660$, $p < 0.01$, La Défense: $\beta = 0.226$, $p < 0.1$) positively affected PA in both sites. However, the relationship between CA and PA is significant only in Lancry ($\beta = 0.124$, $p < 0.1$).

SCI and SCQ did not have significant relationships with CA in either site, but they did affect PA in the Lancry model (SCI: $\beta = 0.146$, $p < 0.1$, SCQ: $\beta = 0.161$, $p < 0.05$). On the contrary SNF ($\beta = 0.253$, $p < 0.05$) affected PA only in the La Défense model.

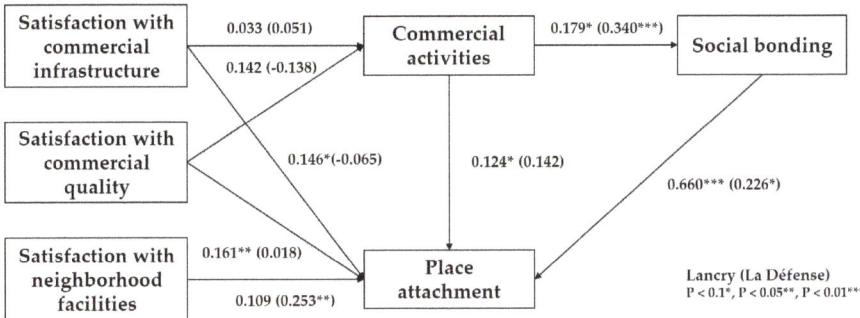

Figure 4. Path analysis of model 1.

Figure 5 presents a path analysis of model 2 for the two sites ($p = 0.003$, CMIN/df = 3.275, RMR = 0.025, RMSEA = 0.118, GFI = 0.968, AGFI = 0.849, NFI = 0.883, CFI = 0.907). The results show what factors affect CA. In this model, only CT ($\beta = 0.215$, $p < 0.01$) had positive effects on CA, while commercial satisfaction (e.g., SCI and SCQ) did not significantly affect CA. Further, CA affected SB ($\beta = 0.272$, $p < 0.01$), while PA ($\beta = 0.186$, $p < 0.01$) and SB ($\beta = 0.305$, $p < 0.01$) significantly affected PA. Especially, the relationship between CA and PA here is stronger than that in model 1; the reason for this is the variability of CA was expanded by combining the data of the two sites.

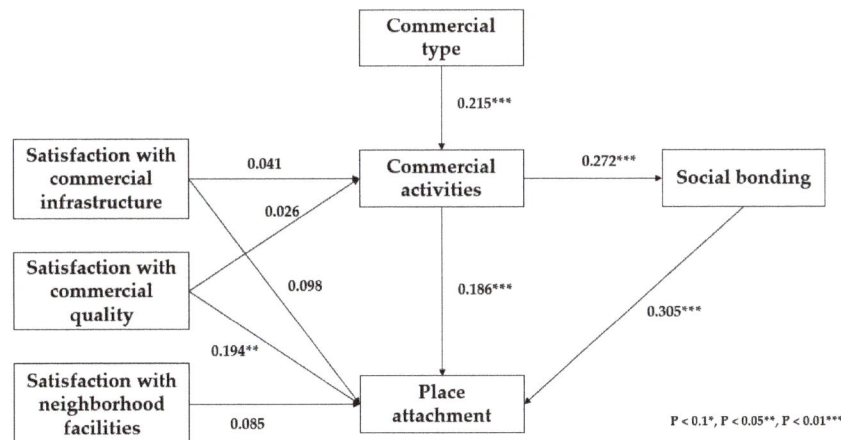

Figure 5. Path analysis of model 2.

Table 5 shows the indirect effects of CA through SB on PA. Analysis was conducted using two-tailed significance in Amos. First, in the Lancry case, the indirect effect of the path CA → SB → PA was 0.118 ($p < 0.05$), which was significant. In the La Défense, the indirect effect of CA → SB → PA was 0.076 ($p < 0.05$). Last, in the combined model, the indirect effect of CT → SB → PA was 0.083 ($p < 0.01$), which was statistically significant.

Table 5. Analysis of indirect effects.

Model	Site	Path	Direct Effects	Indirect Effects	Total Effects
1	Lancry	CA → SB → PA	0.124	0.118 **	0.242
	La Défense	CA → SB → PA	0.142	0.076 **	0.218
2	Two sites	CA → SB → PA	0.186	0.083 ***	0.269

* $p < 0.1$, ** $p < 0.05$, *** $p < 0.01$.

5. Discussion

Previous studies have mainly focused on the satisfaction of residents with a specific place in terms of the relationship between the neighborhood environment and PA. In contrast, this study finds that CT, which is the physical form of certain distinguishing characteristics in a neighborhood, affects the activities of people (such as CA) and that CA positively affects SB and PA. Place attachment (PA) refers to the emotions that a person has in connection to a specific place and his or her experience of activities therein. Accordingly, it was necessary to discuss not only people's primary satisfaction with the physical space, but also whether the characteristics of the physical environment led to certain types of human activity, thereby affecting level of PA. Therefore, this study analyzed the CT and CA, as well as SCI and SCQ as influencing factors of place attachment. The results are as follows.

First, PA was influenced by the satisfaction factors of the neighborhood environment, and PA differed in each case according to the neighborhood characteristics. Model 1 showed that Lancry and La Défense had different factors that influenced PA. Further, note that PA is a complicated emotion that is influenced by various factors. Therefore, high satisfaction with neighborhood environments can affect PA. In the case of Lancry, satisfaction with the commercial environment (SCI and SCQ) had a positive effect on PA because of the high density of shops and high accessibility. On the other hand, La Défense is a new city with appropriate walking areas (i.e., without cars), parks, green areas, and cultural facilities that have been well developed, thus, the neighborhood environment satisfaction (SNF) positively influenced PA.

Second, analysis of the neighborhood CT considering street shops and malls supports our hypothesis presented herein. It was confirmed that CT affects the CA of residents, however commercial satisfaction (SCI and SCQ) did not affect CA; this indicates that commercial satisfaction can improve the PA of people through emotional means, especially through the use of spaces. However, PA does not alter the pattern of shopping in daily life because people typically only buy products when they are needed. This finding confirmed that, in part, the higher value of CA of residents was found in the neighborhood with street shops than in the neighborhood with malls. From the standpoint of neighborhood design, this result shows that different types of commercial structures can produce different patterns of daily life in neighborhood residents. In a neighborhood with street shops, with a high shopping frequency, CT acts as a factor that increases opportunities among residents to meet with one another relative to these opportunities in neighborhoods with other commercial features. Therefore, it is expected that SB will increase in neighborhoods with street shops as well as the promotion of local affection among residents. This expectation is supported by the finding that CA has a positive effect on SB. Increased levels of activity lead to the accumulation of neighborhood experiences with other residents, which leads to the formation of bonds between locals and place attachment.

Finally, the most important factor in this study is the "activities" element. The results show that increased SB of residents is due to the CA of residents, which depends on the CT of neighborhoods. As a result of this study, we see that SB ultimately positively affects residents' PA to a neighborhood. The reason why SB is important in PA is not only the feelings of attachment that people have to a specific place, but also because the emotions felt by people identifying as members of a community through their relationships with neighbors is a factor in the formation of PA. The results show that SB has the greatest direct effect on PA ($\beta = 0.278$). Further, the mediating effects between CA and PA show the importance of SB in PA.

6. Conclusions

The results of this study confirm that the CA among people in neighborhoods is an influencing factor with respect to the SB and PA of local residents, and that CA can be varied by changing CT. Based on the results of the study, two measures could be proposed as measures to increase place attachment in locals for sustainable neighborhood environments. The accumulation of experiences in places and bonding among local people are the methods whereby place attachment is formed. These are items that can be promoted based on the activities of residents in a neighborhood. In particular, this study analyzes the daily activities of residents in terms of their commercial activities, which provides implications for community and neighborhood design with regard to commercial features. The commercial environments of neighborhoods and the commercial activities occurring therein are basic activities in the daily lives of people. Until now, the focus on commercial features in urban design and neighborhood environments has been solely on physical quality and satisfaction. However, the results of the study demonstrate that the commercial environment is not simply a place to facilitate purchase transactions among residents. The commercial environment is an important factor that increases bonding among the people in neighborhoods and strengthens the attachment of locals to their place.

In order to promote the commercial activities of residents in neighborhoods, it is necessary to design neighborhoods that consist of small-scale stores rather than large-scale mall-type stores. However, current commercial spaces are becoming larger than ever. As a result of the decline of small-scale stores—a potential place for interaction and communication among locals—opportunities for exchange activities among residents have decreased. In addition, when small-scale merchants are replaced by large shopping malls, independent stores with diverse personalities tend to disappear. Over time, the unique characteristics of each trade will be replaced by the uniformity of commercial franchises. From the point of view of residents, favorite shops are disappearing, and there are fewer places where they want to go. This can have a negative impact on the formation of place attachment, as the appealing features of neighborhood local commerce are lost.

In order to create neighborhoods that people want to live in, place attachment via social bonding should be strengthened, and urban design that increases the activities of residents in neighborhoods should be reflected. From this point of view, this study suggests that small commercial spaces, such as street shops, are more effective than large malls in enhancing local attachment.

Author Contributions: B.K. conceived the idea and designed this research. J.P. developed the idea for academic research and reviewed the manuscript as a corresponding author. All authors carried out data collection and analysis, and drafted the manuscript.

Acknowledgments: This research was supported by Basic Science Research Program through the National Research Foundation of Korea (NRF) funded by the Ministry of Education (NRF-2017R1D1A1A09000940) and modified based on "Effects of Commercial Environment on Community and Place Attachment in Neighborhoods", 2016 Spring Congress Korea Planning Association (proceeding), Seoul, Korea.

Conflicts of Interest: The authors declare no conflict of interest.

Abbreviations

PA	Place attachment
SB	Social bonding
CA	Commercial activities
CT	Commercial type
SCI	Satisfaction with commercial infrastructure
SCQ	Satisfaction with commercial quality
SNF	Satisfaction with neighborhood facilities

References

1. Low, S.M.; Altman, I. Place attachment. In *Place Attachment*; Springer: Boston, MA, USA, 1992; pp. 1–12.
2. Hidalgo, M.C.; Hernandez, B. Place attachment: Conceptual and empirical questions. *J. Environ. Psychol.* **2001**, *21*, 273–281. [CrossRef]
3. Bailey, N.; Kearns, A.; Livingston, M. Place attachment in deprived neighbourhoods: The impacts of population turnover and social mix. *Hous. Stud.* **2012**, *27*, 208–231. [CrossRef]
4. Liu, Y.Q.; Wu, F.L.; Liu, Y.; Li, Z.G. Changing neighbourhood cohesion under the impact of urban redevelopment: A case study of Guangzhou, China. *Urban Geogr.* **2017**, *38*, 266–290. [CrossRef]
5. Kim, D. A Study on the Place Attachment of Residential Environment-Focused on Multilevel Analysis Using SEM and HLM. Ph.D. Thesis, Seoul National University, Seoul, Korea, 2008.
6. Lewicka, M. Place attachment: How far have we come in the last 40 years? *J. Environ. Psychol.* **2011**, *31*, 207–230. [CrossRef]
7. Kaltenborn, B.P.; Bjerke, T. Associations between landscape preferences and place attachment: A study in Røros, Southern Norway. *Landsc. Res.* **2002**, *27*, 381–396. [CrossRef]
8. Brown, G.; Brown, B.B.; Perkins, D.D. New housing as neighborhood revitalization-place attachment and confidence among residents. *Environ. Behav.* **2004**, *36*, 749–775. [CrossRef]
9. Choi, E.; Yim, H. The perception and the determinants of place attachment. *J. Korea Plan. Assoc.* **2005**, *40*, 53–64.
10. Zhu, Y.S.; Breitung, W.; Li, S.M. The changing meaning of neighbourhood attachment in Chinese commodity housing estates: Evidence from Guangzhou. *Urban Stud.* **2012**, *49*, 2439–2457. [CrossRef]
11. Park, J.; Kim, B. Influence of residents' attachment to the community on participation in community-building activities-focusing on hannam 1 dong. *J. Urban Des. Insit. Korea* **2014**, *15*, 215–226.
12. Hernandez, B.; Hidalgo, M.C.; Salazar-Laplace, M.E.; Hess, S. Place attachment and place identity in natives and non-natives. *J. Environ. Psychol.* **2007**, *27*, 310–319. [CrossRef]
13. Nielsen-Pincus, M.; Hall, T.; Force, J.E.; Wulfhorst, J.D. Sociodemographic effects on place bonding. *J. Environ. Psychol.* **2010**, *30*, 443–454. [CrossRef]
14. Handy, S.L.; Boarnet, M.G.; Ewing, R.; Killingsworth, R.E. How the built environment affects physical activity-views from urban planning. *Am. J. Prev. Med.* **2002**, *23*, 64–73. [CrossRef]
15. Saelens, B.E.; Sallis, J.F.; Black, J.B.; Chen, D. Neighborhood-based differences in physical activity: An environment scale evaluation. *Am. J. Public Health* **2003**, *93*, 1552–1558. [CrossRef] [PubMed]
16. Rosenberg, D.; Ding, D.; Sallis, J.F.; Kerr, J.; Norman, G.J.; Durant, N.; Harris, S.K.; Saelens, B.E. Neighborhood environment walkability scale for youth (news-y): Reliability and relationship with physical activity. *Prev. Med.* **2009**, *49*, 213–218. [CrossRef] [PubMed]
17. Van Dyck, D.; Cardon, G.; Deforche, B.; Sallis, J.F.; Owen, N.; De Bourdeaudhuij, I. Neighborhood SES and walkability are related to physical activity behavior in belgian adults. *Prev. Med.* **2010**, *50* (Suppl. 1), S74–S79. [CrossRef] [PubMed]
18. Ding, D.; Sallis, J.F.; Kerr, J.; Lee, S.; Rosenberg, D.E. Neighborhood environment and physical activity among youth a review. *Am. J. Prev. Med.* **2011**, *41*, 442–455. [CrossRef] [PubMed]
19. Cerin, E.; Macfarlane, D.J.; Ko, H.H.; Chan, K.C.A. Measuring perceived neighbourhood walkability in Hong Kong. *Cities* **2007**, *24*, 209–217. [CrossRef]
20. Shigematsu, R.; Sallis, J.F.; Conway, T.L.; Saelens, B.E.; Frank, L.D.; Cain, K.L.; Chapman, J.E.; King, A.C. Age differences in the relation of perceived neighborhood environment to walking. *Med. Sci. Sports Exerc.* **2009**, *41*, 314–321. [CrossRef] [PubMed]
21. Williams, D.R.; Roggenbuck, J.W. Measuring Place Attachment: Some Preliminary Results. Available online: https://www.fs.fed.us/rm/value/docs/nrpa89.pdf (accessed on 18 May 2018).
22. Proshansky, H.M. The city and self-identity. *Environ. Behav.* **1978**, *10*, 147–169. [CrossRef]
23. Yoon, Y.; Kwock, Y. Residents' place attachment in evaluating tourism destination. *J. Hotel Resort* **2005**, *4*, 511–525.
24. Lee, E. The intrinsic attributes of place attachment: In case of a poem Dashi Elle ege (to Elle again). *J. Cult. Hist. Geogr.* **2006**, *18*, 1–10.

25. Li, X.; Kleinhans, R.; van Ham, M. *Ambivalence in Place Attachment: The Lived Experiences of Residents in Declining Neighbourhoods Facing Demolition in Shenyang, China*; IZA Institute of Labor Economics: Bonn, Germany, 2017.
26. Lewicka, M. What makes neighborhood different from home and city? Effects of place scale on place attachment. *J. Environ. Psychol.* **2010**, *30*, 35–51. [CrossRef]
27. Bonaiuto, M.; Aiello, A.; Perugini, M.; Bonnes, M.; Ercolani, A.P. Multidimensional perception of residential environment quality and neighbourhood attachment in the urban environment. *J. Environ. Psychol.* **1999**, *19*, 331–352. [CrossRef]
28. Pretty, G.H.; Chipuer, H.M.; Bramston, P. Sense of place amongst adolescents and adults in two rural Australian towns: The discriminating features of place attachment, sense of community and place dependence in relation to place identity. *J. Environ. Psychol.* **2003**, *23*, 273–287. [CrossRef]
29. Norman, G.J.; Nutter, S.K.; Ryan, S.; Sallis, J.F.; Calfas, K.J.; Patrick, K. Community design and access to recreational facilities as correlates of adolescent physical activity and body-mass index. *J. Phys. Act. Health* **2006**, *3*, S118–S128. [CrossRef] [PubMed]
30. Pate, R.R.; Colabianchi, N.; Porter, D.; Almeida, M.J.; Lobelo, F.; Dowda, M. Physical activity and neighborhood resources in high school girls. *Am. J. Prev. Med.* **2008**, *34*, 413–419. [CrossRef] [PubMed]
31. Nagel, C.L.; Carlson, N.E.; Bosworth, M.; Michael, Y.L. The relation between neighborhood built environment and walking activity among older adults. *Am. J. Epidemiol.* **2008**, *168*, 461–468. [CrossRef] [PubMed]
32. Wood, L.; Frank, L.D.; Giles-Corti, B. Sense of community and its relationship with walking and neighborhood design. *Soc. Sci. Med.* **2010**, *70*, 1381–1390. [CrossRef] [PubMed]
33. Leyden, K.M. Social capital and the built environment: The importance of walkable neighborhoods. *Am. J. Public Health* **2003**, *93*, 1546–1551. [CrossRef] [PubMed]
34. Kim, J.; Kaplan, R. Physical and psychological factors in sense of community-new urbanist kentlands and nearby orchard village. *Environ. Behav.* **2004**, *36*, 313–340. [CrossRef]
35. KelQuartier. Available online: http://www.kelquartier.com/ile_de_france_paris_paris_10_quartier_chateau_d_eau_lancry_75010-q100518/revenu_moyen.html (accessed on 18 May 2018). (In France)
36. KelQuartier. Available online: http://www.kelquartier.com/ile_de_france_hauts_de_seine_puteaux_quartier_la_defense_92800-q100352/revenu_moyen.html (accessed on 18 May 2018). (In France)
37. Kang, S. *The Impact of Community Attachment on the Attitudes toward Tourism Development*; Hanyang: Seoul, Korea, 2001.

© 2018 by the authors. Licensee MDPI, Basel, Switzerland. This article is an open access article distributed under the terms and conditions of the Creative Commons Attribution (CC BY) license (http://creativecommons.org/licenses/by/4.0/).

Article

Analysis of Embodied Environmental Impacts of Korean Apartment Buildings Considering Major Building Materials

Seungjun Roh [1], Sungho Tae [1,2,*] and Rakhyun Kim [3,*]

1. Sustainable Building Research Center, Hanyang University, 55 Hanyangdaehak-ro, Sangnok-gu, Ansan 15588, Korea; roh.seungjun@gmail.com
2. Department of Architecture & Architectural Engineering, Hanyang University, 55 Hanyangdaehak-ro, Sangnok-gu, Ansan 15588, Korea
3. Architectural Engineering, Hanyang University, 55 Hanyangdaehak-ro, Sangnok-gu, Ansan 15588, Korea
* Correspondence: jnb55@hanyang.ac.kr (S.T.); redwow6@hanyang.ac.kr (R.K.); Tel.: +82-31-400-5187 (S.T.); +82-31-436-8076 (R.K.)

Received: 7 April 2018; Accepted: 12 May 2018; Published: 23 May 2018

Abstract: Because the reduction in environmental impacts (EIs) of buildings using life-cycle assessment (LCA) has been emphasized as a practical strategy for the sustainable development of the construction industry, studies are required to analyze not only the operational environmental impacts (OEIs) of buildings, but also the embodied environmental impacts (EEIs) of building materials. This study aims to analyze the EEIs of Korean apartment buildings on the basis of major building materials as part of research with the goal of reducing the EIs of buildings. For this purpose, six types of building materials (ready-mixed concrete, reinforcement steel, concrete bricks, glass, insulation, and gypsum) for apartment buildings were selected as major building materials, and their inputs per unit area according to the structure types and plans of apartment buildings were derived by analyzing the design and bills of materials of 443 apartment buildings constructed in South Korea. In addition, a life-cycle scenario including the production, construction, maintenance, and end-of-life stage was constructed for each major building material. The EEIs of the apartment buildings were quantitatively assessed by applying the life-cycle inventory database (LCI DB) and the Korean life-cycle impact assessment (LCIA) method based on damage-oriented modeling (KOLID), and the results were analyzed.

Keywords: embodied environmental impact; apartment building; major building material; life-cycle assessment

1. Introduction

With the rising importance of sustainable development, efforts have been made in all industrial areas to reduce environmental impacts (EIs) [1–5]. In line with this, the construction industry has focused its research on cutting-edge technologies (e.g., highly efficient insulating materials, high-performance glass, high-air-tightness windows, and renewable energy systems) capable of dramatically reducing the energy consumption of a building during its operation stage in order to decrease operational environmental impacts (OEIs), which account for over 70% of the EIs of conventional buildings [6–10]. As a result, zero-energy buildings—energy-efficient buildings that use little energy during their operation stage—have been developed and successfully constructed in many countries [11–15].

As technologies to reduce the OEIs of buildings have been commercialized, research on the life-cycle assessment (LCA) of buildings—which considers the reduction in the OEIs of buildings as well as in the embodied environmental impacts (EEIs) caused by the production, construction, maintenance, and end-of-life stages of the building materials used—has been emphasized recently [16–22]. This is because additional building materials may be necessary for energy-efficient buildings compared with conventional buildings, thus increasing EEIs, but decreasing OEIs [17]. Results of previous LCAs of energy-efficient buildings showed that EEIs were higher than OEIs [23,24]. Hence, more research is necessary to assess and reduce the EEIs of buildings, as the importance and influence of EEIs have gradually increased [25,26].

Some of the previous studies on the analysis of EEIs are important in terms of their approach, methodology, and case studies [27–33]. Because they mostly analyzed only carbon emissions during the production stage of the building materials, their use has been limited. Therefore, for a study's results to be used as basic data for reducing the EEIs of buildings, these impacts must be analyzed by considering the following:

- The EEIs of a number of buildings must be analyzed according to the characteristics of those buildings. This is because the results of analyzing the EEIs for one or more buildings cannot be generalized as the EEI characteristics of all buildings.
- The assessment target must be expanded from carbon emissions to other EI categories. To achieve sustainable development, it is necessary to address not only global warming due to carbon emissions but also various other global environmental problems [34].
- The scope of assessment must be extended from the building material's production stage to a life-cycle perspective. This is because the overall EEIs of buildings must be examined quantitatively to be reduced [35].
- The EEI assessment results of buildings must be analyzed from a building-material perspective. In this way, EEIs can be reduced by identifying building materials that have the greatest influence on these EEIs.
- EIs must be assessed not only for EI categories, but also for safety guards. This is because the end-point-level damage to humans and ecosystems by each EI must be identified.

Therefore, the aim of this study is to analyze the EEIs of Korean apartment buildings on the basis of major building materials as part of research with the goal of reducing the life-cycle environmental impacts (LCEIs) of buildings.

2. Background

2.1. Embodied Environmental Impact

The LCEIs of buildings can be divided into EEIs and OEIs [24,27]. The EEIs of buildings correspond to the LCEIs excluding the EIs caused by energy consumption (e.g., heating, cooling, hot water, lighting, and ventilation). In other words, EEIs include EIs that arise from the building-material production stage and the building construction, maintenance, and end-of-life stages. EEIs for a building are calculated using Equation (1):

$$EEI = EI_{PS} + EI_{CS} + EI_{MP} + EI_{ES},\tag{1}$$

where EEI denotes the life-cycle embodied environmental impact (LCEEI) of the building. EI_{PS}, EI_{CS}, EI_{MP}, and EI_{ES} are the EEIs of the building-material production stage, and the building construction, maintenance, and end-of-life stages.

2.2. Environmental Impact Categories

EI categories represent global environmental changes caused by human behavior or technology. Global warming potential (GWP), acidification potential (AP), eutrophication potential (EP), ozone layer depletion potential (ODP), photochemical ozone creation potential (POCP), and abiotic depletion potential (ADP) are representative EI categories, which can be assessed quantitatively through various life-cycle impact assessment (LCIA) methodologies [36,37].

GWP represents climate change, that is, the rise in average temperatures of the earth's atmosphere, and causes environmental problems because of changing ecosystems in soil or water, or because of rising sea levels. AP represents the acidification of water and soil, mainly by the circulation of pollutants, threatening the survival of living organisms such as fish, plants, and animals. EP represents the harmful impacts on the marine environment, such as red tides resulting from the amount of nutrients abnormally increasing through the introduction of chemical fertilizers or sewage. ODP is a phenomenon in which the ozone in the ozone layer—located in the stratosphere 15–30 km above the ground—is destroyed and its density decreases. It can lead to diseases such as skin cancer because of the increase in ultraviolet radiation. POCP is a reaction between air pollutants and sunlight in which chemical compounds such as ozone (O_3) are created, in turn causing damage to ecosystems and human health and inhibiting the growth of crops. ADP represents the cause behind the destruction of ecosystem balance and environmental pollution caused by the excessive collection and consumption of resources.

2.3. Safety Guard and Damage Index

From an environmental ethics perspective, the "safety guard" represents the environment that the human race must protect. It can be classified into human and ecosystem items. The human items are divided into human health, which is required for humans to live a healthy life, and social assets, which support human society. The ecosystem items can be subdivided into biodiversity, which refers to the preservation of animals and plants, and primary production, which is essential for maintaining biodiversity [38].

The damage index quantifies the damage to the aforementioned safety guard (human health, social assets, biodiversity, and primary production) caused by EIs. For assessing damages to human health, disability-adjusted life years (DALY) are used. DALY is a damage index representing the number of years of healthy life lost as a result of EIs. For social assets, the mean economic cost (USD) for the suppression and depletion of crops; fossil fuels; and fishery, forest, and mineral resources is used. In addition, biodiversity is assessed through the expected increase in the number of extinct species (EINES) damage index, that is, the expected number of extinct species of vascular and aquatic plants. For primary production, the net primary production (NPP) is used as a damage index, assessing the amount ($kg/m^2 \cdot y$) of organic matter created by the photosynthesis of land plants and marine plankton. The damage index for each safety guard can be assessed through the end-point-level LCIA methodology, which systematizes damage indexes for each safety guard using research results from natural sciences. Figure 1 is an example of the LCIA method at the end-point level [39]. It shows the structures and degrees of the impacts of GWP caused by 1 ton of CO_2 emission on human health and social assets as safety guards. According to Figure 1, GWP caused by 1 ton of CO_2 emission adversely affects heat stress, exposure to infectious diseases, malnutrition, disaster damage, energy consumption, and agricultural production at the end-point level and ultimately causes a damage of 1.23×10^{-4} DALY and 2.5 USD to human health and social assets, respectively.

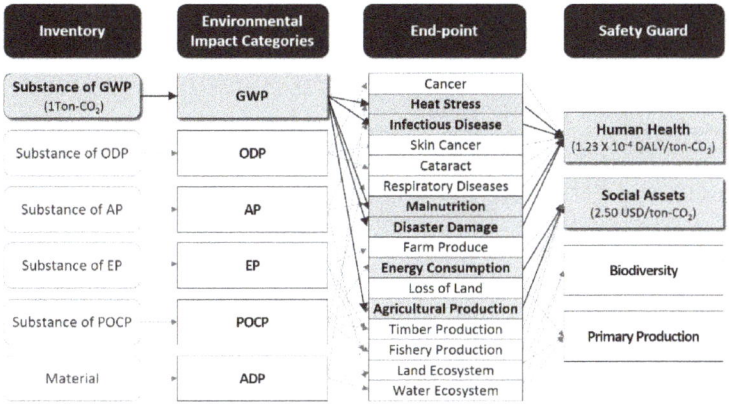

Figure 1. Example of the evaluation method for life-cycle environmental impacts (LCEIs) at the end-point level [38].

3. Materials and Methods

The section details the assessment of the EEIs of Korean apartment buildings on the basis of the major building materials using the sequential LCA methodology. For this purpose, in the goal and scope definition stage, the purpose of LCA and the scope of the system were defined. In the life-cycle inventory (LCI) analysis stage, the average inputs per unit area of major building materials were derived according to the structure types and plans of apartment buildings by analyzing the design and bills of materials of apartment buildings constructed in South Korea. In addition, a life-cycle scenario including the production, construction, maintenance, and end-of-life stage was constructed for each of the major building materials. In the LCIA stage, the EEIs of the six impact categories and damage indexes for each safety guard were quantitatively assessed by applying the life-cycle inventory database (LCI DB) and the Korean LCIA method based on damage-oriented modeling (KOLID) [38], an end-point-level LCEI assessment methodology.

3.1. Goal and Scope Definition

The purpose of performing LCA in this study was to analyze the EEIs of Korean apartment buildings on the basis of major building materials. As for a system boundary, building material production, the building construction, maintenance, and end-of-life stages were included, and six EI categories (GWP, AP, EP, ODP, POCP, and ADP) and four safety guards (human health, social assets, biodiversity, and primary production) were evaluated. Gross floor area (m^2) was established as the functional unit. The criteria used to determine the quality of the LCA results were classified into temporal, regional, and technical ranges, as described in Table 1. Furthermore, the building material inputs were analyzed on the basis of the building material quantities applied to the ground floor of the apartment buildings, and it was assumed that the total quantities of building materials specified in the bills of materials were used in the buildings.

Table 1. Data quality criteria.

Classification	Temporal Ranges	Regional Ranges	Technical Ranges
Internal data (Bills of materials)	Bills of materials prepared at the commencement of the work	Bills of materials prepared in South Korea	Six major building materials listed in the bills of materials
External data (LCI DB)	Latest LCI DB	LCI DB constructed in South Korea and Germany	LCI DB of same or similar building materials

3.2. Life-Cycle Inventory Analysis

3.2.1. Selection of Major Building Materials

To analyze the EEIs of buildings more efficiently, it is necessary to select building materials with the highest EIs. The construction of buildings includes more complex procedures than the production processes for general products, and the EEI analysis requires excessive time and labor, as more than 1000 building materials can be used in a construction project.

Therefore, this study analyzed the EEIs of apartment buildings using results from previous research [39] that derived six major building materials (ready-mixed concrete, reinforcement steel, concrete bricks, glass, insulation, and gypsum) accounting for over 95% of the six EI categories (GWP, AP, EP, ODP, POCP, and ADP) in accordance with the cut-off criteria of ISO 14040, an international standard for LCA.

3.2.2. Analysis of Major Building Material Inputs

A total of 443 apartment buildings in South Korea were selected as samples, and the inputs per unit area of the six major building materials were analyzed according to the structure types and plans of the apartment buildings. In this case, the structure types were divided into wall structures, frame structures, and flat plate structures, and the plans were classified into plate, tower, and mixed types. Table 2 lists the number of samples, and Table 3 represents the average input quantities per unit area of the major building materials according to the structure types and plans of the apartment buildings.

Table 2. Number of samples.

Classification	Wall Structure	Rigid Frame Structure	Flat Plate Structure
Plate type	118	22	6
Tower type	101	40	22
Mixed type	60	64	10

Table 3. Average input quantity of building materials by structure types and plans of apartment building.

Classification	Unit	Wall Structure			Rigid Frame Structure			Flat Plate Structure		
		Plate Type	Tower Type	Mixed Type	Plate Type	Tower Type	Mixed Type	Plate Type	Tower Type	Mixed Type
Ready-mixed concrete	m^3/m^2	0.77	0.73	0.75	0.71	0.60	0.70	0.68	0.58	0.68
Rebar	kg/m^2	98.13	101.35	99.92	118.34	145.26	131.55	127.52	158.40	149.24
Concrete brick	kg/m^2	90.87	90.81	86.52	90.52	89.98	86.89	89.77	88.85	85.54
Glass	kg/m^2	5.87	5.99	5.99	5.87	6.12	5.61	5.74	5.74	5.99
Insulation	kg/m^2	1.56	1.48	1.58	1.62	1.57	1.60	1.44	1.40	1.49
Gypsum board	kg/m^2	2.63	2.68	2.67	2.66	2.72	2.65	2.50	2.69	2.62

3.2.3. Construction of the Life-Cycle Scenario

For the analysis of LCEEIs, the EEIs of the six major building materials with the highest EIs were assessed in the production stage, and a life-cycle scenario was constructed so that the EEIs could be assessed in the construction, maintenance, and end-of-life stages on the basis of the major building material inputs in the production stage [40]. Figure 2 shows the case of ready-mixed concrete as an example of the EEI assessment on the basis of the constructed life-cycle scenario.

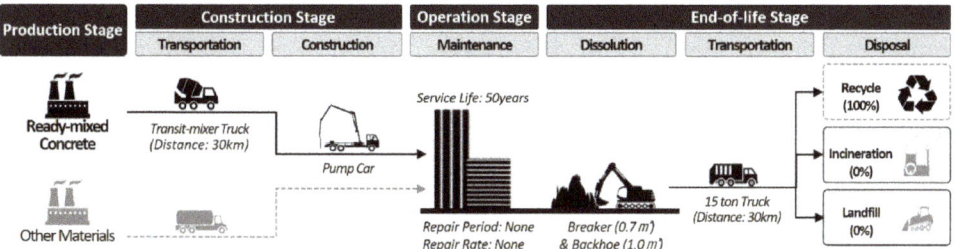

Figure 2. Example of scenario-based embodied environmental impact (EEI) evaluation.

(1) Production stage:

In the production stage, the EIs arising from the production of building materials are assessed. In this study, the EIs of the production stage were assessed using the average inputs per unit area of the six major building materials (ready-mixed concrete, reinforcement steel, glass, concrete bricks, insulation, and gypsum) derived from previous research in accordance with the cut-off criteria of LCA.

(2) Construction stage:

The construction stage is divided into the transportation process of building materials and the construction process of buildings.

In the transportation process, building materials are transported from their production sites to the construction site. In this study, freight vehicles for each of the major building materials were selected as shown in Table 4 on the basis of the standard estimation system for construction works [41]. In addition, the transport distance was assumed to be 30 km for all of the major building materials.

Table 4. Freight vehicles.

Classification	Ready-Mixed Concrete	Rebar	Others
Freight vehicle	Transit-mixer truck	20 ton truck	8 ton truck

The construction stage represents the EIs caused by the use of equipment during construction, and it was assessed using the LCI DB for the unit of construction work for each building material.

(3) Maintenance stage:

In the maintenance stage, the EIs arising from the production and transport of building materials that are periodically replaced in order to recover the status of aging buildings during their service life are assessed. In this study, the service life of buildings was set to 50 years, in accordance with the upper limit of the standard service life of the Enforcement Regulations of the Corporate Tax Act of Korea [42]. In addition, the EIs of the maintenance stage were assessed using the repair period and rate for each building material suggested by the standards for the formulation of the long-term repair plan in the Enforcement Regulations of the Multi-Family Housing Management Act of Korea. In other words, it was assumed that ready-mixed concrete, reinforcement steel, concrete bricks, glass, and insulation, among the selected six major building materials, were not replaced during the service life of the buildings and that 100% of the gypsum boards were replaced every 20 years.

(4) End-of-life stage:

The end-of-life stage is divided into the demolition process, the transportation process of waste building materials, the incineration process, and the landfill process.

In the dissolution process, the EIs of the equipment and machinery used for building demolition are assessed through fuel efficiency (diesel consumption per unit of work) information of the demolition machines after the number of machines is calculated on the basis of the amount of waste material generated in the demolition process. In this study, it was assumed that both crushers (0.7 m^3) and backhoes (1.0 m^3) were used as demolition equipment [19] and that the amount of waste material generated in the demolition process was the same as the input quantities of the six major building materials in the production stage.

In the transportation process, the EIs arising from transporting the waste materials generated in the demolition process to recycling centers, incineration plants, or landfills are assessed. In this case, it was assumed that the waste building materials were transported using 15 ton trucks in accordance with the standard estimation system for construction works [41] and that the distances from the demolition site to recycling centers, incineration plants, and landfills were 30 km.

In the end-of-life process, the EIs arising from incinerating or landfilling waste materials are assessed. In this study, the cut-off method imposed on recycling companies was applied to the EIs of the waste material recycling process, and only the EIs of the incineration and landfilling of non-recycled waste materials were assessed. For this, the construction waste processing data from waste statistics [43] published by the Korean Environmental Industry and Technology Institute were investigated, and the recycling, incineration, and landfill rates of each major building material were applied as shown in Table 5.

Table 5. Processing ratios of waste building materials.

Classification	Recycle Ratio (%)	Incineration Ratio (%)	Landfill Ratio (%)
Waste concrete	100.0	0.0	0.0
Waste rebar	100.0	0.0	0.0
Waste concrete brick	100.0	0.0	0.0
Waste glass	79.0	0.0	21.0
Waste insulation	46.7	53.3	0.0
Waste gypsum board	62.7	0.2	37.1

3.3. Life-Cycle Impact Assessment

3.3.1. Application of the LCI DB

For the assessment of LCEEIs, the LCI DBs for building materials used in buildings, freight vehicles for transporting building materials and waste materials, unit construction work for each building material, and incineration and landfill processes of waste materials must be applied.

In this study, LCI DBs were applied in the order of the Korean LCI DB [44] of the Ministry of Trade, Industry and Energy and the Ministry of Environment (ME) of South Korea; the National Database on Environmental Information of Building Materials of the Korean Institute of Civil Engineering and Building Technology [45]; and Oekobaudat [46] of Germany, considering regional, temporal, and technical correlations, which are the LCI DB selection criteria for LCA suggested by ISO 14040 (refer to Table 6). Furthermore, the EEIs of the six EI categories were assessed through the multiplication of the activity and the EI factor, as shown in Equation (2):

$$EI_i = \sum_{j=1}^{n} (A_j \times EF_{i,j}), \qquad (2)$$

where EI_i are the EEIs of EI category (i), A_j are the building material and energy input quantity for activity (j), and $EF_{i,j}$ is the EI factor of EI category (i) for activity (j).

Table 6. Environmental impact (EI) factors.

Classification		Unit	GWP kg CO$_{2eq}$/ Unit	AP kg SO$_{2eq}$/ Unit	EP kg PO$_4^{3-}$$_{eq}$/ Unit	ODP kg CFC11$_{eq}$/ Unit	POCP kg C$_2$H$_{4eq}$/ Unit	ADP kg Sb$_{eq}$/ Unit	Ref.
Production stage	Ready-mixed concrete	m^3	4.09×10^2	6.82×10^{-1}	7.96×10^{-2}	4.65×10^{-5}	1.10×10^0	2.04×10^0	Ⓐ
	Rebar	kg	4.38×10^{-1}	1.40×10^{-3}	1.79×10^{-4}	1.04×10^{-8}	3.41×10^{-4}	2.79×10^{-3}	Ⓐ
	Concrete brick	kg	1.23×10^{-1}	1.56×10^{-4}	2.26×10^{-5}	4.71×10^{-9}	3.82×10^{-5}	3.02×10^{-4}	Ⓑ
Construction stage	Transit-mixer truck	m^3 × km	6.74×10^{-1}	6.50×10^{-3}	1.03×10^{-3}	2.44×10^{-7}	1.12×10^{-3}	4.47×10^{-3}	Ⓑ
	8 ton truck	kg × km	2.88×10^{-6}	2.17×10^{-8}	3.86×10^{-9}	1.06×10^{-12}	6.45×10^{-9}	1.94×10^{-8}	Ⓐ
End-of-life stage	Diesel	kg	6.82×10^{-2}	1.40×10^{-4}	9.55×10^{-6}	1.26×10^{-10}	1.18×10^{-5}	2.16×10^{-2}	Ⓐ
	Construction waste dumping	kg	6.05×10^{-2}	8.52×10^{-5}	1.31×10^{-5}	1.23×10^{-11}	2.21×10^{-5}	5.02×10^{-9}	Ⓒ

Ⓐ: Korean life-cycle inventory database; Ⓑ: National Database on Environmental Information of Building Materials; Ⓒ: Oekobaudat.

3.3.2. Application of KOLID

To calculate direct impacts on humans and ecosystems using the EEI assessment results derived for each EI category, the end-point-level LCIA methodology, which systematizes the damage index for each safety guard using results from natural science research, is required.

In this study, KOLID [38] was applied. KOLID is an end-point-level damage-calculation LCEI assessment methodology developed by the Korean ME in 2009 to better understand the damage caused by environmental issues and to expand the distribution of environmentally friendly products. This methodology quantifies 16 end-point damages, including cancer, infectious disease, and cataract, attributable to the six EI categories (GWP, AP, EP, ODP, POCP, and ADP) triggered by products and services, and it evaluates the four safety guards (human health, social assets, biodiversity, and primary production) (refer to Figure 1). Regional correlations were considered, and the damage index for safety guard objects was quantitatively calculated using the LCEEIs of the apartment buildings. Table 7 shows the safety guards and damage indexes of KOLID, and Equation (3) represents the damage-index calculation formula for each safety guard using KOLID:

$$SI_i = \sum_{j=1}^{n} (EI_j \times DF_{i,j}), \quad (3)$$

where SI_i is the damage index of safety guard (i), EI_j are the EEIs of EI category (j), and $DF_{i,j}$ is the damage factor of safety guard (i) for EI category (j).

Table 7. Safety guard and damage index of Korean life-cycle impact assessment method based on damage-oriented modeling (KOLID).

Classification	Safety Guard	End Point	Indicator	Damage Factor
GWP	Human health	Mortality damages caused by heat/cold stress, infections, natural disaster damage, and malnutrition	Lost life	1.23×10^{-7} DALY/kg CO$_2$
	Social assets	Decreases in agricultural production output	Agricultural production output	2.54×10^{-3} USD/kg CO$_2$
		Changes in energy consumption due to increases in cooling and decreases in heating	Energy consumption quantity	
		Sea-level rising	Land prices	
AP	Human health	Damages caused by asthma and respiratory diseases	Lost life	2.38×10^{-4} DALY/kg SO$_2$
	Social assets	Decreases in wood production output	Wood production output	4.76×10^0 USD/kg SO$_2$
	Primary production	Decreases in primary production output of land plants	Primary production output	2.27×10^1 kg/kg SO$_2$

Table 7. Cont.

Classification	Safety Guard	End Point	Indicator	Damage Factor
EP	Social assets	Decreases in fishery production output	Fishery production output	2.16×10^0 USD/kg PO_4^{3-}
ODP	Human health	Damages caused by malignant melanoma, basal cell carcinoma, and spinocellular carcinoma	Lost life	1.35×10^{-3} DALY/kg CFC-11
ODP	Social assets	Decreases in agricultural and wood production output	Agricultural and wood production output	1.21×10^0 USD/kg CFC-11
ODP	Primary production	Decreases in primary production output of land plants and phytoplankton	Primary production output	2.79×10^2 kg/kg CFC-11
POCP	Human health	Damages caused by sudden death, asthma, and respiratory diseases	Lost life	3.22×10^{-5} DALY/kg C_2H_4
POCP	Social assets	Decreases in agricultural and wood production output	Agricultural and wood production output	0.77×10^0 USD/kg C_2H_4
POCP	Primary production	Decreases in primary production output of land plants	Primary production output	2.64×10^1 kg/kg C_2H_4
ADP	Social assets	Decreases in resource deposits	Users' costs	1.42×10^1 kg/kg Sb
ADP	Biodiversity	Changes in the composition of plant species	Species changes	1.53×10^{-1} EINES/kg Sb
ADP	Primary production	Land changes, and potential NPP decreases in land use	Primary production output	8.90×10^{-14} kg/kg Sb

4. Results and Discussion

This section describes the assessment of the LCEEIs of the apartment buildings with different structure types and plans, as well as the analysis of the assessment results and characteristics from the perspectives of total EIs, building life-cycle stages, major building materials, and safety guards.

4.1. Analysis of Total Environmental Impacts

Figure 3 shows the results of the LCEEI assessment of the apartments analyzed in this study. According to Figure 3, the EIs of tower-type apartment buildings with a flat plate structure were the lowest for all EI categories, while those of plate-type apartment buildings with a wall structure were the highest. If the reduction in EIs is considered during the apartment building design stage using such characteristics, planning only tower-type apartment buildings with a flat plate structure instead of plate-type buildings with wall structures will reduce the potential EIs of each EI category by between 10.74% and 21.67% (refer to Table 8).

Table 8. Reduction ratio of environmental impacts (EIs).

Classification	Wall Structure, Plate Type	Flat Plate Structure, Tower Type	Reduction Ratio
GWP (kg CO_{2eq}/m^2)	4.18×10^2	3.57×10^2	14.59%
AP (kg SO_{2eq}/m^2)	9.33×10^{-1}	8.32×10^{-1}	10.83%
EP (kg $PO_4^{3-}{}_{eq}/m^2$)	1.21×10^{-1}	1.08×10^{-1}	10.74%
ODP (kg $CFC11_{eq}/m^2$)	5.26×10^{-5}	4.12×10^{-5}	21.67%
POCP (kg C_2H_{4eq}/m^2)	1.01×10^0	7.96×10^{-1}	21.19%
ADP (kg Sb_{eq}/m^2)	2.26×10^0	1.96×10^0	13.27%

The increase or decrease in EIs according to the structure types and plans of the apartment buildings tended to vary relatively regularly for all EI categories. In other words, it was found that the EIs tended to decrease as the structure type changed from a wall structure to frame structures and flat plate structures. Within the same structure type, EIs also varied regularly according to the change in plan. In other words, in wall structures, the EIs of all EI categories decreased as the plan changed from plate to mixed and to tower type. In the flat plate structure, EIs decreased as the plan changed from mixed to plate and to tower type. In the frame structure, the tower type exhibited the lowest EIs despite that changes in some EIs depended on the EI category.

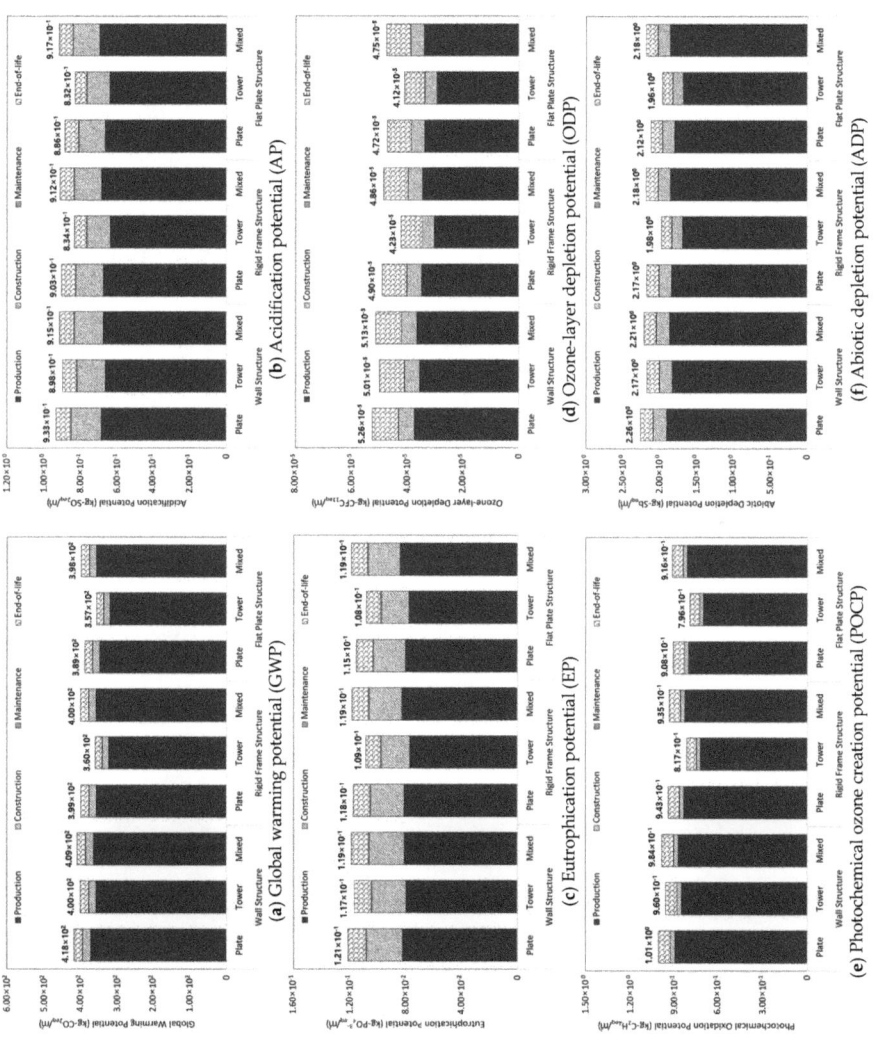

Figure 3. Results of life-cycle embodied environmental impacts (LCEEIs) by life-cycle stages.

4.2. Analysis by Life-Cycle Stage

According to Figure 3, among the LCEEIs of the apartment buildings, the impacts of the production stage were the highest for all EI categories, while those of the maintenance stage were the lowest. In particular, among the overall EEIs assessed in this study, the percentages of EEIs caused by the production stage ranged from 67.96% (EP, wall structure, plate type) to 90.04% (GWP, flat plate structure, tower type), indicating that reducing EEIs during the production stage is imperative for decreasing LCEEIs of apartment buildings.

The percentages of the EEIs caused by the construction stage ranged from 2.94% (POCP, flat plate structure, tower type) to 21.04% (EP, wall structure, plate type) depending on the EI category, and the percentages caused by the end-of-life stage ranged from 5.23% (GWP, flat plate structure, tower type) to 18.29% (ODP, flat plate structure, tower type). In particular, the proportions of EEIs caused by the construction and end-of-life stages were generally higher for the ODP, AP, and EP impact categories. This indicates that the GWP analysis focused on the production stage, which was mainly performed in previous studies, as well as that the EEI analysis, which considers various EI categories, is necessary for reducing EEIs caused by buildings. This confirms that EI reduction strategies in terms of a building's entire life cycle, including production, construction, and end-of-life stages, are indispensable.

4.3. Analysis by Major Building Materials

Figure 4 shows the results of the LCEEI assessment for the major building materials. As shown in the figure, the impacts of ready-mixed concrete were the highest for all EI categories, while those of glass were the lowest. In particular, among the overall EEIs assessed in this study, the percentages of those caused by ready-mixed concrete ranged from 68.13% (EP, flat plate structure, tower type) to 94.75% (ODP, wall structure, plate type). This indicates that the development and application of concrete with reduced EIs, which considerably replaces conventional concrete with supplementary cementitious materials (SCMs), must be performed to reduce the LCEEIs of apartment buildings.

The percentages of EEIs caused by reinforcement steel ranged from 2.83% (ODP, wall structure, plate type) to 27.52% (AP, flat plate structure, tower type) depending on the EI category. In particular, the EEIs of reinforcement steel were inversely proportional to those of ready-mixed concrete. This is because reinforcement steel and ready-mixed concrete are materials that largely constitute the structures of buildings, and thus the input quantity of reinforcement steel relatively decreased as that of ready-mixed concrete increased, depending on the structure types and plans. As such, from the perspective of EI reduction for apartment buildings, it is necessary to design structural materials considering the balance between EEIs of ready-mixed concrete and reinforcement steel.

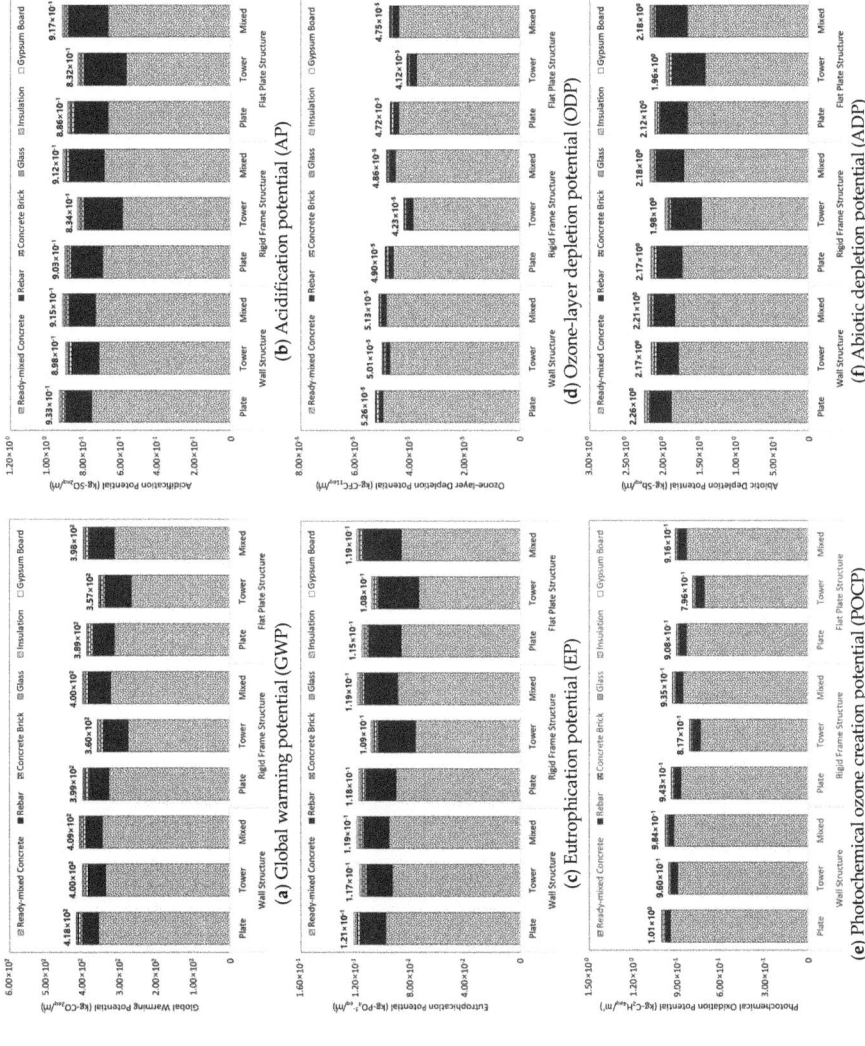

Figure 4. Results of life-cycle embodied environmental impacts (LCEEIs) by major building materials.

4.4. Analysis by Safety Guards

Table 9 shows the results of the damage index assessment by safety guards according to the structure types and plans of the apartment buildings. As can be seen from the figure, the tower-type buildings with a flat plate structure exhibited the lowest damage indexes for all safety guards, while the plate-type buildings with a wall structure showed the highest values. In addition, the increase or decrease in the damage indexes by safety guards tended to vary relatively similarly to how the structure types and plans of the apartment buildings changed regularly for all items. In other words, the damage index by safety guard tended to decrease as the structure type changed from wall structures to frame structures and flat plate structures in the same way as the characteristics of the EI categories changed, as mentioned earlier in Section 4.1: Analysis of Total Environmental Impacts. Furthermore, for wall structures, the damage indexes for all safety guards decreased as the plan changed from plate to mixed and to tower type. In flat plate structures, the damage indexes for safety guards tended to decrease as the plan changed from mixed to plate and to tower type.

Table 9. Results of damage index by safety guard.

Classification	Unit	Wall Structure			Rigid Frame Structure			Flat Plate Structure		
		Plate Type	Tower Type	Mixed Type	Plate Type	Tower Type	Mixed Type	Plate Type	Tower Type	Mixed Type
Human health	DALY/m^2	3.06×10^{-4}	2.94×10^{-4}	3.00×10^{-4}	2.94×10^{-4}	2.69×10^{-4}	2.96×10^{-4}	2.88×10^{-4}	2.68×10^{-4}	2.97×10^{-4}
Social assets	USD/m^2	6.11×10^{0}	5.87×10^{0}	5.99×10^{0}	5.88×10^{0}	5.38×10^{0}	5.92×10^{0}	5.75×10^{0}	5.34×10^{0}	5.93×10^{0}
Biodiversity	EINES/m^2	3.46×10^{-1}	3.32×10^{-1}	3.39×10^{-1}	3.32×10^{-1}	3.03×10^{-1}	3.34×10^{-1}	3.25×10^{-1}	3.01×10^{-1}	3.34×10^{-1}
Primary Production	kg/m^2	5.11×10^{-2}	4.89×10^{-2}	5.00×10^{-2}	4.86×10^{-2}	4.35×10^{-2}	4.86×10^{-2}	4.72×10^{-2}	4.29×10^{-2}	4.83×10^{-2}

4.5. Discussion

As the reduction in the LCEEIs of buildings has been emphasized recently, studies should be carried out to analyze the EEIs of building materials. This is because quantitative values of the LCEEIs of buildings and their major causes must be analyzed first in order to reduce these impacts.

This study provides a significant contribution towards this goal because it presents basic data for reducing the EEIs of buildings by selecting 443 Korean apartment buildings as samples and analyzing their EEIs in terms of six impact categories and damage indexes by safety guards. In particular, the EEIs analyzed in this study according to the structure types and plans of the apartment buildings can be used as factors for easily identifying the EEIs of apartment buildings in construction practice. Furthermore, it appears that the improvement in the environmental performance of ready-mixed concrete, which was found to be the main cause of EEIs, can be utilized as basic data for reducing the EEIs of apartment buildings.

On the other hand, plate-type apartment buildings with a wall structure produced the highest results for all EI categories within the scope of this study, because plate-type apartment buildings with a wall structure used the highest quantity of ready-mixed concrete, which was the most influential in all the EI categories compared to the apartment buildings with other structure types and plans. Therefore, in order to effectively reduce the EEIs of plate-type apartment buildings with a wall structure, it would be effective to apply high-strength concrete to the vertical structural member to reduce the input quantity of ready-mixed concrete and rebar by way of reducing the cross-section. In addition, it is necessary to actively use low-EI concrete that replaces cement, which causes high EIs, with industrial by-products such as fly ash (FA) and ground-granulated blast-furnace slag (GGBS) as a binder for concrete.

This study, however, conducted research only for apartment buildings, not considering various other building types, and the numbers of samples for each structure type and plan were not even a result of difficulty in data collection. In the future, it is necessary to extend the analysis to other building types and improve the reliability and significance of analysis results by securing additional sample data. Moreover, further studies are required to conduct deterministic analyses of EEIs of buildings in combination with probabilistic analysis methods. Research to facilitate the decision-making of stakeholders by integrating EEI assessment results composed of various impact categories and damage indexes for each safety guard into a single index is also required.

5. Conclusions

The purpose of this study was to analyze the EEIs of Korean apartment buildings on the basis of major building materials as part of research with the goal of reducing the LCEIs of buildings. The results are summarized as follows:

1. The LCEEIs of apartment buildings according to structure types and plans were assessed using 443 apartment buildings in South Korea, and the results were analyzed from the perspectives of total EIs, building life-cycle stages, major construction materials, and safety guards.
2. The analysis results showed that the tower-type apartment buildings with a flat plate structure exhibited the lowest EIs for all EI categories (GWP, AP, EP, ODP, POCP, and ADP) on the basis of total EIs, whereas the plate-type apartment buildings with a wall structure showed the highest EIs.
3. In particular, the percentage of EEIs caused by the production stage was the highest for all EI categories; for example, the maximum proportion of 90.04% was found for the tower-type apartment buildings with a flat plate structure for GWP. In addition, the percentages of EEIs of the construction and end-of-life stages reached 21.04% and 18.29%, respectively, depending on the EI category.
4. It was confirmed that ready-mixed concrete and reinforcement steel, both of which constitute the structures of apartment buildings, are major construction materials that cause such EEIs and that the EEIs of ready-mixed concrete are inversely proportional to those of reinforcement steel. In particular, the percentage of the EEIs caused by ready-mixed concrete reached 94.75% for ODP in plate-type apartment buildings with a wall structure, whereas that caused by reinforcement steel reached 27.52% for AP in tower-type apartment buildings with a flat plate structure.
5. The damage index by safety guard was the lowest in the tower-type apartment buildings with a flat plate structure, similarly to total EIs, and was the highest in the plate-type apartment buildings with a wall structure.

Author Contributions: All authors contributed substantially to all aspects of this article.

Acknowledgments: This research was supported by a grant (16CTAP-C114806-01) from the Technology Advancement Research Program (TARP), funded by the Ministry of Land, Infrastructure, and Transport of the Korean government.

Conflicts of Interest: The authors declare no conflict of interest.

Abbreviations

The following abbreviations are used in this manuscript.

LCA	Life-cycle assessment
LCI	Life-cycle inventory analysis
LCIA	Life-cycle impact assessment
LCI DB	Life-cycle inventory database
KOLID	Korean life-cycle impact assessment method based on damage-oriented modeling
LCEI	Life-cycle environmental impact
LCEEI	Life-cycle embodied environmental impact
EI	Environmental impact
EEI	Embodied environmental impact
OEI	Operational environmental impact
GWP	Global warming potential
AP	Acidification potential
EP	Eutrophication potential
ODP	Ozone-layer depletion potential

POCP	Photochemical ozone creation potential
ADP	Abiotic depletion potential
DALY	Disability-adjusted life years
EINES	Expected increase in number of extinct species
NPP	Net primary production
SCM	Supplementary cementitious material
FA	Fly ash
GGBS	Ground-granulated blast-furnace slag

References

1. Simpson, N.P.; Basta, C. Sufficiently capable for effective participation in environmental impact assessment? *Environ. Impact Assess. Rev.* **2018**, *70*, 57–70. [CrossRef]
2. Gorobets, A. Eco-centric policy for sustainable development. *J. Clean. Prod.* **2014**, *64*, 654–655. [CrossRef]
3. Vrieze, R.; Moll, H. An analytical approach towards sustainability-centered guidelines for Dutch primary school building design. *Int. J. Sustain. Build. Technol. Urban Dev.* **2017**, *8*, 93–124.
4. Vonka, M.; Hajek, P.; Lupisek, A. SBToolCZ: Sustainability rating system in the Czech Republic. *Int. J. Sustain. Build. Technol. Urban Dev.* **2013**, *1*, 46–52. [CrossRef]
5. Braganca, L.; Mateus, R.; Koukkari, H. Building Sustainability Assessment. *Sustainability* **2010**, *2*, 2010–2023. [CrossRef]
6. Simona, P.L.; Spiru, P.; Ion, I.V. Increasing the energy efficiency of buildings by thermal insulation. *Energy Procedia* **2017**, *128*, 393–399. [CrossRef]
7. Li, J.; Colombier, M. Managing carbon emissions in China through building energy efficiency. *J. Environ. Manag.* **2009**, *90*, 2436–2447. [CrossRef] [PubMed]
8. Pierucci, A.; Cannavale, A.; Martellotta, F.; Fiorito, F. Smart windows for carbon neutral buildings: A life cycle approach. *Energy Build.* **2018**, *165*, 160–171. [CrossRef]
9. Dong, L.; Wang, Y.; Li, H.X.; Jiang, B.; Al-Hussein, M. Carbon Reduction Measures-Based LCA of Prefabricated Temporary Housing with Renewable Energy Systems. *Sustainability* **2018**, *10*, 718. [CrossRef]
10. Huedo, P.; Mulet, E.; Lopez-Mesa, B. A model for the sustainable selection of building envelope assemblies. *Environ. Impact Assess. Rev.* **2016**, *57*, 63–77. [CrossRef]
11. Cellura, M.; Guarino, F.; Longo, S.; Mistretta, M. Different energy balances for the redesign of nearly net zero energy buildings: An Italian case study. *Renew. Sustain. Energy Rev.* **2015**, *45*, 100–112. [CrossRef]
12. Haase, M.; Andresen, I.; Gustavsen, A.; Dokka, T.H.; Hestnes, A.G. Zero Emission Building Concepts in Office Buildings in Norway. *Int. J. Sustain. Build. Technol. Urban Dev.* **2011**, *2*, 150–156. [CrossRef]
13. Ferrante, A.; Mochi, G.; Predari, G.; Badini, L.; Fotopoulou, A.; Gulli, R.; Semprini, G. A European Project for Safer and Energy Efficient Buildings: Pro-GET-onE (Proactive Synergy of inteGrated Efficient Technologies on Buildings' Envelopes). *Sustainability* **2018**, *10*, 812. [CrossRef]
14. Harkouss, F.; Fardoun, F.; Biwole, F.H. Multi-objective optimization methodology for net zero energy buildings. *J. Build. Eng.* **2018**, *16*, 57–71. [CrossRef]
15. Annunziata, E.; Frey, M.; Rizzi, F. Towards nearly zero-energy buildings: The state-of-art of national regulations in Europe. *Energy* **2013**, *57*, 125–133. [CrossRef]
16. AIA. *AIA Guide to Building Life Cycle Assessment in Practice*; AIA: Washington, DC, USA, 2010.
17. Zuo, J.; Zhao, Z.Y. Green building research–Current status and future agenda: A review. *Renew. Sustain. Energy Rev.* **2014**, *30*, 271–281. [CrossRef]
18. Oliveira, Z.L.; David, V.L.; Serra, S.M.B.; Barreto, D. Decision making process assisted by life cycle assessment: Greenhouse gas emission. *Int. J. Sustain. Build. Technol. Urban Dev.* **2017**, *8*, 244–253.
19. Roh, S.; Tae, S. Building Simplified Life Cycle CO_2 Emissions Assessment Tool (B-SCAT) to Support Low-Carbon Building Design in South Korea. *Sustainability* **2016**, *8*, 567. [CrossRef]
20. Rovers, R. Material-Neutral Building: Closed Cycle Accounting for Buildings Construction a new practical way to measure improvements in creating a balanced resource use for construction. *Int. J. Sustain. Build. Technol. Urban Dev.* **2010**, *1*, 152–159. [CrossRef]
21. Geng, S.; Wang, Y.; Zuo, J.; Zhou, Z.; Du, H.; Mao, G. Building life cycle assessment research: A review by bibliometric analysis. *Renew. Sustain. Energy Rev.* **2017**, *76*, 176–184. [CrossRef]

22. Rashid, A.F.A.; Yusoff, S. A review of life cycle assessment method for building industry. *Renew. Sustain. Energy Rev.* **2015**, *45*, 244–248. [CrossRef]
23. Iban-Mohammed, T.; Greenough, R.; Taylor, S.; Ozawa-Meida, L.; Acquaye, A. Operational vs. embodied emissions in buildings—A review of current trends. *Energy Build.* **2013**, *66*, 232–245. [CrossRef]
24. Chastas, P.; Theodosiou, T.; Bikas, D.; Kontoleon, K. Embodied Energy and Nearly Zero Energy Buildings: A Review in Residential Buildings. *Procedia Environ. Sci.* **2017**, *38*, 554–561. [CrossRef]
25. Basbagill, J.; Flager, F.; Lepech, M.; Fischer, M. Application of life-cycle assessment to early stage building design for reduced embodied environmental impacts. *Build. Environ.* **2013**, *60*, 81–92. [CrossRef]
26. Luo, Z.; Yang, L.; Liu, J. Embodied carbon emissions of office building: A case study of China's 78 office buildings. *Build. Environ.* **2016**, *95*, 365–371. [CrossRef]
27. Roh, S.; Tae, S.; Shin, S. Development of building materials embodied greenhouse gases assessment criteria and system (BEGAS) in the newly revised Korea Green Building Certification System (G-SEED). *Renew. Sustain. Energy Rev.* **2014**, *35*, 410–421. [CrossRef]
28. Li, X.; Yang, F.; Zhu, Y.; Gao, Y. An assessment framework for analyzing the embodied carbon impacts of residential buildings in China. *Energy Build.* **2014**, *85*, 400–409. [CrossRef]
29. Moncaster, A.M.; Song, J.Y. A comparative review of existing data and methodologies for calculating embodied energy and carbon of buildings. *Int. J. Sustain. Build. Technol. Urban Dev.* **2012**, *1*, 26–36. [CrossRef]
30. Zeng, R.; Chini, A. A review of research on embodied energy of buildings using bibliometric analysis. *Energy Build.* **2017**, *155*, 172–184. [CrossRef]
31. Shi, Q.; Yu, T.; Zuo, J. What leads to low-carbon buildings? A China study. *Renew. Sustain. Energy Rev.* **2015**, *50*, 726–734. [CrossRef]
32. Wen, T.J.; Siong, H.C.; Noor, Z.Z. Assessment of embodied energy and global warming potential of building construction using life cycle analysis approach: Case studies of residential buildings in Iskandar Malaysia. *Energy Build.* **2015**, *93*, 295–302. [CrossRef]
33. Roh, S.; Tae, S.; Shin, S.; Woo, J. Development of an optimum design program (SUSB-OPTIMUM) for the life cycle CO_2 assessment of an apartment house in Korea. *Build. Environ.* **2014**, *73*, 40–54. [CrossRef]
34. Heinonen, J.; Säynäjoki, A.; Junnonen, J.; Pöyry, A.; Junnila, S. Pre-use phase LCA of a multi-story residential building: Can greenhouse gas emissions be used as a more general environmental performance indicator? *Build. Environ.* **2016**, *95*, 116–125. [CrossRef]
35. Dixit, M.K. Life cycle embodied energy analysis of residential buildings: A review of literature to investigate embodied energy parameters. *Renew. Sustain. Energy Rev.* **2017**, *79*, 390–413. [CrossRef]
36. Guinee, J.B. *Handbook on Life Cycle Assessment Operational Guide to the ISO Standards, CML*; Leiden University: Leiden, The Netherlands, 2002.
37. ISO. *ISO 14025: Environmental Labels and Declarations—Type III Environmental Declarations—Principles and Procedures*; ISO: Geneva, Switzerland, 2006.
38. Korea Environmental Industry & Technology Institute. *Development of Integrated Evaluation Technology on Product Value for Dissemination of Environmentally Preferable Products*; Korea Ministry of Environment: Sejong, Korea, 2009.
39. Roh, S.; Tae, S.; Suk, S.J.; Ford, G. Evaluating the embodied environmental impacts of major building tasks and materials of apartment buildings in Korea. *Renew. Sustain. Energy Rev.* **2017**, *73*, 135–144. [CrossRef]
40. Roh, S.; Tae, S.; Suk, S.J.; Ford, G.; Shin, S. Development of a building life cycle carbon emissions assessment program (BEGAS 2.0) for Korea's green building index certification system. *Renew. Sustain. Energy Rev.* **2016**, *53*, 954–965. [CrossRef]
41. Korea Institute of Civil Engineering and Building Technology (KICT). *Standard Estimating System of the Construction Work*; KICT: Goyang, Korea, 2017.
42. Korea Legislation Research Institute (KLRI). Korea Corporate Tax Act: Korea Ministry of Strategy and Finance. 2013. Available online: http://elaw.klri.re.kr/kor_service/lawView.do?hseq=28577&lang=ENG (accessed on 31 March 2018).
43. Korea Environmental Industry & Technology Institute (KEITI). *Waste Statistics*; KEITI: Seoul, Korea, 2017.
44. Korea Environmental Industry & Technology Institute (KEITI). Korea Life Cycle Inventory Database. 2004. Available online: http://www.edp.or.kr/lci/lci_db.asp (accessed on 12 March 2018).

45. Korea Institute of Civil Engineering and Building Technology (KICT). *The Final Report of National DB on Environmental Information of Building Materials*; KICT: Goyang, Korea, 2008.
46. Germany Federal Ministry of the Interior, Building and Community. Oekobaudat. 2017. Available online: http://www.oekobaudat.de/en/database/database-oekobaudat.html (accessed on 31 March 2018).

© 2018 by the authors. Licensee MDPI, Basel, Switzerland. This article is an open access article distributed under the terms and conditions of the Creative Commons Attribution (CC BY) license (http://creativecommons.org/licenses/by/4.0/).

Article

Building Ownership, Renovation Investments, and Energy Performance—A Study of Multi-Family Dwellings in Gothenburg

Mikael Mangold [1,*], **Magnus Österbring** [2], **Conny Overland** [3], **Tim Johansson** [4] **and Holger Wallbaum** [2]

1. City Development, Research Institutes of Sweden, 41261 Gothenburg, Sweden
2. Architecture and Civil Engineering, Chalmers University of Technology, 41296 Gothenburg, Sweden; magost@chalmers.se (M.Ö.); holger.wallbaum@chalmers.se (H.W.)
3. Department of Business Administration, University of Gothenburg, 40530 Gothenburg, Sweden; conny.overland@handels.gu.se
4. Department of Civil, Environmental and Natural Resources Engineering, Luleå University of Technology, 97187 Luleå, Sweden; tim.johansson@ltu.se
* Correspondence: mikael.mangold@ri.se; Tel.: +46-70-297-9778

Received: 28 March 2018; Accepted: 17 May 2018; Published: 22 May 2018

Abstract: The European building stock was renewed at a rapid pace during the period 1950–1975. In many European countries, the building stock from this time needs to be renovated, and there are opportunities to introduce energy efficiency measures in the renovation process. Information availability and increasingly available analysis tools make it possible to assess the impact of policy and regulation. This article describes methods developed for analyzing investments in renovation and energy performance based on building ownership and inhabitant socio-economic information developed for Swedish authorities, to be used for the Swedish national renovations strategy in 2019. This was done by analyzing measured energy usage and renovation investments made during the last 30 years, coupled with building specific official information of buildings and resident area characteristics, for multi-family dwellings in Gothenburg (N = 6319). The statistical analyses show that more costly renovations lead to decreasing energy usage for heating, but buildings that have been renovated during the last decades have a higher energy usage when accounting for current heating system, ownership, and resident socio-economic background. It is appropriate to include an affordability aspect in larger renovation projects since economically disadvantaged groups are over-represented in buildings with poorer energy performance.

Keywords: renovation extent; energy retrofitting; rent affordability; tenure; energy performance certificate; decision support

1. Introduction

In 2010, buildings accounted for 32% of total global final energy usage [1]. The European Directive 2012/27/EU [2] requires member states to have a strategy for renovation of the building stock with the target of reducing energy usage by 20% by 2020 compared with 1990. In many European countries, the building stock increased at a rapid pace during the period 1950–1975 [3]. This aging building stock needs to be renovated, and there is a possibility to introduce energy efficiency measures in the renovation process [4,5].

Building condition, building ownership, and rent affordability determine what kind of renovation measures are possible and optimal [6]. Furthermore, tenure types, building regulations, construction practices etc. vary between different contexts, which increases the need for country-specific studies on renovation investments and energy performance [7–9]. The authors have developed methods for analyzing investments in renovation and building energy performance based on comprehensive building-specific information that can be used to analyze subgroups of building owners and specific socio-economic inhabitant groups. This was done for the Swedish case and to be used by the Swedish authorities. The paper tries to answer the question how can merged official comprehensive building specific information be used to describe the developments in the multi-family dwelling stock?

There are several studies on how ownership and socio-economic area conditions affect investment and energy performance. For rental multi-family dwellings, the landlord–tenant problem can be a barrier to energy retrofitting and renovation [10–14]. The multi-family dwellings that are resident-owned also have internal barriers to larger investments in energy retrofitting [15–17]. There is also not a consensus on which parameters should be included in an optimization of renovation projects. For instance, renovations financed by increased rents in socio-economically disadvantaged areas risk aggravating societal inequities [18–20].

The purpose of this article is to describe quantitative methods developed for analyzing how overall renovation progress and improved energy performance of buildings is related to ownership structures. The methods were developed for the multi-family dwellings in Gothenburg, Sweden, with the intention of being applicable for the entire national building stock in the update of the national strategy of energy efficient renovation in 2019. This article also contributes within the field of analysis of building specific information coupled with *measured* energy usage, also presented as Building Energy Epidemiology by the International Energy Agency, Annex 70. Countries have different registers and specific ways of storing and working with information. This article presents the Swedish case and methods developed for working with Swedish building specific information.

Building specific energy usage for all Gothenburg multi-family dwellings was gathered from Swedish Energy Performance Certificates (EPC). In most other counties in Europe the EPCs contain estimated energy usage, while in Sweden the EPCs should be issued by an certified energy expert that registers measured energy for heating from the 13 most commonly used heating sources, heating for tap water, and domestic electricity usage [21,22]. Renovation investments were deduced from official property taxation records. By aligning these datasets with building-specific information on ownership, tenure, and living-area socio-economic information, it is possible to analyze the progress toward the goals of reduced energy usage and decreased segregation. Michelsen et al. [23] also used measured energy usage from EPC data for German the building stock and made similar analyses focusing mainly on the difference between smaller and larger real estate owners. Michelsen et al. [23] found that larger real estate companies outperform smaller companies in terms of extensive renovations, but smaller companies can be better at continuous maintenance. This article compares real estate companies with other types of ownership. Albatici et al. [24] and Šijanec Zavrl et al. [25] merged mainly energy-related databases for the Italian and Slovenian building stock to model scenarios of developments toward energy efficiency. Our additional contribution is to work toward the development of methods that also include socio-economic parameters of building stock development.

Matschoss et al. [17] made a larger European summary of renovation policy that include tenure and found that incentives for condominium owners to energy retrofit buildings are needed in Europe to reach the 20-20-20 targets. This article sheds light on the renovation activities of multi-family dwellings in Sweden in two main ways. First, it describes methods developed for analyzing energy usage and investment in renovation in different ownership groups intended to be applied future analysis of impact of building regulations and subsidies. Second, we add to the literature on decision-making by presenting an empirical analysis of a large sample of homogeneous organizations, where one can distinguish between individual and social decision making.

Real-Estate Ownership in Sweden

To analyze the impact of regulations and subsidy schemes, it is necessary to distinguish between different ownership for which decision-making structures, taxation rules, subsidies, and the availability of funds are different. In addition, there is a variation between the ownership types in living area per person and in the resident profit or increased cost from improved living standards and energy performance.

Swedish multi-family dwellings are primarily of two tenure types: rental apartments or resident owned apartments, which is a type of tenure comparable with condominiums. Residents own a tradable share in a resident's organization that gives them the right to inhabit a specific apartment of the building. Member of the resident's organization are required to pay a membership fee toward maintenance and capital costs.

The rental apartments can be divided in two larger groups: municipally owned and privately owned [26]. From 1960 to 1975, a national initiative was initiated with the aim to build one million dwellings to cover an urgent housing need in Sweden [26,27]. The multi-family dwellings constructed during this period were constructed as privately owned rental apartments (20%), municipal rental apartments (50%), and resident-owned apartments (30%) [26]. Details about the building age typology can be found in Table A1. The main construction types in the Gothenburg multi-family building stock are slab block, enclosed block, point block, and gallery access block.

Private real estate firms are for-profit organizations that typically are family controlled, though investment funds are growing in importance within this group. Private real estate firms span from private individuals that own one building to large firms that own properties nationally as well as internationally. The modus operandi of private building owners is expected to be that of for-profits in general—to maximize asset value over time. For this reason, renovation as investments in general by private building owners are more likely to take place in attractive areas where risk is lower. With lower risk, the value added to money invested is higher. Some variation should also be found between different private tenants in how they make decisions. For instance, large firms possess superior organizational capacity and have better access to external finance. This means that they are likely to be better equipped to make investments in general and thereby are more inclined to follow through with investments [23].

Municipality owned multi-family dwellings in Gothenburg consist of four large companies. Municipally owned apartments are located in both areas with higher and lower income. They differ from private firms as they, historically, have not had profits as an organizational goal. In fact, they have not even been allowed to make profits. Instead, their focus has been to provide sufficient and affordable housing. The pro-social goal of these firms indicates that they are more inclined to make investments in lower income areas. Renovation can also be more common in municipally owned building due the fact that these companies have had access to cheaper finance through municipal bails.

Resident-owned apartments differ from both private and municipal housing companies. To start with, they are not subject to rental regulation. That means that they can increase their member fees for supplying value-adding features to their members in ways that rental housing companies cannot. For instance, if a private real estate company invests in reduced energy usage and improved indoor climate, this investment has to be carried by the cost reductions from the lower energy usage only. They are not allowed to charge a higher rent for the improved indoor climate and area branding. The group of resident-owned apartments, by contrast, can internalize such features in their member fees. This should increase the scope for more far-reaching renovation.

However, investments in resident-owned buildings are dampened by obstacles associated with social choice. In particular, if the membership fee is increased, the market value of the apartment decreases, making it difficult for all tenants to agree on a time when renovation should happen. Mastschoss et al. [17] studied barriers to energy retrofitting of owner-occupied multi-family dwellings in nine European countries, not including Sweden, finding that collective decision problems is a challenge in every country and that lack of professional experience of real estate management is another challenge. Indeed, suboptimal decisions in co-ops can be costly [28].

As part of the strategy to tackle the housing need that existed before the 1960–1975 era, two cooperatives for groups of resident owned apartments were founded, HSB and Riksbyggen.

These cooperatives constructed buildings that were sold to their member groups. The cooperatives also manage and maintain the resident-owned buildings. The cooperatives promote maintenance plans and certain types of renovations. The economic model may include a larger maintenance plan with more frequent visits and thus require less renovation projects.

After the Swedish bank crisis in 1991, privately owned rental apartments were converted into resident-owned apartments at an increased pace. It is more common with resident-owned buildings in attractive parts of the city, and the price increase of apartments in resident owned buildings has had a strong positive development during the past 20 years. During the same period, municipally owned multi-family dwellings have been sold to private capital funds in order to finance renovation and maintenance [29].

2. Materials

In this chapter, the materials from several Swedish authorities and institutions are described. Sampling was not necessary because 82% of the Gothenburg multi-family dwellings are covered by all matched datasets. Buildings were grouped based on the described ownership types and conversions.

2.1. Matched Datasets

Some datasets used for the analyses in this article were previously described in an article on the data quality of Swedish EPC data [30] and an article describing the renovation and energy retrofitting need in Gothenburg [31]. In these articles analyses of aspects related to the work behind this article such as data uncertainties can be found. Data on real-estate ownership was extracted from the Swedish National Land Survey and Retriever Business was matched with the EPC and base area socio-economic information. Base areas are the smallest demographical statistics area in Sweden containing 50 to 4000 residents. The combined geocoded datasets that describe the Gothenburg building stock are presented in Table 1.

Table 1. Details on datasets and data providers for multi-family dwellings in Gothenburg.

	The National Board of Housing Building and Planning	Swedish Land Survey	Gothenburg City Executive Office	Retriever Business
Aggregation level	Building	Building	Base area [1]	Organization
N	6320	64,600	434	1140
Information used in this article	Heating and energy usage, identified energy saving measures, number of apartments, heated floor area [3]	Construction year, renovation year, value year [2], building owner, corporate form	Average income, number of inhabitants	Organization establishment year

[1] Base areas are the smallest demographical statistics area in Sweden containing 50 to 4000 residents; [2] Value year is further explained in 3.1 Renovation extent; [3] Heated floor area is a measure specifically developed for EPC in Sweden (A_{temp}). Heated floor area is defined as the heated floor space including shared spaces and footprints of walls but not including garages.

The Swedish Land Survey information contains the corporate form under which the building is owned. This makes it possible to separate the resident-owned buildings and the municipal housing companies. Due to legal limitations, socio-economic data had to be aggregated to base area level. For studies in 2019, it is recommended to aggregate sensitive socio-economic inhabitant information to the multi-family dwelling level.

2.2. Characteristics of Building Ownership Groups

The three tenure types are described in 1.1. Real estate ownership in Sweden was further subdivided to reflect differences in regard to the distinction between individual and social decision-making and area attractivity. The privately owned group was further subdivided into buildings owned by: private persons, private companies, or owned by foundations based on the distinction between individual and social decision making. To be noted is that student housing with a total heated floor area of 200,000 m^2 is owned by foundations.

Income was previously used as an indicator to geographically separate the city into more and less attractive areas in Gothenburg by Mangold et al. [31]. The threshold of average income of 200,000 SEK/year before tax was used to divide the ownership groups municipally owned and private company owned into dummy groups of more disadvantaged suburban areas or more attractive rental areas (1\$ = 8.26 SEK as of 1 March 2018). It should be noted that this separation does not reflect the official low-income definition, which is 60% of median income, 148,000 SEK/year. The resident-owned buildings were divided into buildings that were built by the cooperatives for resident owned-buildings (HSB and Riksbyggen) and buildings that have been converted into resident-owned buildings. The details for the ownership groups can be seen in Table A2.

2.3. Buildings Converted into Resident-Owned Buildings

Retriever Business is a register of all companies in Sweden including organizations of resident-owned buildings, which contains establishment year. Most of the buildings registered as resident-owned in Retriever Business are registered as resident-owned in the Swedish National Land Survey—see Table A3—but there are overlaps with other ownership groups due to a few different verified reasons: change of ownership between the records were made, split ownership of the building, and a company can own a share of a resident owned building. From 2000 to 2009 the buildings registered as resident owned increased from 901,000 to 2,280,000 m² heated floor area. During the same time the group "private company owned, higher income" decreased from 3,380,000 to 2,550,000 m² heated floor area. This made it possible to identify under which tenure renovations have taken place by comparing establishment year and reconstruction year.

3. Methods

The regressions were made to find and compare patterns of renovation investments and energy usage. The intention is to use these types of regressions for the assessment of building regulations and subsidy schemes in 2019. However, there are several other data analysis methods and visualization tools that are needed to communicate results with decision makers. The categorization of the building owners in this paper is an illustration of how to make the comprehensive building specific information more understandable. Timelines is another way to illustrate development that make it easy for practitioners, academics and decision makers to work together in the development of policy.

Regressions were made in R to explain variance in renovation investments and energy performance, see Table 2. Renovation investments and energy performance were derived from value year and renovation year in the register of the Swedish National Land Survey and from the Energy Performance Certificates (EPC). The value year is initially the year of construction, but as the building is renovated, the value year will be changed depending on the cost of the renovation. This is further described in Table A4. The purpose of recording a value year is to have an official record of anticipated remaining service life of buildings [32].

Table 2. Description of models in the regression analyses.

	Type	Dependent	Unit/Value	N
Model 1	Ordinal	Renovation group	0, 1, 2, 3	6244
Model 2	Linear	Renovation extent	%	6244
Model 3	Linear	Energy performance	kWh/m².year	5725
Model 4	Linear	Energy performance excluding energy for heating domestic hot water	kWh/m².year	5697

3.1. Renovation Investments

In Sweden, renovations are registered on a municipal level. The Swedish Tax Office require that a renovation is registered as a change in value year depending on the cost of the renovation in relation to new building cost as described in Table A4 and Equation (1). New building cost is revised yearly by the Swedish tax authority to reflect inflation and changes in construction costs.

$$\text{Renovation extent} = \frac{\text{Renovation cost}}{\text{New building cost}} = \frac{\text{Value year} - \text{Construction year}}{\text{Renovation year} - \text{Construction year}} \quad (1)$$

The groups in Table A4 were used in the ordinal (logit) regression in model 1. The groups are further described in Table A5. Using Equation (1), it was possible to calculate the renovation extent and equivalent cost of previous renovation projects from the value year and reconstruction year for group 2. The renovation extent was used as a dependent in model 2.

3.2. Energy Performance

Most EPCs for multi-family dwellings in Gothenburg were issued in 2008 and 2009 [30]. Since only one measurement of energy use is available from the EPC the statistical analysis for the relationship between building energy performance and investment in renovation the buildings were divided using renovation year of 2009 as a watershed. Energy performance of buildings renovated before 2009 include the impact of renovations. Only these buildings were used to analyze the importance of renovation investment for building energy performance, $N = 5697$. The buildings renovated after 2009 were used to analyze the importance of energy savings advice provided in the EPC for renovation investments.

The EPC will be renewed in 2019. This will make it possible to replace energy performance with change in energy performance as a dependent in the future analyses of success and impact of changed building regulations and subsidy schemes.

Energy for heating domestic hot water is included in the energy performance presented in the EPC. However, regression analyses have been made for both energy performance including and excluding energy for heating domestic hot water. There is a large difference in registered heating for domestic hot water between the building ownership groups in Gothenburg (see Figure 1). Removal of outliers was conducted with the criterion 2.5 standard deviation above average buildings energy performance (223 and 182 kWh/m^2.year for energy performance with and without energy for heating domestic hot water).

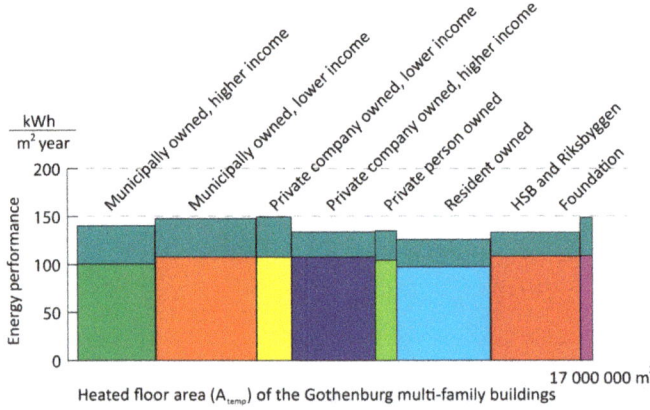

Figure 1. Energy usage in the Gothenburg multi-family dwellings separated into ownership groups. The energy for heating domestic hot water is teal.

Some studies suggest that water usage can be used as a proxy for number of inhabitants and time spent at home [33]. In Figure 1, the energy for heating of domestic hot water has been separated from the energy performance to illustrate the differences between the ownership groups. Living area per person has impact on both the need of renovation and on energy usage in buildings.

3.3. Independent Variables in Regression Analyses

Nominal, ordinal, and scalar variables were used in the regression analyses. The independent variables building age, heating system, and ownership group are nominal variables that were converted into nominal groups of variables for the linear regression analyses. One variable in each group was kept outside of the linear regression analyses as a reference category; these are indicated with asterisk in Table A6. In the regression analyses, it was also possible to use base area average income as an independent variable instead of separating private company and municipal ownership into subgroups. For the statistical analyses of building energy performance, the importance of the time period in which the renovation happened was studied by creation an ordinal variable in segments 1979–1989, 1990–1999, and 2000–2009. This variable was excluded in the analyses of renovation investment because of the strong interdependence of the two variables both derived from renovation year.

3.4. Robustness Checks

In order to estimate renovation investments made in the entire building stock and be able to make a linear regression for the renovation costs, costs for renovations for group 1 and group 3 in Table A4 were assumed. The robustness of these assumptions was checked through the comparison between model 1 and 2, as well as an additional linear regression for only group 2. The assumptions caused no larger differences in the predictions.

Data on building ownership from Swedish National Land Survey only describe the building ownership in 2014. The changes in ownership were included by the addition of conversions registered in Retriever Business. The analysis of investment in relation to conversion was done only for the buildings registered in Retriever Business.

4. Results

Analyses of total costs and estimates of future cost of renovations and energy retrofitting were conducted by Mangold et al. [31]. In Figure 2, past renovation costs have been separated further in the ownership groups. A difference in how consistently the members in ownership groups invest in renovations can be seen and the economic crises in 1991 and 2000 are observable.

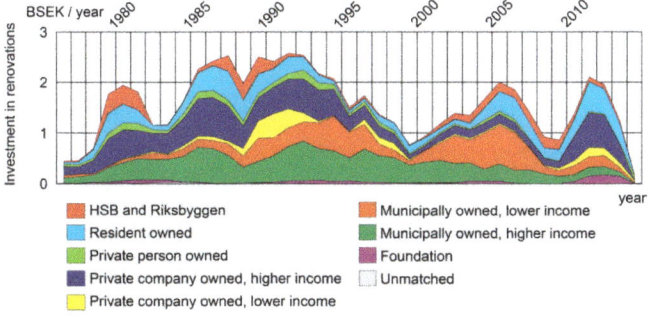

Figure 2. Costs of renovation registered for ownership groups.

The division of areas into higher and lower income is based on income statistics from 2015 which means that changes in income is not reflected in Figure 2. Some of the renovation costs in the municipally owned, higher income areas in the 1980ties and 1990ties were made in central areas that had a lower income level when they were renovated but are now higher income areas.

The renovation costs in Figure 2 include buildings from all construction periods. In the two largest ownership groups of multi-family dwellings constructed during 1960–1975—HSB and Riksbyggen and municipally owned, lower income—there is a considerable amount of multi-family dwellings that has not been renovated (see Table 3). The group resident-owned contain the most renovations that have been conducted with a cost of less than 20% of new building cost in Table 3. It is easier for resident that own their buildings to agree on and carry out smaller renovation investments than it is for other ownership groups.

For the discussion on municipal or private ownership of rental buildings in lower income areas, it is relevant to notice that private companies, and especially private persons, have registered fewer investments through renovations than municipal companies in Table 3.

Table 3. Share of heated floor area (10^3 m^2) in the renovation cost groups of the buildings built during 1960–1975.

	No Renovation	Less than 20%	20–70%	More than 70%	Grand Total
Municipally owned, higher income	44%	20%	9%	26%	398
Municipally owned, lower income	50%	23%	20%	6%	2390
Private company owned, lower income	50%	38%	9%	3%	745
Private company owned, higher income	37%	41%	18%	3%	1019
Private person owned	92%	8%	0%	0%	119
Resident owned	30%	55%	12%	2%	570
HSB and Riksbyggen	68%	26%	6%	0%	1358
Foundation owned	74%	12%	3%	12%	167
Unmatched	100%	0%	0%	0%	30

In Figure 3, the renovation costs in relation to conversion to resident ownership have been illustrated. Right after the conversion renovation projects are the most frequent. Fewer buildings were renovated and then converted. In connection with the analysis of shared ownership (such as resident owned buildings) it was hypothesized that larger multi-family dwellings would be less renovated. However, a separation of the building stock in regard to building heated floor area did not show any significant relationship between resident owned buildings' size and renovation cost.

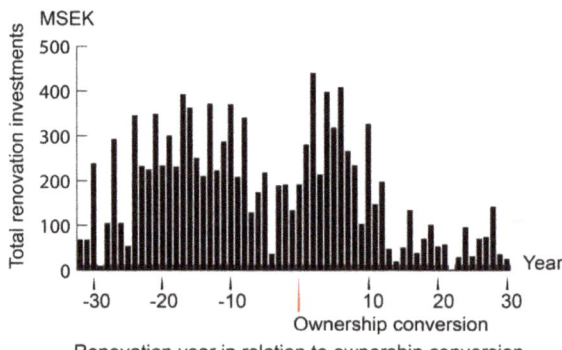

Figure 3. Costs of renovations of resident-owned building summed by the years between renovation year and the year the building tenure was converted into resident ownership, for only buildings registered as resident-owned in Retriever Business.

Regression Results

The coefficients in Table 4 mostly reflect expected patterns in the building stock of Gothenburg. This is a promising result for the later analysis of changes in energy usage connect using the same methods for the national building stock in 2019. However, there are also some important considerations to be made when applying these methods.

Table 4. Ordinal (logit) and linear regression results.

Variable	Model 1 Estimate	Sig.	Model 2 Coefficient	Sig.	Model 3 Coefficient	Sig.	Model 4 Coefficient	Sig.
Constant			73.594 ***	0.000	167.775 ***	0.000	132.131 ***	0.000
Renovation extent [%]					−0.084 ***	0.000	−0.092 ***	0.000
Energy performance [kWh/m^2.year]	−0.007 ***	0.000	−0.156 ***	0.000				
Private company owned	0							
Municipality owned	0.335 ***	0.000	3.018 **	0.010	8.089 ***	0.000	−1.656	0.063
Private person owned	−1.080 ***	0.000	−16.991 ***	0.000	−0.116	0.935	−1.56	0.094
Resident owned	−0.475 ***	0.000	−11.061 ***	0.000	−1.129	0.264	1.8	0.079
HSB and Riksbyggen	−0.483 ***	0.000	−6.947 ***	0.000	−2.36 *	0.034	−0.222	0.866
Foundation owned	0.106	0.560	5.331	0.065	19.369 ***	0.000	13.48 ***	0.000
Base area share of university degree [%]	0.094	0.572	0.005	0.437	7.503 ***	0.000	11.803 ***	0.000
Base area average income [KSEK]	−0.001 **	0.005	−0.037 ***	0.000	−0.047 ***	0.000	−0.037 ***	0.000
Was recently renovated					1.86 ***	0.000	1.27 **	0.005
FTX or other heat recovery from ventilation	0.643 ***	0.000	6.041 ***	0.000	−2.865 **	0.007	−3.362 ***	0.001
District heating	0							
Heat pumps	−0.453 ***	0.000	1.521	0.502	−68.443 ***	0.000	−51.217 ***	0.000
Electricity	−1.187 ***	0.000	−9.632 **	0.003	−32.891 ***	0.000	−29.621 ***	0.000
Boiler	−0.335 ***	0.000	−7.85 *	0.028	5.982 *	0.028	6.413 *	0.010
Constructed before 1945	0							
Constructed 1945–1960	−0.754 ***	0.000	−16.415 ***	0.000	0.511	0.628	−1.049	0.279
Constructed 1960–1975	−1.631 ***	0.000	−35.937 ***	0.000	1.732	0.157	−1.424	0.206
Constructed after 1975	−3.583 ***	0.000	−49.154 ***	0.000	−27.724 ***	0.000	−26.676 ***	0.000
Heated floor area [10^3 m^2]	0.014	0.356	−0.039	0.868	−0.852 ***	0.000	−0.564 ***	0.001
Sides with other buildings	−0.085	0.033	−0.78	0.237	−5.325 ***	0.000	−4.3 ***	0.000
Number of floors	−0.026	0.107	−1.047 ***	0.000	−1.074 ***	0.000	−0.847 ***	0.000
Number of staircases	−0.006	0.559	−0.061	0.701	−0.301 *	0.015	−0.258 *	0.024
Heated basements	0.232 ***	0.000	0.463	0.621	−6.591 ***	0.000	−5.582 ***	0.000
Heated garage ratio to building [%]	−2.115 **	0.004	−31.77 **	0.001	1.096	0.832	6.279	0.191
N	6235		6235		5725		5697	
Unit			%		kWh/m^2.year		kWh/m^2.year	
R^2 adjusted			0.255		0.484		0.400	

* Coefficient is significant at the 0.05 level (2-tailed); ** Coefficient is significant at the 0.01 level (2-tailed); *** Coefficient is significant at the 0.001 level (2-tailed).

Models 3 and 4 explain variance in energy performance including and excluding heating for domestic hot water. This separation is important to make in order to understand why municipally owned multi-family dwellings have worse energy performance registered in the EPC than other ownership groups.

More costly renovations have resulted in lower energy usage, but buildings that have been renovated during the last decades have a higher energy usage when accounting for current heating system, ownership, and resident socio-economic background. As seen in Table 4, parts of the variance in energy performance are explainable by variables describing the energy systems of the buildings. The types of energy systems also differ between the ownership groups, as seen in Table 5. This should also be considered when comparing the energy performance of the different ownership groups. The group heat pump stands out in the regression analysis because the energy supplied to the heat pump is recorded in the EPC, and not the energy provided by the heat pump to the building. Installation of ventilation with heat recovery and changing heating systems from electricity or boilers to heat pumps or district heating have been registered as renovations. Buildings constructed after the 1975 and after the oil crises were built with better energy performance.

Table 5 also illustrates how different ownership results in different types of energy savings measures. Installing heat pumps have been a profitable energy savings measure for especially private persons and residents owning their buildings. The price of district heating is close to the price of electric heating directly which incentivizes building owners to install heat pump. However, for municipal housing companies, buying district heat from waste heat and more carbon-neutral heating distributed by the municipal energy company is the most prevalent solution. Other commonly used energy savings measures in Sweden have been roof insulation, ventilation heat recovery, wall insulation, replacement of windows, and balcony heat bridge mitigation.

Table 4 also illustrates that municipally owned real estate companies make most of the larger investments in renovations. Fewer renovations are registered for resident owned buildings. Contributing factors to this pattern are that: parts of the renovations cost are shifted from the building owner to the apartment owner, ownership of the apartment reduces needs of maintenance, it is more difficult to agree on larger investments in larger groups of owners, HSB and Riksbyggen sell maintenance services which reduce the need for more costly renovations, and finally one unverified interpretation is that buildings built by HSB and Riksbyggen are in larger need of renovations.

Multi-family dwellings in base areas inhabited by economically disadvantaged people have a worse energy performance and fewer renovations have been made in these areas. This emphasizes the importance of including affordability aspects in the sustainability analysis of renovations of multi-family dwellings.

Table 5. Multi-family dwellings with different heat sources and energy efficient ventilation (FTX or heat recovery) in building ownership groups.

	Boiler	District Heating	Electricity	Heat Pump	Total	Energy Efficient Ventilation
Municipally owned, high income	10	1020	13	26	1069	169
Municipally owned, low income	0	810	9	1	820	86
Private company owned, low income	1	293	2	8	304	40
Private company owned, high income	13	975	14	38	1040	111
Private person owned	5	368	9	26	408	15
Resident owned	40	1188	67	165	1460	204
HSB and Riksbyggen	18	908	0	65	991	104
Foundation owned	0	140	1	11	152	25
Unmatched	0	49	0	26	75	37
Grand Total	87	5751	115	366	6319	791

Separate statistical analyses were also made for the buildings renovated between 2009 and 2014 (N = 485). The EPCs were issued in 2009, so for this analysis, the recorded energy performance was instead seen as a predictor of renovation. It was found that buildings with a poor energy performance were prioritized for larger renovations after 2009. The average building energy performance (excluding heating for domestic hot water) was 161 kWh/m^2.year for renovation investment cost Group 3 (N = 17); compared with total average energy performance of 131 kWh/m^2.year. However, it was not possible to establish a significant relationship between suggested energy saving measures and renovation extent.

For the statistical analysis of the socio-economic information, we found that using base area as a level of aggregation for socio-economic information introduces error sources and limitations for the interpretation of results. Existence of other building types made it difficult to establish clear links between number of residents in multi-family dwellings and base areas. Analyses that explain socio-economic status of areas would need to include information aggregated to the multi-family dwelling level.

5. Discussion

In this article methods of analyzing building-specific investment and energy performance have been described. The methods are intended to be used to present decision makers with analyses of developments in the multi-family dwelling stock. The interpretation of results is highly context specific and is also a political matter. Different subsidy schemes apply for the separate ownership groups and priorities in housing policy are based on political decisions.

Hsu [34] also used comprehensive building energy usage and engineering audits and found that variance in energy performance of buildings is not sufficiently explained by building and heating system characteristics. In this article we demonstrate that variance in building energy performance can be further described when adding building ownership and area socio-economic information. Huber et al. [20] studied the challenges of renovating and energy retrofitting multi-family dwellings in socio-economically disadvantaged areas. In this article, we find that the municipally owned real estate companies in Gothenburg are most exposed to those challenges. Trade-offs need to be made between energy usage reductions, rent increases, and increased living standards. Curtis et al. [35] used EPC data matched with census data for the Irish building stock, and found building energy performance to be worse in buildings in socio-economically disadvantaged areas.

Michelsen et al. [23] found that larger real estate companies outperform smaller companies in terms of extensive renovations, but that smaller companies can be better at continuous maintenance. Company size was not part of the analysis in this article. However, we would like to add that the least number of renovations were registered for private person owned rental buildings. The municipally owned real estate companies are large, and they have executed most large renovation projects.

Matschoss et al. [17] compared energy efficiency renovations in multi-family dwellings in eight European countries with regard to building ownership, not including Sweden. They found that resident joint ownership may have internal barriers to making larger investments in energy retrofitting. In Sweden, the costs of renovating the interior of the resident owned apartments are not included in renovation cost of the entire building. Interior apartment renovations are frequent in resident owned buildings, due to the strong market development of resident owned apartments. Including these renovations costs and associated environmental impact would be necessary to make a more complete evaluation of multi-family dwelling resident ownership as a tenure type. Furthermore, other studies have shown that it is not possible to establish that energy usage cost calculations in the EPC affect the price of the resident owned apartments [36–39].

For rental apartments, the landlord–tenant problem can be a barrier to energy retrofitting and renovation [10,12–14]. It was difficult to isolate these aspects in the analysis of the Gothenburg multi-family dwellings. In Sweden, the real estate owner pays the heating cost. In the past decades, individual volumetric metering billing of electricity and water have been increasingly installed in rental apartments in Gothenburg [40].

Central for the regression analyses with building energy performance as dependent was the removal of domestic hot water heating. Only a minor share of the heat in domestic hot water remains in the building after usage. Most of the heat exits the building in the waste water [41]. Excluding heating for water usage from energy performance, which is most strongly predicted by living area per person [33,40], is one manner of separating the impact of user behavior. When excluding energy for heating domestic hot water, no significant association between energy performance and building ownership could be made. This finding is relevant in the context that European countries are reissuing EPCs in which more countries will include measured, instead of calculated, energy and water usage. Because of the differences in water usage between the ownership groups, there are risks that groups of multi-family dwellings that are more crowded appear as have comparatively poorer energy performances depending on how the energy performance certification with measured energy usage is implemented in European counties.

6. Conclusions

This article analyzed variance in investments in renovations and building energy performance of multi-family dwellings, based on building ownership, location and building characteristics. We found that both building ownership and socio-economic area characteristics are useful in explaining variance. Central findings are that energy used for heating domestic hot water and heating systems varies between ownership groups and that municipally owned real estate companies face challenges in the renovation of building in socio-economically disadvantaged areas. Including affordability aspects in the sustainability analyses of larger renovations in such areas is necessary since economically disadvantaged groups are overrepresented in multi-family dwellings with poor energy performance. Buildings which have had higher energy usage are overrepresented in the group of buildings having gone through larger renovations in recent years. Large renovation is a predictor of lower energy usage. However, when accounting for current heating system, ownership, and resident socio-economic background, buildings that have been renovated during the last decades have a higher energy usage.

Building ownership is context specific, but the overall renovation progress and energy performance of buildings in different types of ownership groups can be described using official and publicly accessible databases. Comprehensive building specific information, separated in ownership groups, made it possible to demonstrate the differences in renovations and building energy performance between ownership groups. Furthermore, statistical analyses of the data can be used to reinforce conclusions.

Future Studies

The EPCs were designed to be renewed every 10 years. This will make it possible to describe and explain the changes in energy usage in Sweden in 2019. These studies will investigate predictors of energy retrofitting projects and changes in socio-economic inhabitation characteristics. Ideally, casual relationships can be made between these changes and changes in building regulation and subsidy schemes. In this study, sensitive socio-economic inhabitant information should be aggregated to the multi-family dwelling level during the time of the two waves of EPCs.

In this article, information about the Gothenburg multi-family dwelling stock has been studied. Johansson et al. [42] also used the information sources described in this paper to develop methods for connecting the entire Swedish national databases. This will make it possible to provide decision support to the national authorities which are responsible for building regulation and subsidy schemes.

Adding building-specific energy models would enable comprehensive strategic advice for building owners [43,44]. Using the records of renovation cost and energy performance linked with scenarios will make it possible to point out which building owners should be in line for energy retrofitting.

Author Contributions: M.M. wrote the article and did the statistical data analysis. M.Ö. provided support in writing and analysis of data. C.O. provided support in real estate economics and wrote the segments on 1.1 Real estate ownership in Sweden. T.J. did the merging of different data sources and helped in the writing of the paper. H.W. supervised the project, provided support in the writing and framed the paper as part of the research field.

Acknowledgments: The authors would like to thank Martin Storm at the National Board of Housing Building and Planning for sharing EPC data, Lutz Ewert at Gothenburg CEO for all discussions and for sharing socio-economic data and Eric Jeansson at CPA for giving us access to the Gothenburg spatial city plan. This work was financially supported by Chalmers Infrastructure Engineering, SIRen, and the Swedish Research Council for Environment, Agricultural Sciences and Spatial Planning-Formas. The Formas funds includes a budget post open access payment which will be used for this paper.

Conflicts of Interest: The authors declare no conflict of interest. The founding sponsors had no role in the design of the study; in the collection, analyses, or interpretation of data; in the writing of the manuscript, and in the decision to publish the results.

Appendix A Tables with Descriptions of Ownership Groups and Variables for the Regression Analyses

Table A1. Heated floor area of multi-family dwellings in Gothenburg divided by construction periods and ownership groups [10^3 m^2].

	Built before 1931	1931–1945	1946–1960	1961–1975	1976–1990	1991–2005	Built after 2005
Municipally owned, higher income	512	383	788	398	320	169	76
Municipally owned, lower income	34	31	746	2387	94	7	7
Private company owned, lower income	21	21	351	745	1	1	17
Private company owned, higher income	389	624	383	999	208	65	62
Private person owned	180	275	123	119	12	6	2
Resident owned	919	467	402	590	194	235	252
HSB and Riksbyggen	87	159	755	1358	346	183	51
Foundation owned	58	19	43	167	9	66	41
Unmatched	11	1	4	31	22	1	151
Total	**2211**	**1980**	**3595**	**3049**	**1206**	**733**	**659**

Table A2. Descriptions of the ownership groups. Geographical analyses were conducted to verify the validity of the groups.

	Number of Buildings	Heated Floor Area [10^3 m^2]	Average Income [SEK/Person.Year]	Higher Education [1]
Municipally owned, higher income	1069	2646	241,000	44%
Municipally owned, lower income	820	3305	153,000	27%
Private company owned, lower income	304	1156	156,000	30%
Private company owned, higher income	1040	2729	263,000	48%
Private person owned	408	717	264,000	54%
Resident owned	1460	3059	283,000	52%
HSB and Riksbyggen	991	2939	243,000	38%
Foundation owned	152	403	238,000	68%
Unmatched	75	223	286,000	19%

Table A3. Retriever Business data description and overlap with data from Swedish National Land Survey.

	Buildings Registered as Resident Owned	Heated Floor Area in Retriever Business [10^3 m^2]	Overlap of the Datasets	Average Construction Year	Average Organization Registration Year
Municipally owned, higher income	53	147	6%	1944	2004
Municipally owned, lower income	1	1	0%	1977	2005
Private company owned, lower income	5	20	2%	1931	2006
Private company owned, higher income	185	578	21%	1941	2004
Private person owned	98	214	30%	1937	2005
Resident owned	696	1561	51%	1944	1990
HSB and Riksbyggen	94	481	16%	1957	1956
Foundation owned	7	10	3%	1958	2006

Table A4. Methods for setting a value year based on renovation costs according to Swedish Tax Agency [32].

	Renovation Cost	Method of Setting the Value Year
Group 1	Less than 20% of the new building cost [1]. Assumed to be 10% in the linear regression model.	No change in Value year
Group 2	20–70% of the new building cost [1]	The value year is set based on the cost of the renovation compared with the cost of constructing a comparable building using Equation (1).
Group 3	More than 70% of the new building cost [1]. Assumed to be 90% in the linear regression model.	Value year is set to the year of the renovation

[1] New building cost is set and updated by the Swedish tax authorities. In order to calculate and compare renovation costs between years New building cost of 2012 [32] is be used in this article, which was 15,300 SEK/m^2 (1$ = 8.26 SEK as of 1 March 2018). The new building cost is increasing based on Inflation, changes in construction costs and property value.

Table A5. Description of renovation cost groups. Renovation cost is presented as a percentage of New building cost as specified by the Swedish Tax Agency [32].

	Number of Buildings	Heated Floor Area [10^3 m^2]	Average Building Size [m^2]	Average Construction Year
No renovation	3090	8050	2610	1966
Group 1	984	3770	3830	1958
Group 2	1040	2750	2640	1943
Group 3	1210	2610	2160	1937

Table A6. Independents in the regression analyses.

Type	Variable	Unit	Average	Std. dev.
Nominal	Private company owned [1]	Yes/no	0.198	-
Nominal	Municipality owned	Yes/no	0.308	-
Nominal	Private person owned	Yes/no	0.068	-
Nominal	Resident owned	Yes/no	0.219	-
Nominal	HSB and Riksbyggen	Yes/no	0.170	-
Nominal	Foundation owned/student housing	Yes/no	0.025	-
Scalar	Base area share of university degree	%	0.433	0.196
Scalar	Base area average income	KSEK	238	64.7
Ordinal	Was recently renovated (decades from 1988)	Ordinal	0.924	1.18
Nominal	FTX or other heat recovery from ventilation	Yes/no	0.123	-
Nominal	District heating [1]	Yes/no	0.903	-
Nominal	Heat pumps	Yes/no	0.062	-
Nominal	Electricity	Yes/no	0.020	-
Nominal	Boiler	Yes/no	0.015	-
Nominal	Constructed before 1945 [1]	Yes/no	0.353	-
Nominal	Constructed 1945–1960	Yes/no	0.215	-
Nominal	Constructed 1960–1975	Yes/no	0.224	-
Nominal	Constructed after 1975	Yes/no	0.208	-
Scalar	Heated floor area	10^3 m^2	2.63	2.97
Ordinal	Sides shared with other buildings	Integer	0.500	0.807
Ordinal	Number of floors	Integer	4.17	2.26
Ordinal	Number of staircases	Integer	3.58	3.89
Ordinal	Number of heated basements	Integer	0.700	0.494
Scalar	Heated garage ratio to building	%	0.090	0.044

[1] Variable used as reference category in the regression analyses.

References

1. IPCC. Working Group III—Mitigation of Climate Change, Chapter 9 Buildings. 2014. Available online: https://www.ipcc.ch/pdf/assessment-report/ar5/wg3/ipcc_wg3_ar5_frontmatter.pdf (accessed on 28 April 2018).
2. EU (The European Union). *Directive 2012/27/EU on Energy Efficiency*; EU: Brussels, Belgium, 2012.
3. Meijer, F.; Itard, L.; Sunikka-Blank, M. Comparing European residential building stocks: Performance, renovation and policy opportunities. *Build. Res. Inf.* **2009**, *37*, 533–551. [CrossRef]
4. Ball, E. Holistic strategies for energy efficient refurbishment of the housing stock and renewal of the related energy supply system. In *WP Transnational Manual*; German Association for Housing, Urban and Spatial Development (DV): Berlin, Germany, 2011.
5. Heeren, N.; Jakob, M.; Martius, G.; Gross, N.; Wallbaum, H. A component based bottom-up building stock model for comprehensive environmental impact assessment and target control. *Renew. Sustain. Energy Rev.* **2013**, *20*, 45–56. [CrossRef]
6. Ostermeyer, Y.; Wallbaum, H.; Reuter, F. Multidimensional Pareto optimization as an approach for site-specific building refurbishment solutions applicable for life cycle sustainability assessment. *Int. J. Life Cycle Assess.* **2013**, *18*, 1762–1779. [CrossRef]
7. Pombo, O.; Rivela, B.; Neila, J. The challenge of sustainable building renovation: Assessment of current criteria and future outlook. *J. Clean. Prod.* **2016**, *123*, 88–100. [CrossRef]
8. Grossmann, K.; Kabisch, N.; Kabisch, S. Understanding the social development of a post-socialist large housing estate: The case of Leipzig-Grünau in eastern Germany in long-term perspective. *Eur. Urban Reg. Stud.* **2017**, *24*, 142–161. [CrossRef]
9. Corvacho, H.; Alves, F.B.; Rocha, C. A Reflection on Low Energy Renovation of Residential Complexes in Southern Europe. *Sustainability* **2016**, *8*, 987. [CrossRef]
10. Charlier, D. Energy efficiency investments in the context of split incentives among French households. *Energy Policy* **2015**, *87*, 465–479. [CrossRef]
11. Deng, Y.; Wu, J. Economic returns to residential green building investment: The developers' perspective. *Reg. Sci. Urban Econ.* **2014**, *47*, 35–44. [CrossRef]
12. Hope, A.J.; Booth, A. Attitudes and behaviours of private sector landlords towards the energy efficiency of tenanted homes. *Energy Policy* **2014**, *75*, 369–378. [CrossRef]

13. Kholodilin, K.A.; Michelsen, C. The Market Value of Energy Efficiency in Buildings and the Mode of Tenure. *Urban Stud.* **2017**, *54*, 3218–3238. [CrossRef]
14. Pivo, G. Unequal access to energy efficiency in US multifamily rental housing: Opportunities to improve. *Build. Res. Inf.* **2014**, *42*, 551–573. [CrossRef]
15. Bardhan, A.; Jaffee, D.; Kroll, C.; Wallace, N. Energy efficiency retrofits for U.S. housing: Removing the bottlenecks. *Reg. Sci. Urban Econ.* **2014**, *47*, 45–60. [CrossRef]
16. Genus, A.; Theobald, K. Creating low-carbon neighbourhoods: A critical discourse analysis. *Eur. Urban Reg. Stud.* **2016**, *23*, 782–797. [CrossRef]
17. Matschoss, K.; Heiskanen, E.; Atanasiu, B.; Kranzl, L. Energy renovations of EU multifamily buildings: Do current policies target the real problems. In *Rethink, Renew, Restart*; IEEE: Piscataway Township, NJ, USA, 2013.
18. Ahlfeldt, G.M. Blessing or curse? Appreciation, amenities and resistance to urban renewal. *Reg. Sci. Urban Econ.* **2011**, *41*, 32–45. [CrossRef]
19. Arbaci, S.; Tapada-Berteli, T. Social inequality and urban regeneration in Barcelona city centre: Reconsidering success. *Eur. Urban Reg. Stud.* **2012**, *19*, 287–311. [CrossRef]
20. Huber, A.; Mayer, I.; Beillan, V.; Goater, A.; Trotignon, R.; Battaglini, E. Refurbishing residential buildings: A socio-economic analysis of retrofitting projects in five European countries. In Proceedings of the World Sustainable Energy Days, Wels, Austria, 2–4 March 2011; pp. 2–4.
21. Pagliaro, F.; Cellucci, L.; Burattini, C.; Bisegna, F.; Gugliermetti, F.; de Lieto Vollaro, A.; Salata, F.; Golasi, I. A methodological comparison between energy and environmental performance evaluation. *Sustainability* **2015**, *7*, 10324–10342. [CrossRef]
22. Hjortling, C.; Björk, F.; Berg, M.; Af Klintberg, T. Energy mapping of existing building stock in Sweden-Analysis of data from Energy Performance Certificates. *Energy Build.* **2017**, *153*, 341–355. [CrossRef]
23. Michelsen, C.; Rosenschon, S.; Schulz, C. Small might be beautiful, but bigger performs better: Scale economies in "green" refurbishments of apartment housing. *Energy Econ.* **2015**, *50*, 240–250. [CrossRef]
24. Albatici, R.; Gadotti, A.; Baldessari, C.; Chiogna, M. A decision making tool for a comprehensive evaluation of building retrofitting actions at the regional scale. *Sustainability* **2016**, *8*, 990. [CrossRef]
25. Šijanec Zavrl, M.; Stegnar, G.; Rakuščk, A.; Gjerkeš, H. A bottom-up building stock model for tracking regional energy targets—A case study of kočevje. *Sustainability* **2016**, *8*, 1063. [CrossRef]
26. Hall, T.; Vidén, S. The Million Homes Programme: A review of the great Swedish planning project. *Plan. Perspect.* **2005**, *20*, 301–328. [CrossRef]
27. Castell, P. *The Swedish Suburb as Myth and Reality*; Chalmers University of Technology: Göteborg, Sweden, 2010.
28. Almenberg, J.; Karapetyan, A. Hidden costs of hidden debt. *Rev. Financ.* **2013**, *18*, 2247–2281. [CrossRef]
29. Lind, H. Does the Law on Rental Appartments Lead to Correct Renovations? (Leder Hyreslagens Regler Till Rätt Renoveringar: Analys Och Förslag). Available online: http://www.renoveringscentrum.lth.se/fileadmin/renoveringscentrum/SIRen/Publikationer/Leder_hyreslagens_regler_till_ratt_renovering.pdf (accessed on 28 April 2018).
30. Mangold, M.; Österbring, M.; Wallbaum, H. Handling data uncertainties when using Swedish energy performance certificate data to describe energy usage in the building stock. *Energy Build.* **2015**, *102*, 328–336. [CrossRef]
31. Mangold, M.; Österbring, M.; Wallbaum, H.; Thuvander, L.; Femenias, P. Socio-economic impact of renovation and energy retrofitting of the Gothenburg building stock. *Energy Build.* **2016**, *123*, 41–49. [CrossRef]
32. Swedish Tax Agency. *The Tax Office General Advice (Skatteverkets Allmänna Råd)*; Swedish Tax Agency: Solna, Sweden, 2015.
33. Pullinger, M.; Browne, A.; Anderson, B.; Medd, W. *Patterns of Water: The Water Related Practices of Households in Southern England, and Their Influence on Water Consumption and Demand Management*; Lancaster University: Lancaster, UK, 2013; Available online: https://www.escholar.manchester.ac.uk/uk-ac-man-scw:187780 (accessed on 28 April 2018).
34. Hsu, D. How much information disclosure of building energy performance is necessary? *Energy Policy* **2014**, *64*, 263–272. [CrossRef]

35. Curtis, J.; Devitt, N.; Whelan, A. Using census and administrative records to identify the location and occupancy type of energy inefficient residential properties. *Sustain. Cities Soc.* **2015**, *18*, 56–65. [CrossRef]
36. Bonde, M.; Song, H.-S. Is energy performance capitalized in office building appraisals? *Prop. Manag.* **2013**, *31*, 200–215. [CrossRef]
37. Bruegge, C.; Carrión-Flores, C.; Pope, J.C. Does the housing market value energy efficient homes? Evidence from the energy star program. *Reg. Sci. Urban Econ.* **2016**, *57*, 63–76. [CrossRef]
38. Cerin, P.; Hassel, L.; Semenova, N. Energy performance and housing prices. *Sustain. Dev.* **2012**, *22*, 404–419. [CrossRef]
39. Fuerst, F.; McAllister, P.; Nanda, A.; Wyatt, P. Energy performance ratings and house prices in Wales: An empirical study. *Energy Policy* **2016**, *92*, 20–33. [CrossRef]
40. Mangold, M.; Morrison, G.; Harder, R.; Hagbert, P.; Rauch, S. The transformative effect of the introduction of water volumetric billing in a disadvantaged housing area in Sweden. *Water Policy* **2014**, *16*, 973–990. [CrossRef]
41. McNabola, A.; Shields, K. Efficient drain water heat recovery in horizontal domestic shower drains. *Energy Build.* **2013**, *59*, 44–49. [CrossRef]
42. Johansson, T.; Olofsson, T.; Mangold, M. Development of an energy atlas for renovation of the multifamily building stock in Sweden. *Appl. Energy* **2017**, *203*, 723–736. [CrossRef]
43. Österbring, M.; Mata, É.; Thuvander, L.; Mangold, M.; Johnsson, F.; Wallbaum, H. A differentiated description of building-stocks for a georeferenced urban bottom-up building-stock model. *Energy Build.* **2016**, *120*, 78–84. [CrossRef]
44. Mastrucci, A.; Baume, O.; Stazi, F.; Leopold, U. Estimating energy savings for the residential building stock of an entire city: A GIS-based statistical downscaling approach applied to Rotterdam. *Energy Build.* **2014**, *75*, 358–367. [CrossRef]

© 2018 by the authors. Licensee MDPI, Basel, Switzerland. This article is an open access article distributed under the terms and conditions of the Creative Commons Attribution (CC BY) license (http://creativecommons.org/licenses/by/4.0/).

Article

How does the Ecological Well-Being of Urban and Rural Residents Change with Rural-Urban Land Conversion? The Case of Hubei, China

Min Song [1,2,*], **Lynn Huntsinger [2]** and **Manman Han [3]**

1. School of Business Administration, Zhongnan University of Ecomomics and Law, Wuhan 430073, China
2. Department of Environmental Science, Policy and Management, University of California at Berkeley, Berkeley, CA 94720-3110, USA; huntsinger@berkeley.edu
3. Department of Land Management, Zhejiang University, Hangzhou 310058, China; hanman1016@163.com
* Correspondence: songmin0211@hotmail.com; Tel.: +86-189-862-43230

Received: 4 January 2018; Accepted: 14 February 2018; Published: 15 February 2018

Abstract: Human well-being can be affected by the loss of ecosystem services from conversion of agricultural lands. Uncovering negative ecological consequences of rural-urban conversion is important for regulating rural-urban land conversion. This paper evaluates the impacts of rural-urban land conversion on the ecological well-being of different interest groups in China and makes policy recommendations for mitigating them. This research empirically quantifies and compares changes in the ecological well-being of rural and urban residents due to rural-urban land conversion and examines how transformation factors affect such changes in Hubei, China using the Fuzzy Synthetic Evaluation Model. Results show that compared with urban residents, rural resident ecological well-being level declines more obviously with rural-urban land conversion. Two socio-demographic characteristics, age and education level, as well as zoning characteristics, influence both rural and urban resident well-being changes. It is argued that there is a need for quantitative measurement of agricultural ecosystem services changes and that the construction of ecological compensation policies in areas undergoing rural-urban land conversion is essential for regulating rural-urban land conversion and for maintaining resident ecological well-being.

Keywords: ecological well-being changes; rural-urban land conversion; transformation factors; urban residents; rural residents; China

1. Introduction

Humans are dependent upon the services provided by nature and unless we effectively account for the range of values from ecosystems in our efforts to protect the environment, we cannot sustain human well-being [1,2]. Land use and land cover change (LUCC) is a driver of global change that directly influent the status and integrity of ecosystems and in last term its capacity to supply ecosystem services [3]. While human well-being, as an endpoint and central yardstick for sustainability, is widely recognized as an important issue but is difficult to be studied empirically [4–6]. One of the outcomes of urbanization in China, "rural-urban land conversion" is the change from agricultural land in rural areas to developed urban land which is a kind of LUCC [7,8]. Rural-urban land conversion provides land element for urban sprawl, industrial development and economic growth. But it also presents a challenge to the ecological system because many ecosystem services provided by agricultural land are lost in the process of conversion, which can be described as the negative external ecological effects of rural-urban land conversion [9–13]. Fortunately, such losses in human well-being have received increasing attention in economic analysis and public policy making, based on the ecosystem services functions as well as human well-being indicators made in the Millennium Ecosystem Assessment

(MA) [14–17]. Currently, some empirical studies aimed to uncover the relationship between land use change or ecosystem services and human well-being have been reported in many countries, including the U.S. [18], Brazil [19], South Africa [20], China [6], Spanish [3] and the U.K. [21]. However, the issue of how to quantify the changes in ecological well-being of rural and urban residents due to rural-urban land conversion and how conversion factors affect such changes has not been adequately studied [22–24]. In view of this, this paper constructs an ecological well-being index for residents, quantifies the changes in ecological well-being of residents caused by rural-urban land conversion and examines the impact of transformation factors (socio-demographic characteristics and zoning characteristics) on these changes for rural and urban residents.

The agricultural social-ecological system includes agricultural land, the natural environment and human interventions. It is commonly characterized as semi-natural and semi-artificial [25–27]. The agroecosystem provides not only agricultural products including grain, fiber, vegetable, wood and fishery products but also multiple ecological services such as air purification, climate regulation, soil stability and eco-landscapes [14,28,29]. With the acceleration of land-based urbanization in China since the 1980s, the population has been migrating to urban areas triggering unprecedented, fierce land competition and tremendous land use changes. The character of rural-urban land conversion in China in the last 30 years has been described as 'accelerating urban expansion, establishing the eastern part of China as the center of gravity and extending out to the Midwest of China' [16]. Rural-urban land conversion transfers the semi-natural landscapes to a landscape of buildings and roads. The biological community, soils, water flow and surface structure of agricultural lands is destroyed in this process which means that the land's original ecosystem degrades [30–32]. The capability of agricultural land to providing various ecological services is diminishing and declining accordingly [33]. Therefore, rural-urban land conversion which is a common social-economic phenomenon in developing countries has significant negative externalities from the ecological point of view [13,31,34,35]. Such land use transitions had negative effects on local ecological systems and human well-being may decline when these negative externalities are ignored and incorrectly treated [3,36,37]. In 2005, the MA presented a conceptual framework which revealed the interactions between ecosystem services, human well-being and the driving forces of change and it pointed out that ecosystem services (including provision services, regulating services, cultural services and supporting services) are closely related to human well-being. Human well-being was explained as the basic material for a good life, health, good social relations, security and freedom of choice and action [38,39]. In view of these, the human well-being discussed in this paper is limited to those factors closely associated with ecosystem services and termed "human ecological well-being".

The existing research usually focuses on the evaluation of a certain kind of ecological service from the agricultural ecosystem, such as supply services that underpin basic livelihood capacity [40], coordinating services that can prevent extreme weather, conserve water and purify the air [31,41], support services that provide habitat for plants and animals and maintain the potential for sustainable use of agricultural land resources in the future [42,43], as well as cultural services that include educational opportunities, inspiration, entertainment and local identity [44–46]. Scholars in China have paid close attention to the changes in the overall welfare of some stakeholders in the process of rural-urban land conversion [11,47,48], especially the overall welfare of farmers [49–51]. Most of these papers used the theories and methods of welfare economics to calculate the economic or material welfare changes caused by rural-urban land conversion, while neglecting the impact of ecosystem change on human well-being from the externalities of land use change [40]. In recent years, some scholars have begun to link human ecological well-being loss with ecosystem change from the perspective of the relationship between ecosystem services and human well-being, based on the framework for ecosystem services as well as human well-being constituents from the MA [15,17,52]. Such research calls for the investigation of resident ecological well-being changes in the process of rural-urban land conversion. Of quantitative measurement methods, the most commonly used methods are stated-preference methods including the Contingent Valuation Method (CVM), Choice

Experiment Method (CE), etc. [44–46]. However, these methods have some inherent defects, to be more specific, CVM can only estimate a well-being change caused by one certain state of environmental change and the accuracy of its result is affected by more than ten kinds of inherent latent deviations such as hypothetical deviation, partial-global deviation, strategic deviation, etc. [53,54]. Although CE is able to calculate the economic value of different attributes of resources, its estimation accuracy depends on the authenticity and reliability of the survey data and is influenced by embedding bias, hypothetical deviation, etc. [55]. Therefore, the systematic research on the relationship between agroecosystem services and human well-being needs to be further advanced in terms of content and method.

Moreover, for the sake of protecting limited agricultural land, China implemented the world's most stringent Land Use Control System starting in the 1980s, upgrading to Land Use Spatial Control in recent ten years. It can be seen as a conversion from reserving the quantity of agricultural land to regulating agricultural land protection spatially [13,56,57]. Specifically, Land Use Spatial Control in China divides the entire land space into four major function oriented zones including the Optimizing Development Zone, Key Development Zone, Restricted Development Zone as well as the Forbidden Development Zone according to resource and environment carrying capacity, existing development density and development potential, natural environmental constituents, the level of socio-economic development, ecosystem characteristics and the spatial differentiation of human activities in different areas. This raises some questions, notably, when rural-urban land conversion occurs in the different zones, are there different changes in the ecological well-being of residents? Investigation of this question can provide a basis for the implementation of spatially heterogeneous ecological compensation mechanisms for agricultural land protection.

The goal of this paper, therefore, is to examine the specific human well-being constituents that closely relate to the ecological services from agricultural lands and which foster "human ecological well-being". We construct an evaluation index for resident ecological well-being, quantify the changes in ecological well-being of residents due to rural-urban land conversion and analyze the effect of individual and zoning transformation factors. We examine impacts on both rural and urban residents: we hypothesize that both are affected by the agroecosystem because of spillover effects but that they are affected differently owing to their different relationship with and distance from agricultural lands. The research findings from this study provide a foundation for evaluating impacts of conversion on the ecological well-being of different interest groups, for regulating rural-urban land conversion by zoning and for establishing an efficient ecological compensation mechanism for rural-urban land conversion.

2. Constituents of Resident Ecological Well-Being

The Millennium Ecosystem Assessment (MA) was called for by United Nations Secretary-General Kofi Annan in 2000 and had lasted for 4 years since it was initiated in 2001. The objective of the MA was to assess the consequences of ecosystem change for human well-being and to establish the scientific basis for actions needed to enhance the conservation and sustainable use of ecosystems and their contributions to human well-being. Two of the four core questions of MA are "Who caused the ecosystem and their services change and how these changes affected human well-being?" In 2005, the MA presented a conceptual framework which revealed the interactions between ecosystem services and human well-being and it pointed out that ecosystem services (including provision services, regulating services, cultural services and supporting services) are closely related to human well-being (including the basic material needs for a good life, health, good social relations, security and freedom of choice and action) [38]. On this basis, series indices are selected in this paper to set up the evaluation index system for resident ecological well-being based on the classification for human well-being constituents of MA, combined with the characteristics of agroecosystem and the existing literature about human well-being (Figure 1 and Table 1). In view of the different ecological impacts on rural and urban residents from rural-urban land conversion, the specific indices of ecological well-being for the two groups are slightly different (Table 1).

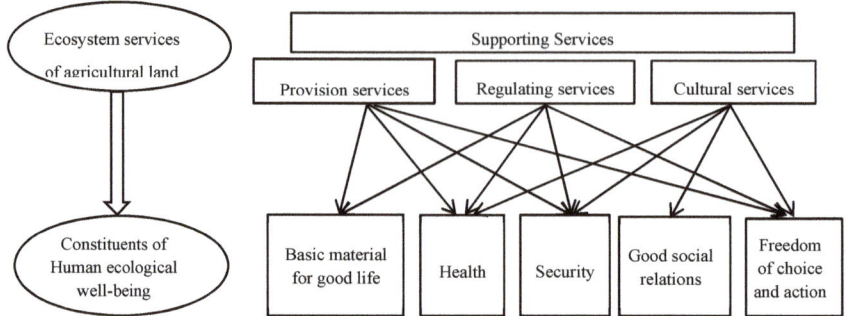

Figure 1. Relationship between agro ecosystem services and human ecological well-being.

Table 1. Index system of resident ecological well-being.

Constituents of Ecological Well-Being	Criterion	Indices	
		Rural residents	Urban residents
Security X_1	Access to clean and secure living spaces	Waste recycling capability X_{11}	
	Reduction of environmental attack and threats	Frequency of agroecosystem-related meteorological disasters (such as drought, floods, soil erosion and desertification) X_{12}	
Basic material for a good life X_2	Access to resources for income and livelihood	Obtain daily staple food X_{21} Obtain daily vegetables X_{22} Obtain daily meat X_{23}	
Health X_3	Access to clean air	Satisfaction with air quality X_{31}	
	Access to adequate and clean water	Satisfaction with water quality X_{32}	
	Obtain adequate nutrition	Safety of food, vegetables and meat consumption X_{33}	
	Avoidance of preventable diseases	Pollution-related diseases X_{34}	
Good social relations X_4	Opportunities to express cultural and spiritual values associated with the ecosystem	Rural life nostalgia X_{41} Children's rural experiences X_{42}	
	Opportunities to experience the aesthetic and recreational values associated with the ecosystem		Frequency of ecotourism X_{43} Satisfaction with the natural landscape X_{44}
Freedom of choice and action X_5	Achieving the status of valuable survival state	Livelihood choices X_{51}	

2.1. Security

The MA recognizes that the security refer to safety of person and possessions, secure access to necessary resources and security from natural and human-made disasters [38]. Rural-urban land conversion weakens the land's capacity for water conservation, weather moderation and waste recycling [31,40]. The recycling of surface runoff and flood regulation are significantly affected [31]. This paper selects the indices of waste recycling and the frequency of meteorological disasters (such as drought, floods, soil erosion and desertification etc.), which are related to the agricultural ecosystem, to characterize resident well-being level for the security constituent. What needs to be explained is that the index of waste recycling capability is only for rural resident survey and analysis, because waste needs ecosystems to degrade in many rural areas where still short of complete waste disposal system.

2.2. Basic Materials for a Good Life

Basic material for a good life refers to the ability to have a secure and adequate livelihood, including income and assets, enough food and water at all times, shelter, ability to have energy to keep warm and cool and access to goods [38]. Agroecosystem synthesize organic compounds by solar energy and artificial auxiliary energy, providing the fundamental materials for human life [58]. Rural-urban land conversion changes the basic material conditions of rural residents as the direct users of agricultural land. Therefore, the existing compensation institutions for agricultural land acquisition should not only take economic compensation into account but also compensation for degradation of access to ecosystem resources [40]. In the assessment's conceptual framework of interactions between ecosystem services and human well-being constructed by MA, one of the services—the provisioning service—refers to the products obtained from the ecosystem. Changes in provisioning services such as food, water and fuelwood have very strong influences on the adequacy of material for a good life. And it is also stated that the access to these materials is heavily mediated by socioeconomic circumstances [38]. This research focuses on agroecosystem which produce food mainly. Besides, in many underdeveloped rural areas in China, farming is still the main way for farmers to get food to meet the basic need of life. Therefore, this paper uses changes in the way that rural residents obtain daily ingredients (staple food, vegetables and meat) to represent the change in basic materials for a good life. It should be noted that the relationship between an urban resident's basic materials for a good life and the agroecosystem is not significant, because on the one hand, they have no direct connection with the agricultural land (they do not farm), on the other hand, as a relatively wealthy group, local changes in ecosystems may not cause a significant change in their access to necessary material goods [38]. Accordingly, this paper considers that the basic materials for an urban resident's good life remain unchanged in rural-urban land conversion. So the indices of obtained daily staple food, obtained daily vegetables and obtained daily meat, which are employed to characterize the constituents of basic materials for a good life, are only used for the rural resident survey and analysis.

2.3. Health

The MA states that the constituent of health is the ability of an individual to feel well and be strong, or in other words to be adequately nourished and free from disease, to have access to adequate and clean drinking water and clean air and to have the ability to have energy to keep warm and cool [38]. Rural-urban conversion destroys the nutrient recycling capacity of agroecosystem [59–62]. Therefore, this paper investigates the satisfaction of rural and urban residents with their daily consumption of staple food, vegetables and meat to measure the change in their capability of obtaining enough nutrition. Existing studies also suggest that agroecosystem have the function of blocking and absorbing a certain amount of particulate waste [63], preventing the deterioration of river water [64] and controlling the spread of epidemic diseases [58]. In view of these, in this paper, the capability of avoiding preventable diseases is measured by the indices of number of ecological pollution-related diseases. Access to adequate and clean water as well as clean air is measured by the indices of resident satisfaction with water quality and air quality separately.

2.4. Good Social Relations

Good social relations can be explained as the presence of social cohesion, mutual respect and the ability to help others and provide for children [38]. The capability of providing aesthetic values and interactions between human and the ecological environment offered by agricultural lands have been recognized as vital service functions [17,65]. Rural-urban land conversion typically triggers aesthetic damage [66]. For urban residents, the decline of agricultural land reduces their opportunities for rural tourism, agritainment, etc. In this paper, the constituents of good social relations for urban residents are measured by indices of frequency of ecotourism and satisfaction with the natural landscape. Agroecosystem fosters specific agricultural cultures which include agricultural production activities and rural lifestyles [67] and also the link between humans and nature as well as humans and society [44]. In China, farmers are strongly attached to their agricultural land. A deep and irreversible ruin of the cultural and spiritual heritage of rural areas takes place when rural-urban land conversion proceeds [68].Therefore, the constituent of good social relations for rural residents is characterized by indices of rural life nostalgia and children's rural experiences.

2.5. Freedom of Choice and Action

Freedom of choice and action refers to the ability of individuals to control what happens to them and to be able to achieve what they value doing or being [38]. The basic livelihood of rural populations depends mainly on the services provided by agroecosystem [40]. Agricultural populations who live in poor areas are influenced by land use change more directly and profoundly [69].

Before rural-urban land conversion, rural residents have the freedom to make choices between farming and working in urban areas. However, after the rural-urban land conversion, their livelihood choices are more limited because they have no other choice but to work in urban areas. Their low level of education and inadequate non-agricultural labor skills exacerbate this limitation. While for urban residents, rural residents entering urban areas to work compete for the available jobs. Therefore, this paper uses livelihood choices as an index to measure changes in resident freedom of choice and action.

3. Materials and Methods

3.1. Study Area

Wuhan City and Shiyan City in Hubei Province were selected as the study area (Figure 2). Wuhan City which is located in a Key Development Zone according to the Main Function Oriented Division (MFOD) is a megalopolis in middle of China. Wuhan City, with an urbanization rate of 71.7% at the end of 2016, significantly higher than any other city in Hubei province, is developing at soaring speed in recent years. It is in the phase of accelerated urbanization which accompanies rapid urban expansion. Wuhan City has 13 districts, including 7 central districts and 6 suburban districts. The latter is composed of Jiaxia district, Caidian district, Huangpi district and so on and the area with the highest rate of rural-urban land conversion in the last few years. However, Shiyan City, located in the northwest border of Hubei Province, is in a Restricted Development Zone. It is not only a poor region of the Qin-ba mountain area but also the Core Water Source Area of the Middle Route of the South-to-North Water Diversion Project. This city, which has adopted a "focus on ecology externally and focus on humanities internally" as its development strategy is now facing the dilemma of balancing ecological protection and economic development.

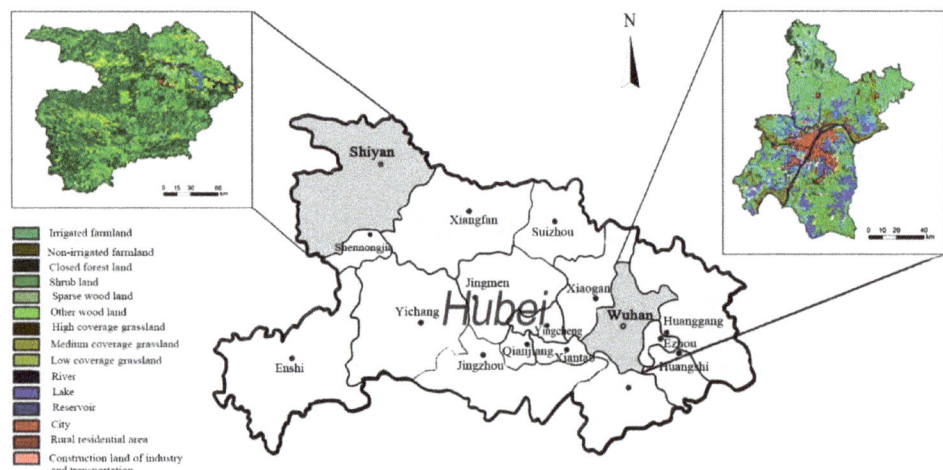

Figure 2. Study areas (Color should be used).

3.2. Questionnaire Design and Data Collection

In general, the questionnaire collected information regarding (1) a brief description about the multifunction of agricultural land and agricultural land loss in the process of rapid urbanization in China (2) respondents' perception of the ecosystem services changes caused by rural-urban land conversion. Questions are designed according to the indices showed in Table 1 (i.e. "Do you think the Frequency of drought/floods/soil erosion/desertification increased?" If "yes," "Do you think it is related to rural-urban land conversion?" The five-point Likert scale (i.e. strongly disagree, mildly disagree, unsure, mildly agree and strongly agree) was used for respondents to choose (3) the socio-demographic information of area residents. It must be noted that, considering the specific indices of ecological well-being for the rural residents and urban residents are slightly different (Table 1), two different questionnaires were developed for these two groups of respondents. Specifically, rural residents were asked to compare the state of each kind of agroecosystem services before and after they lost their agricultural land (either positive or negative). If they think there were changes, they needed to indicate their opinion about whether those changes were related to rural-urban land conversion as well as the degree of relevance. Urban residents were asked to clarify their perception of the changes of each kind of agroecosystem services in the last fifteen years and also needed to indicate their views on whether those changes were related to rural-urban land conversion as well as the degree of relevance. To avoid investigators deviation in survey, we made no effort to force respondents to provide arguments; they were given the option to not provide an argument if they did not recognize any change [3,70].

In this research, respondents are composed of two groups, rural residents and urban residents, defined according to the census registration system in China. "Rural residents" refers to people who are registered in a village and are generally eligible to be assigned agricultural land to work and make a living from. "Urban residents" are those who are registered as urban permanent residents and live in urban areas. Major Function Oriented Zoning Division (MFOZD) of China which is the guideline for optimizing the spatial pattern of regional development in China is put forward in 2011. Based on comprehensive analysis on regional resources and environment capacity, existing development density and potential, MFOZD divides the land space of China into a series of regional units according to a specified major function for each area. Some regions are planned as urbanizing regions (develop functional regions) whose major function is providing industrial products and services. Such areas are named "Priority Development Zone" or "Key Development Zone". Other regions are planned as agricultural regions (ecology functional areas) whose major function is providing ecological services. Such areas are named "Restricted Development Zone" or "Forbidden Development Zone". In order to investigate the impact of location on resident ecological well-being changes, Wuhan City and Shiyan City of Hubei Province are selected as study areas in this study. The two cities belong to a Key Development Zone and a Restricted Development Zone, respectively, according to the MFOZD of China.

Specifically, survey sites for rural residents were selected in the suburbs, where rural-urban land conversion has been frequent in recent years, while for urban residents the sites selected were open spaces or parks with a large flow of people. Ultimately, rural residents were surveyed in five administrative villages including Xingfu Village, Xiangyang Village, Fangliang Village, Liuhe Village and Chunhe Village in the Jiangxia District of Wuhan City, as well as in five administrative villages including Erdaopo Village, Caijialing Village, Wolonggang Village, Qinglongshan Village and Changping Village in Yunyang District of Shiyan City. Urban residents were surveyed in two central districts including the Hongshan District and Hangjiang District in Wuhan City as well as the Zhangwan District in Shiyan City. Sample size was estimated using the following sampling formula proposed by Scheaffer [71].

$$N^* = N/\left[(N-1)\sigma^2\right] + 1$$

where N^* is the needed theoretical sample size, N is the population of study area, σ is the sampling error ($\sigma = 5\%$) [71]. By the end of 2015, the population of Wuhan and Shiyan were 1166.24×10^4 obtained from Wuhan Statistical Yearbook (2016) and Wuhan Statistical Yearbook (2016), including 729.57×10^4 urban residents and 436.67×10^4 rural residents. Accordingly, the sample size required for our research should be at least 402. Considering the possible unresponsive or invalid samples, a total of 500 interviewees including 220 rural residents and 280 urban residents were randomly selected to be surveyed in the study areas mentioned above. Our research team which was composed of 8 postgraduates majoring in land resource management took face-to-face interviews to obtain data throughout the entire data collection periods in May 2015 and May 2016. Interviewers were asked to learn the contents of the questionnaire, especially the key issues and questions, through pre-training and pre-research to avoid investigators deviation. Besides, each interview was limited to 20–25 min to avoid length of residence time bias. After a brief description of the purpose of our research and obtaining permissions, we explained meaning of each index embedded in the questions of the questionnaire. And their answers were faithfully recorded by the interviewers. Ultimately, valid questionnaires were obtained from 209 rural residents and 266 urban residents after discarding invalid samples that included illogical or incomplete information.

3.3. Methods

3.3.1. Fuzzy Synthetic Evaluation Model

The core essence of Fuzzy Mathematics is regarding fuzzy concepts as study objects, employing fuzzy sets to determine imprecise or complex things and adopting the value of membership function to describe the degree to which a certain constituent belongs to a fuzzy set, which is employed to describe imprecision or vagueness [49]. The 'human ecological well-being' which is studied in this paper is one aspect of the concept of 'human well-being.' Researching the impacts of changes in life status and the subjective feelings and psychological characteristics of rural residents and urban residents as they are related to the ecosystem of agricultural land is essential to exploring this issue. Since most of these constituents are subjective and vague, the Fuzzy Synthetic Evaluation Model has a significant advantage for dealing with the subjective evaluation indexes is appropriate for this study.

X is defined as the fuzzy set of resident state of ecological well-being. W is a subset of X which represents the possible change in ecological well-being as a result of rural-urban land conversion. Then, the ecological well-being function $W^{(n)}$ of the nth rural dweller or urban dweller can be expressed as follows:

$$W^{(n)} = \{x, \mu_w(x)\}$$

where $\mu_w(x)$ is the membership value function for x belonging to W, $\mu_w(x) \in [0,1]$ and $x \in X$. It is well accepted that the higher the membership value is, the higher the ecological well-being level is. When the membership value is 1, the ecological well-being of residents is the best; when it is 0 it is the worst and 0.5 is the medium state, neither good nor bad.

3.3.2. Forms of Membership Function

The membership function is one of the critical points of the fuzzy synthetic evaluation method and the membership function is different according to the different types of variables used. Variables are usually divided into three types: virtual dichotomous variables, virtual continuous variables and virtual qualitative variables. Let x_{ij} represent the value of the jth evaluation index of the ith constituent of the resident's ecological well-being. x_i is the ith constituent subset of resident ecological well-being which is determined by the primary index x_{ij}. The initial index of ecological well-being is $x = [x_{11}, \cdots, x_{ij}, \cdots]$.

There are only two cases of virtual dichotomous variables which are commonly described by "true" or "false". For example, the question "Whether you and your family have had diseases that are related to ecosystem?" applies to the virtual dichotomous variable whose membership function can be expressed as follows:

$$\mu(x_{ij}) = \begin{cases} 0 & X_{ij} = 0 \\ 1 & X_{ij} = 1 \end{cases} \quad (1)$$

When the respondent has diseases due to ecological environmental pollution, X_{ij} is 0, otherwise X_{ij} is 1.

Virtual continuous variables are the indices that have continuous values. Their membership function can be expressed as follows. Equation (2) describes the positive relationship between X_{ij} and the state of ecological well-being, while Equation (3) describes the negative relationship between them.

$$\mu(x_{ij}) = \begin{cases} 0 & 0 \leq x_{ij} \leq x_{ij}^{min} \\ \frac{x_{ij} - x_{ij}^{min}}{x_{ij}^{max} - x_{ij}^{min}}, & x_{ij}^{min} \leq x_{ij} \leq x_{ij}^{max} \\ 1 & x_{ij} \geq x_{ij}^{max} \end{cases} \quad (2)$$

$$\mu(x_{ij}) = \begin{cases} 0 & 0 \leq x_{ij} \leq x_{ij}^{min} \\ \frac{x_{ij}^{max} - x_{ij}}{x_{ij}^{max} - x_{ij}^{min}}, & x_{ij}^{min} \leq x_{ij} \leq x_{ij}^{max} \\ 1 & x_{ij} \geq x_{ij}^{max} \end{cases} \quad (3)$$

where x_{ij}^{min} and x_{ij}^{max} represent the minimum and maximum value of X_{ij} respectively.

Quantitative data are generally not available for some ecological well-being indices as some can be described by qualitative words only. Examples like this including the respondents' answers about "How satisfied are you with the air quality," "How about your nostalgia about rural life?" etc. These kinds of indexes are dealt with by a Likert scale which assigns the range of agroecosystem services from 1 through 5 at equidistant intervals. When satisfaction is measured, a larger value indicates a better ecological well-being state. On the contrary, when a negative indicator such as the frequency of agroecosystem-related meteorological disasters is measured, a larger value indicates a worse state. The membership function $\mu(x_{ij})$ of the virtual qualitative variable is as follows:

$$\mu(x_{ij}) = \begin{cases} 0 & 0 \leq x_{ij} \leq x_{ij}^{min} \\ \frac{x_{ij} - x_{ij}^{min}}{x_{ij}^{max} - x_{ij}^{min}}, & x_{ij}^{min} \leq x_{ij} \leq x_{ij}^{max} \\ 1 & x_{ij} \geq x_{ij}^{max} \end{cases} \quad (4)$$

$$\mu(x_{ij}) = \begin{cases} 0 & 0 \leq x_{ij} \leq x_{ij}^{min} \\ \frac{x_{ij}^{max} - x_{ij}}{x_{ij}^{max} - x_{ij}^{min}}, & x_{ij}^{min} \leq x_{ij} \leq x_{ij}^{max} \\ 1 & x_{ij} \geq x_{ij}^{max} \end{cases} \quad (5)$$

where x_{ij}^{min} and x_{ij}^{max} represent the minimum and maximum value of X_{ij} respectively.

3.3.3. Weight Determination

The Delphi method and the Analytic Hierarchy Process are commonly used in the calculation of weights but these are questioned due to the subjectivity of their results. This paper uses the weight definition proposed by [72] and employs the weight function as modified by [49,73].

$$\omega_{ij} = \ln\left[1/\overline{\mu(x_{ij})}\right] \quad (6)$$

3.3.4. Aggregate the Membership Values of Indices

According to the different types of the constituents described above in Section 3.3.2, corresponding membership functions are selected to calculate the indices membership value. And then the constituents' membership value is calculated according to the weight function. Therefore, the membership value of resident ecological well-being is as follows:

$$W^{(n)} = \sum_{i=1}^{I} \mu(x_{ij})^n \cdot \omega_{ij}^{(n)'} / \sum_{i=1}^{I} \omega_i \tag{7}$$

3.3.5. Transformation Factors

Well-being can be described as an individual's satisfaction with some aspects of his/her life as determined by his/her individual life states, physiological conditions and psychological factors [74]. It is recognized as the freedom and capability to choose different kinds of lifestyles in order to get access to a beautiful life, a healthy state, rich experiences, good social relations, cultural identity, a deep sense of belonging, respect and self-realization etc. in the process of natural ecosystem utilization and development [75,76]. Capability approach was put forward by Sen and he discussed the relationship between an individual's functioning, capability and values (or utilities which are defined in the usual terms of pleasures, happiness, or desire fulfillment) [77–79]. On this basis, the changes of farmers' economic and non-economic well-being in the process of rural-urban land conversion have been researched [49,51,73]. It figured out that the conversion degree and efficiency by which the goods or service changing to well-being are significantly different for each individual due to the differences in personal, social and environmental conditions which are conceptualized as "transformation factors". For example, different age groups have different anticipations when facing same environmental changes, more highly educated groups have better ability for livelihood selection and further environmental cognition [51,80]. In addition, an individual's preference for ecosystem services is affected by local environmental resources endowments to some extent [29]. Transformation factors could explain that why different people in same state have different levels of well-being. Sen (1999) studied five categories of conversion factors including the heterogeneity of individual, the diversity of the environment, the difference of social atmosphere, the difference of interpersonal relationship and the distribution of the family. Considering that the social atmosphere is stable in certain period of time in China, interpersonal relationship mainly reflect different social differences and ecological well-being is closely related to personal perception not the family internal distribution, this paper uses 'transformation factors' which influence the impact of rural-urban land conversion on an individual's ecological well-being to further investigate the effects of rural-urban land conversion. Socio-demographic characteristics including gender (T_1), age (T_2) and education level (T_3), as well as zoning characteristic which is represented by location (T_4) are selected as transformation factors which are employed to analyze the differences of ecological well-being change between rural residents and urban residents.

4. Results

4.1. Changes in Resident Ecological Well-Being

According to the actual situation obtained through the survey, the membership values of rural resident ecological well-being indices including satisfaction with air quality, satisfaction with water quality, rural life nostalgia and children's rural experiences are all set as 0.5 which means that the state of rural resident ecological well-being mentioned above is neither good nor bad before rural-urban land conversion and the membership values of their ecological pollution-related diseases is set as 1 which means they did not get diseases due to ecological environment pollution. The membership values of urban residents' each ecological well-being index and overall ecological well-being are all set as 0.5 which means that the state of their ecological well-being is neither good nor bad before rural-urban land conversion. Resident ecological well-being changes are displayed in Table 2.

Table 2. Fuzzy evaluation results of resident ecological well-being change due to urban-rural land conversion.

Ecological Well-Being Constituents	Rural Residents				Urban Residents			
	Membership Value		Weight		Membership Value		Weight	
	Before	After	Before	After	Before	After	Before	After
X_1	0.553	0.509	0.593	0.676	0.500	0.437	0.657	0.828
X_{11}	0.542	0.518	0.565	0.487	—	—	—	—
X_{12}	0.564	0.500	0.573	0.693	0.500	0.437	0.657	0.828
X_2	0.677	0.417	0.390	0.876	—	—	—	—
X_{21}	0.770	0.397	0.261	0.924	—	—	—	—
X_{22}	0.692	0.417	0.368	0.875	—	—	—	—
X_{23}	0.615	0.438	0.486	0.826	—	—	—	—
X_3	0.500	0.456	0.693	0.786	0.500	0.377	0.739	0.976
X_{31}	0.500	0.420	0.693	0.868	0.500	0.373	0.739	0.986
X_{32}	0.500	0.479	0.693	0.736	0.500	0.389	0.747	0.944
X_{33}	0.500	0.380	0.693	0.968	0.500	0.437	0.715	0.828
X_{34}	1.000	0.708	0.000	0.345	0.500	0.325	0.627	0.724
X_4	0.500	0.250	0.693	1.385	0.500	0.520	0.723	0.654
X_{41}	0.500	0.211	0.693	1.556	—	—	—	—
X_{42}	0.500	0.301	0.693	1.201	—	—	—	—
X_{43}	—	—	—	—	0.500	0.585	0.615	0.536
X_{44}	—	—	—	—	0.500	0.473	0.763	0.749
X_5	0.413	0.202	0.884	1.599	0.500	0.487	0.597	0.719
X_{51}	0.413	0.202	0.884	1.599	0.500	0.487	0.597	0.719
Overall	0.507	0.326			0.500	0.447		

4.1.1. Changes in the Ecological Well-Being of Rural Residents

Rural resident overall level of ecological well-being in Wuhan City and Shiyan City declines from 0.507 to 0.326 with urban-rural land conversion, a decrease of 35.67% (Table 2; Figure 3). In fact, levels of all the five well-being constituents decline.

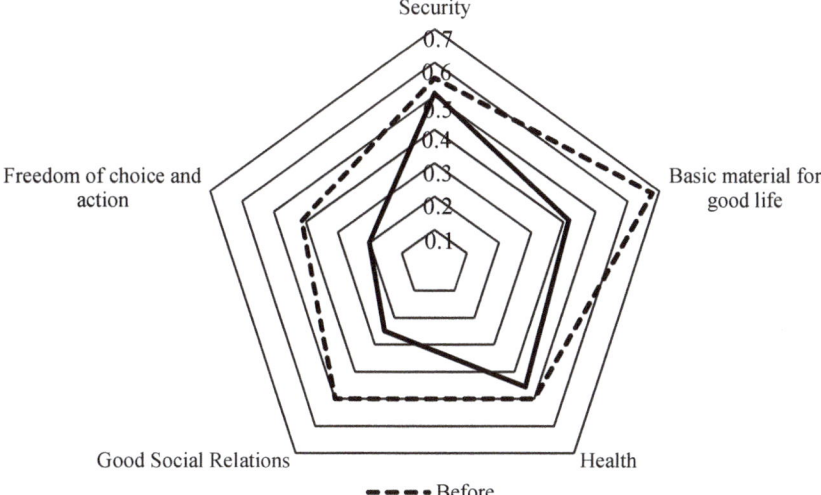

Figure 3. Membership value changes with the constituents of rural resident ecological well-being due to rural-urban land conversion.

- Security

    Before rural-urban land conversion, waste in rural areas was normally disposed of using traditional methods (landfill, incineration, natural piling, etc.). Most livestock manure, litter, food waste and other production and household waste can be degraded by agroecosystem automatically. The survey found that although garbage collection boxes were provided in the study area after rural-urban land conversion, the inadequate infrastructure of garbage disposal and untimely recycling of garbage made the environment worse and some of the wastes that are still disposed of using traditional methods cannot actually be degraded by the environment any more. Consequently, the level of rural resident ecological well-being due to recycling capacities declines (from 0.542 to 0.518). In addition, rural-urban land conversion promotes the transfer of a huge amount of agricultural land to urban construction land in China. Changes in the surface environment trigger off declines in soil water holding capacity, which further weakens the local environment's capacity for moderating floods, droughts and other extreme weather. Results show that the level of rural resident ecological well-being about the frequency of agroecosystem-related meteorological disasters declines by more than 10% (from 0.564 to 0.500). In general, the overall level of the rural resident ecological well-being as related to security declines by 7.96% (from 0.553 to 0.509).

- Basic materials for a good life

    With the rural-urban land conversion, the ability to have a secure and adequate livelihood of rural resident turns from a good status (0.677) in a negative direction (0.417), which breaks through the fuzzy state of neither good nor bad (0.500) and changes toward a bad direction (Table 2). The way that rural residents get access to daily necessities is changed directly by rural-urban land conversion. Before rural-urban land conversion, rural residents basically obtain daily staple food (rice, flour and beans), vegetables and meats from their own farms. A small amount of supplementary vegetables and meat products are purchased from the market. When farmers lose their agricultural land after rural-urban land conversion, they are forced to buy daily foods in the market. Our survey reveals that this change greatly increases their living expenses and makes their lives difficult.

- Health

    Due to rural-urban land conversion, agricultural land's original ecosystem services such as atmospheric regulation and air purification are weakened. In addition, air pollution, construction dust and sewage disposal soar early in land development. The two effects are superimposed so that the indices of air quality satisfaction and water quality satisfaction fall by 16.00% (from 0.500 to 0.420) and 4.20% (from 0.500 to 0.479) separately. Further, rural residents have to purchase for the original materials of food from the market which was previously acquired from farming. This raises the risk of food safety issues. The index of safety satisfaction with staples, vegetables and meat falls by 24.00% (from 0.500 to 0.380). Moreover, the membership value for the index of ecological pollution-related diseases rises (0.708 > 0.500) after rural-urban land conversion but still has a certain degree of decline (from 1.000 to 0.708). On the whole, rural resident ecological well-being for the constituent of health is declines by 8.86% (from 0.500 to 0.456).

- Good social relations

    This constituent is greatly influenced by rural-urban land conversion. Agricultural land is the material carrier of farming culture. Rural residents in China have a strong sense of dependence on agricultural land which is the basis of maintaining good social relations. Rural-urban land conversion cuts off their access to agroecosystem and cultures, which makes the indices of rural life nostalgia and children's rural experiences fall by 57.8% (from 0.500 to 0.211) and 39.8% (from 0.500 to 0.301) separately. As a whole, the membership value of the constituent of good social relations declines by 50.00% (from 0.500 to 0.250).

- Freedom of choice and action

Rural residents' freedom of choice and action is constrained by their lower ability to make a living, so it is at a low level before the rural-urban land conversion (0.413 < 0.500). With the loss of agricultural land, this constituent appears to be the constituent that declines most significantly, by 51.09% (from 0.413 to 0.202). The reason is that rural-urban land conversion directly reduces the natural capital of rural residents, which has a critical impact on their livelihoods [81]. In the surveyed area, the young landless rural residents have to work in township enterprises nearby or work in cities but their inadequate vocational skills make them poor. While older residents basically lose their way to make a living by farming and most of them can only stay at home. Considering the irreversibility of rural-urban land conversion, the decline of their freedom of choice and action is permanent.

4.1.2. Changes in the Ecological Well-Being of Urban Residents

Similar to the analysis of changes in rural resident ecological well-being, changes in the ecological well-being of urban residents can be summarized as follows briefly. Table 2 and Figure 4 show that the rural-urban land conversion makes the overall ecological well-being level drop from 0.500 to 0.447, 10.64%. Specifically, the three constituents of urban resident ecosystem well-being including security, health and freedom of choice and action all declines after the rural-urban land conversion. The constituent of health declines most significantly by 24.67% (from 0.500 to 0.377), within which the two indices including satisfaction with water quality and experience of ecological pollution-related diseases drop strikingly. The constituent of security declines by 12.60% (from 0.500 to 0.437) which reveals that urban residents approve of agricultural land's ecosystem services functions to a certain degree. Different from the rural residents, urban residents' freedom of choice and action is not affected by rural-urban land conversion significantly, it only drops by 2.60% (from 0.500 to 0.487) because urban residents are not closely linked with agricultural land and they have pluralistic livelihoods. It should be noted that the membership value of the constituent of good social relations rose slightly (from 0.500 to 0.520). Although the frequency of ecotourism reflects good social relations and has positive impact on resident ecological well-being, we discovered in the survey that the rise in this index's membership value is mainly due to improvement in people's living standards, allowing them to participate in ecological recreational activities more frequently. In other words, it is not closely related to the increase or decrease of agricultural land.

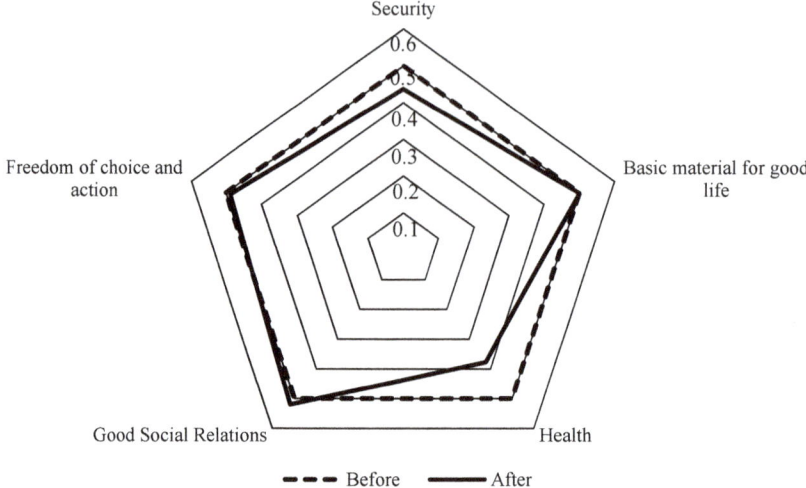

Figure 4. Membership value changes with the constituents of urban resident ecological well-being due to rural-urban land conversion.

4.1.3. Comparison

Comparing the ecological well-being of rural and urban residents, it can be observed that there are obvious differences between these two major stakeholder groups. The changes in rural residents overall ecological well-being and in each constituent of ecological well-being, are all greater than in urban residents. As direct users of agricultural land, rural residents are more closely and complexly interact with the agroecosystem than urban residents [82]. Owing to their high dependence on the agroecosystem, rural residents' ecological well-being unavoidably suffers from rural-urban land conversion. Rural residents are forced to move away from their original residence, lose their agricultural land and directly experience ecological deterioration, etc. Different than for rural residents, the link between urban residents and the agroecosystem is indirect, regardless of the spatial distance to the agricultural land or the intensity of interactions with the ecosystem.

4.2. Impact of Transformation Factors on the Changes in Resident Ecological Well-Being

4.2.1. Rural Residents

Table 3 displays the impact of transformation factors on the changes in rural resident ecological well-being.

- Impact of socio-demographic characteristics

Results show that the ecological well-being of the oldest sample group is the most affected, declining by 37.54%, followed by the least educated sample group, declining by 37.11%. Observing the impact of transformation factors on each constituent of rural resident ecological well-being, it is obvious that their age and education level have a momentous impact on both the ecological well-being constituents of good social relations and freedom of choice and action. The survey which reveals that the older and the less educated the interviewees are, the greater the dependence on their original way of life and the more difficult it is for them to find jobs after rural-urban land conversion. In other words, older and less educated rural residents have less opportunity to express cultural and spiritual values associated with the ecosystem and have less livelihood choices after losing their agricultural land.

- Impact of zoning characteristics

As described above, there are fundamental differences between Wuhan City and Shiyan City in terms of zoning. Specifically, the former is mountainous, while the latter is plains; in view of the MFOZD of China, the former is in a Key Development Zone, while the latter is located in China's Restricted Development Zone; from the ecological perspective, the latter's ecological vulnerability is higher than the former's. The overall change in the rural resident ecological well-being in Shiyan City (declines by 38.19%) is much greater than in Wuhan City (declines by 31.27%) (Table 3). Zoning characteristics have greatest impact on the ecological well-being constituents of freedom of choice and action (declines from 0.426 to 0.211 for Wuhan City and from 0.413 to 0.198 for Shiyan City, both of which declines more than 50%). The decline in the overall ecological well-being levels of rural residents in Wuhan City is mainly due to the decline in two constituents including basic material for a good life and freedom of choice and action. However, in Shiyan City, the dominant constituent in addition to the two above is the constituent of good social relations.

4.2.2. Urban Residents

Table 4 displays the impact of transformation factors on the changes in urban resident ecological well-being.

- Impact of socio-demographic characteristics

There are no significant differences in the impacts of socio-demographic characteristics on the overall levels of ecological well-being for urban residents. This field survey found urban residents' perceptions of changes in the ecological environment is more diverse and that they also have more diverse capabilities for adapting to the changes in the ecological environment brought about by rural-urban land conversion. In comparison, women's ecological well-being declines more than men's when rural-urban land conversion occurs and in the meantime, the older and the less educated the interviewees are, the greater their overall ecological well-being declines.

- Impact of zoning characteristics

Compared with Wuhan City with a higher level of urbanization and higher frequency of rural-urban land conversion, Shiyan City with its rich ecological resources endowment is part of China's key ecological function area and has higher environmental sensitivity. In other words, in comparison with Wuhan City, rural-urban land conversion occurring in Shiyan City will make the marginal benefit of ecosystem services decrease much more, resulting in the fact that urban residents in Shiyan City have more intense perceptions about ecological environment changes.

Table 3. Fuzzy evaluation of rural resident ecological well-being changes using transformation factors.

Transformation Factors			X_1		X_2		X_3		X_4		X_5		Overall		
			Before	After	Before	After	Before	After	Before	After	Before	After	Before	After	Change (%)
	T_1	Male	0.556	0.511	0.678	0.415	0.500	0.459	0.500	0.255	0.416	0.208	0.509	0.331	−35.04
		Female	0.552	0.508	0.675	0.421	0.500	0.452	0.500	0.248	0.409	0.196	0.506	0.323	−36.08
Socio-demographic characteristics	T_2	20–30	0.563	0.512	0.685	0.412	0.500	0.457	0.500	0.292	0.419	0.221	0.511	0.346	−32.25
		31–40	0.565	0.508	0.702	0.421	0.500	0.461	0.500	0.286	0.422	0.223	0.513	0.347	−32.40
		41–50	0.557	0.517	0.674	0.409	0.500	0.452	0.500	0.249	0.414	0.205	0.508	0.326	−35.78
		51–65	0.548	0.499	0.659	0.420	0.500	0.449	0.500	0.231	0.405	0.198	0.503	0.318	−36.85
		≥66	0.550	0.506	0.679	0.416	0.500	0.459	0.500	0.233	0.403	0.187	0.503	0.314	−37.54
	T_3	Elementary school graduate and below	0.561	0.507	0.693	0.422	0.500	0.451	0.500	0.246	0.408	0.188	0.507	0.319	−37.11
		Middle school graduate	0.547	0.512	0.665	0.416	0.500	0.457	0.500	0.253	0.413	0.206	0.506	0.329	−34.93
		High school graduate and above	0.554	0.508	0.681	0.417	0.500	0.462	0.500	0.261	0.421	0.211	0.510	0.334	−34.55
Zoning characteristics	T_4	Wuhan	0.548	0.503	0.667	0.439	0.500	0.445	0.500	0.329	0.426	0.211	0.510	0.351	−31.27
		Shiyan	0.557	0.514	0.691	0.412	0.500	0.469	0.500	0.217	0.413	0.198	0.508	0.314	−38.19

Table 4. Fuzzy evaluation result of urban resident ecological well-being changes using transformation factors.

Transformation Factors			X_1		X_2		X_3		X_4		X_5		Overall		
			Before	After	Before	After	Before	After	Before	After	Before	After	Before	After	Change (%)
	T_1	Male	0.500	0.456	0.500	0.379	0.500	0.523	0.500	0.491	0.500	0.454	−9.21		
		Female	0.500	0.435	0.500	0.374	0.500	0.532	0.500	0.483	0.500	0.446	−10.72		
Socio-demographic characteristics	T_2	20–30	0.500	0.445	0.500	0.386	0.500	0.522	0.500	0.501	0.500	0.456	−8.90		
		31–40	0.500	0.451	0.500	0.389	0.500	0.505	0.500	0.499	0.500	0.455	−9.00		
		41–50	0.500	0.442	0.500	0.375	0.500	0.534	0.500	0.486	0.500	0.450	−10.09		
		51–65	0.500	0.434	0.500	0.369	0.500	0.528	0.500	0.481	0.500	0.443	−11.35		
		≥66	0.500	0.429	0.500	0.367	0.500	0.531	0.500	0.482	0.500	0.442	−11.64		
	T_3	Elementary school graduate and below	0.500	0.503	0.500	0.388	0.500	0.516	0.500	0.423	0.500	0.435	−12.92		
		Middle school graduate	0.500	0.466	0.500	0.376	0.500	0.533	0.500	0.486	0.500	0.448	−10.42		
		High school graduate and above	0.500	0.429	0.500	0.383	0.500	0.512	0.500	0.500	0.500	0.450	−10.05		
Zoning characteristics	T_4	Wuhan	0.500	0.421	0.500	0.373	0.500	0.527	0.500	0.509	0.500	0.450	−9.91		
		Shiyan	0.500	0.443	0.500	0.384	0.500	0.522	0.500	0.489	0.500	0.452	−9.63		

5. Discussion

This study investigated changes in the ecological well-being of rural and urban residents due to rural-urban land conversion, as well as the impact of transformation factors. It confirms what previous studies have indicated: Changes in rural resident overall ecological well-being and in each constituent of ecological well-being are all greater than those of urban resident [29,49,51,80,83]. And transformation factors including age, education level as well as location have significant influence on both rural and urban resident ecological well-being but the degree of influence is different. Rural residents and urban residents have different perception of the ecological services provided by agricultural land because they have a different degree of linkage with agricultural land and different knowledge of agroecosystem functions [60,84,85]. Rural resident ecological well-being experiences greater decline compared to urban residents, because they have closer and more direct contact with agricultural land [82]. Therefore, it is of great practical significance to analyze the ecological well-being changes brought out by rural-urban land conversion from the perspective of different interest groups. In addition, this is the first time location has been used in this way, representing zoning characteristics as a transformation factor for analyzing the effects of rural-urban land conversion on the ecological well-being changes in residents living in different areas with different ecological resources endowments. It could provide a theoretical foundation for regulating rural-urban land conversion by zoning.

This paper reveals the change amplitudes in ecological well-being brought out by rural-urban land conversion using a fuzzy synthetic evaluation model. How to quantify the currency value of the loss of resident ecological well-being is still a big challenge for researchers. The key point is how to transfer natural science field variables employed to quantify ecosystem services change into social science field variables which can be employed to quantify the human ecological well-being and measure changes. An effective linkage in needed. Additionally, different preferences for ecosystem services among interest groups should be identified in decision making and policy formulation, such that the response of interest groups and the priorities for ecosystem services compensation should be set up and decided both based on the revealed preference difference.

Previous studies have fully explained the significant effect of rural-urban land conversion on water, soil, climate and other aspects of ecosystem [38,55,69]. Combining the assessment of ecosystems by MA with the findings in this paper, it is found that the ecosystem functions of agricultural land including supply services, regulatory services, cultural services and support services are degraded or limited due to agricultural land loss which is a result of rural-urban land conversion. Consequently, resident ecological well-being, consisting of security, basic material for a good life, health, good social relations and freedom of choice and action declines. In China, this loss of ecological well-being ca not be compensated due to the lack of an eco-compensation mechanism. In other words, the loss of ecological well-being brought about by rural-urban land conversion exists in the form of a negative externality which has not been accounted for as a part of the cost of rural-urban land conversion. It can be recognized as one of the explanations for why the loss agricultural land in China is excessive. Therefore, it is necessary to establish efficient eco-compensation policies for rural-urban land conversion from the perspective of ecological well-being.

6. Conclusions and Future Work

Rural-urban land conversion in the process of urbanization in China causes a sharp decrease in agricultural land, resulting in the loss of its original ecosystem services. Consequently, people's opportunities for accessing various ecological services from agro-ecosystems are reduced and their freedom of choice and actions is limited. The ecological consequences in the form of negative externalities lead to a decline in people's ecological well-being. This paper constructs ecological well-being indices for rural and urban residents and then estimates the changes in resident ecological well-being caused by rural-urban land conversion using the Fuzzy Synthetic Evaluation Model. In addition, the differences of ecological well-being changes and the impact of transformation factors for ecological well-being changes between rural and urban residents are analyzed. Compared to urban residents, rural resident ecological well-being declines more due to rural-urban land conversion, because of the more serious ecological damage. For rural residents, all the five ecological well-being constituents including security, basic material for a good life, health, good social relations and freedom of choice and action show varying degrees of deterioration and the constituent of freedom of choice and action deteriorates most seriously. For urban residents, after rural-urban land conversion, all the ecological well-being constituents except the constituent of good social relations shows varying degrees of deterioration and the constituent of health deteriorates most seriously. The impact of transformation factors are also investigated in this paper. Two socio-demographic characteristics including age and education level as well as zoning characteristics have influence on both rural and urban resident well-being. Moreover, we argue that there is a need for quantitative measurement of agroecosystem services and for the construction of an ecological compensation mechanism for the process of rural-urban land conversion in China, as it is essential for regulating rural-urban land conversion and maintaining ecological well-being.

Our research is a starting point for quantifying the changes in ecological well-being of residents caused by rural-urban land conversion and examining the impact of transformation factors in this process. Since it was conducted at a provincial level, further research in a larger study area in future research is needed.

Acknowledgments: The authors acknowledge the funding received from the projects "Regulation of rural-urban land conversion based on negative externalities governance (No. 71303260)" and "Differential Eco-Compensation Mechanism for Farmland Protection under the Spatial Control of Land Use: Scale Dependence and Spatial difference (No. 71774174)" of the National Natural Science Foundation of China (NSFC).

Author Contributions: Min Song and Manman Han conceived and designed this research; Manman Han performed the field survey; Min Song and Manman Han analyzed the data; Min Song and Lynn Huntsinger wrote the paper.

Conflicts of Interest: The authors declare no conflict of interest.

References

1. Smith, L.M.; Case, J.L.; Smith, H.M.; Harwell, L.C.; Summers, J.K. Relating ecoystem services to domains of human well-being: Foundation for a U.S. index. *Ecol. Indic.* **2013**, *28*, 79–90. [CrossRef]
2. Leisher, C.; Samberg, L.H.; Buekering, P.V.; Sanjayan, M. Focal Areas for Measuring the Human Well-Being Impacts of a Conservation Initiative. *Sustainability* **2013**, *5*, 997–1010. [CrossRef]
3. Quintas-Soriano, C.; Castro, A.J.; Castro, H.; García-Llorente, M. Impacts of land use change on ecosystem services and implications for human well-being in Spanish drylands. *Land Use Policy* **2016**, *54*, 534–548. [CrossRef]
4. Kazana, V.; Kazaklis, A. Exploring quality of life concerns in the context of sustainable rural development at the local level: A Greek case study. *Reg. Environ. Chang.* **2009**, *9*, 209–219. [CrossRef]
5. Summers, J.K.; Smith, L.M. The Role of Social and Intergenerational Equity in Making Changes in Human Well-Being Sustainable. *Ambio* **2014**, *43*, 718–728. [CrossRef] [PubMed]
6. Wang, B.; Tang, H.; Xu, Y. Perceptions of human well-being across diverse respondents and landscapes in a mountain-basin system, China. *Appl. Geogr.* **2017**, *85*, 176–183. [CrossRef]
7. Zhang, A.; Yang, G.; Lu, H. The influence of rural-urban land conversion on the sustainable development of agriculture. *Theory Mon.* **1999**, *12*, 7–11.
8. Liu, Y.; Tang, W.; He, J.; Liu, Y.; Ai, T.; Liu, D. A land-use spatial optimization model based on genetic optimization and game theory. *Comput. Environ. Urban Syst.* **2015**, *49* (Suppl. C), 1–14. [CrossRef]
9. Stobbe, T. The Economics and Externalities of Agricultural Land in the Urban Fringe. Ph.D. Thesis, University of Victoria, Victoria, BC, Canada, 2008.
10. Eagle, A.J. Threats to Agriculture at the Extensive and Intensive Margins: Economic Analyses of Selected Land-Use Issues in the US West and British Columbia. Ph.D. Thesis, Wageningen University, Wageninge, The Netherlands, 2009.
11. Chen, Z.; Ju, D.; Zhang, A. Measuring external benefits of agricultural land preservation: An application of choice experiment in Wuhan, China. *Acta Ecol. Sin.* **2013**, *33*, 3213–3221. [CrossRef]

12. Racevskis, L.; Ahearn, M.; Alberini, A.; Bergstrom, J.; Boyle, K.B.; Libby, L.; Paterson, R.; Welsch, M. Improved information in support of a national strategy for open land policies: A review of literature and report on research in progress. In Proceedings of the 24th International Conference of Agricultural Economists, Berlin, Germany, 13–18 August 2000; pp. 1–16.
13. Song, M.; Han, M. Compensation mechanism of rural-urban land conversion in the perspective of ecological well-being: Literature review and framework construction. *Issues Agric. Econ.* **2016**, *11*, 94–102.
14. Smith, H.F.; Sullivan, C.A. Ecosystem services within agricultural landscapes—Farmers' perceptions. *Ecol. Econ.* **2014**, *98*, 72–80. [CrossRef]
15. Dai, G.; Na, R.; Dong, X.; Yu, B. The dynamic change of herdsmen well-being and ecosystem services in grassland of Inner Mongolia: Take Xilinguole League as example. *Acta Ecol. Sin.* **2014**, *34*, 2422–2430.
16. Liu, J.; Kuang, W.; Zhang, Z.; Xu, X.; Qin, Y.; Ning, J.; Zhou, W.; Zhang, S.; Li, R.; Yan, C.; et al. Spatiotemporal characteristics, patterns, and causes of land-use changes in China since the late 1980s. *J. Geogr. Sci.* **2014**, *24*, 195–210. [CrossRef]
17. Yang, L.; Zhen, L.; Li, F.; Wei, Y.; Jiang, L.; Cao, X.; Long, X. Impacts of ecosystem services change on human well-being in the Loess Plateau. *Resour. Sci.* **2010**, *32*, 849–855.
18. Poor, P.J.; Brule, R. An Investigation of the Socio-Economic Aspects of Open Space and Agricultural Land Preservation. *J. Sustain. Agric.* **2007**, *30*, 165–176. [CrossRef]
19. Da Silva, J.M.C.; Prasad, S.; Diniz-Filho, J.A.F. The impact of deforestation, urbanization, public investments, and agriculture on human welfare in the Brazilian Amazonia. *Land Use Policy* **2017**, *65*, 135–142. [CrossRef]
20. Hamann, M. Exploring Connections in Social-Ecological Systems: The Links between Biodiversity, Ecosystem Services, and Human Well-Being in South Africa. Ph.D. Thesis, Stockholm Resilience Centre, Stockholm University, Stockholm, Sweden, 2016.
21. Fischer, A.; Eastwood, A. Coproduction of ecosystem services as human–nature interactions—An analytical framework. *Land Use Policy* **2016**, *52*, 41–50. [CrossRef]
22. Offiong, R.A.; Eteng, O.E. Effect of urbanization on greenareas in Calabar metropolis. *Int. J. Eng. Sci.* **2014**, *3*, 71–75.
23. Sallustio, L.; Quatrini, V.; Geneletti, D.; Corona, P.; Marchetti, M. Assessing land take by urban development and its impact on carbon storage: Findings from two case studies in Italy. *Environ. Impact Assess. Rev.* **2015**, *54*, 80–90. [CrossRef]
24. Song, M.; Lei, Y. Negative externalities from different directions of rural-urban land conversion using CE Method. *China Popul. Resour. Environ.* **2017**, *27*, 28–39.
25. Liang, M.; Yang, Z. Research on ecological well-being loss and compensation of non-agricultural land transference. *J. Gansu Sci.* **2014**, *26*, 35–38.
26. Reyers, B.; Biggs, R.; Cumming, G.S.; Elmqvist, T.; Hejnowicz, A.P.; Polasky, S. Getting the measure of ecosystem services: A social–ecological approach. *Front. Ecol. Environ.* **2013**, *11*, 268–273. [CrossRef]
27. Fischer, J.; Hartel, T.; Kuemmerle, T. Conservation policy in traditional farming landscapes. *Conserv. Lett.* **2012**, *5*, 167–175. [CrossRef]
28. Zhang, W.; Ricketts, T.H.; Kremen, C.; Carney, K.; Swinton, S.M. Ecosystem services and dis-services to agriculture. *Ecol. Econ.* **2007**, *64*, 253–260. [CrossRef]
29. Soy-Massoni, E.; Langemeyer, J.; Varga, D.; Sáez, M.; Pintó, J. The importance of ecosystem services in coastal agricultural landscapes: Case study from the Costa Brava, Catalonia. *Ecosyst. Serv.* **2016**, *17*, 43–52. [CrossRef]
30. Qu, F.; Lu, N.; Feng, S. Effects of land use change on carbon emissions. *China Popul. Resour. Environ.* **2011**, *10*, 76–83.
31. Lee, Y.C.; Ahern, J.; Yeh, C.T. Ecosystem services in peri-urban landscapes: The effects of agricultural landscape change on ecosystem services in Taiwan's western coastal plain. *Landsca. Urban Plan.* **2015**, *139*, 137–148. [CrossRef]
32. Kroeger, T.; Casey, F. An assessment of market-based approaches to providing ecosystem services on agricultural lands. *Ecol. Econ.* **2007**, *64*, 321–332. [CrossRef]
33. Plieninger, T.; Bieling, C. *Resilience and the Cultural Landscape: Understanding and Managing Change in Human-Shaped Environments*; Cambridge University Press: Cambridge, UK, 2012; pp. 1–366.
34. Li, S.; Zhang, A. Research of rural-urban land conversion and social loss based on the demonstration study of Wuhan city circle. *Resour. Sci.* **2014**, *36*, 303–310.

35. Chen, Z.; Zhang, A.; Zhang, X.; Song, M. Measurement of external costs in rural-urban land conversion processes. *Resour. Sci.* **2010**, *32*, 1141–1147.
36. Long, H.; Liu, Y.; Hou, X.; Li, T.; Li, Y. Effects of land use transitions due to rapid urbanization on ecosystem services: Implications for urban planning in the new developing area of China. *Habitat Int.* **2014**, *44* (Suppl. C), 536–544. [CrossRef]
37. Chuai, X.; Huang, X.; Wu, C.; Li, J.; Lu, Q.; Qi, X.; Zhang, M.; Zuo, T.; Lu, J. Land use and ecosystems services value changes and ecological land management in coastal Jiangsu, China. *Habitat Int.* **2016**, *57*, 164–174. [CrossRef]
38. Millennium Ecosystem Assessment. *Ecosystems and Human Well-Being: Synthesis*; Island Press: Washington, DC, USA, 2005.
39. Singh, R.; Singh, G. Ecosystem services: A bridging concept of ecology and economics. *Ecol. Quest.* **2017**, *25*. [CrossRef]
40. Sinare, H.; Gordon, L.J.; Enfors Kautsky, E. Assessment of ecosystem services and benefits in village landscapes—A case study from Burkina Faso. *Ecosyst. Serv.* **2016**, *21*, 141–152. [CrossRef]
41. Brander, L.; Brouwer, R.; Wagtendonk, A. Economic valuation of regulating services provided by wetlands in agricultural landscapes: A meta-analysis. *Ecol. Eng.* **2013**, *56* (Suppl. C), 89–96. [CrossRef]
42. Omer, A.; Pascual, U.; Russell, N. A theoretical model of agrobiodiversity as a supporting service for sustainable agricultural intensification. *Ecol. Econ.* **2010**, *69*, 1926–1933. [CrossRef]
43. Landis, D.A. Designing Agricultural Landscapes for Biodiversity-Based Ecosystem Services. *Basic Appl. Ecol.* **2017**, *18*, 1–12. [CrossRef]
44. Barrena, J.; Nahuelhual, L.; Báez, A.; Schiappacasse, I.; Cerda, C. Valuing cultural ecosystem services: Agricultural heritage in Chiloé island, southern Chile. *Ecosyst. Serv.* **2014**, *7* (Suppl. C), 66–75. [CrossRef]
45. Van Zanten, B.T.; Zasada, I.; Koetse, M.J.; Ungaro, F.; Häfner, K.; Verburg, P.H. A comparative approach to assess the contribution of landscape features to aesthetic and recreational values in agricultural landscapes. *Ecosyst. Serv.* **2016**, *17*, 87–98. [CrossRef]
46. Van Berkel, D.B.; Verburg, P.H. Spatial quantification and valuation of cultural ecosystem services in an agricultural landscape. *Ecol. Indic.* **2014**, *37*, 163–174. [CrossRef]
47. Li, H.; Huang, X.; Kwan, M.P.; Bao, H.X. H.; Jefferson, S. Changes in farmers' welfare from land requisition in the process of rapid urbanization. *Land Use Policy* **2015**, *42*, 635–641. [CrossRef]
48. Huang, Y. Study on Estimating the Value of Cultivated Land Ecosystem Services and Economic Compensation of Cultivated Land Protection in Zhuzhou Area. Master Thesis, Hunan Normal University, Changsha, China, 2015.
49. Gao, J.; Qiao, R.; Zhang, A. Fuzzy evaluation of farmers' well-being in rural-urban land conversion based on Sen's capability approach. *Manag. World* **2007**, *6*, 45–56.
50. Peng, K.; Zhang, P.; Zhang, A. Welfare balance of different interest groups during rural-urban land conversion. *Chin. J. Popul. Resour. Environ.* **2009**, *7*, 57–64.
51. Peng, K.; Zhu, H. The Impacts of Rural-Urban Land Conversion on the Welfare of Different Aged Land-lost Farmers. *China Land Sci.* **2015**, *29*, 71–78.
52. Wang, B.; Tang, H. Human well-being and its applications and prospects in Ecology. *J. Ecol. Rural Environ.* **2016**, *32*, 697–702.
53. Venkatachalam, L. The contingent valuation method: A review. *Environ. Impact Assess. Rev.* **2004**, *24*, 89–124. [CrossRef]
54. Freeman, A.M., III; Herriges, J.A.; Kling, C.L. *The Measurement of Environmental and Resource Values: Theory and Methods*, 3rd ed.; RFF Press: New York, NY, USA, 2014.
55. Yang, X.; Burton, M.; Zhang, A. Estimation of farmland eco-compensation criteria based on latent class model: A case of discrete choice experiment. *China Popul. Resour. Environ.* **2016**, *7*, 27–36.
56. Cai, Y.; Zhang, A. Researching progress and trends of agricultural land's ecological compensation under land use planning control. *J. Nat. Resour.* **2010**, *25*, 868–880.
57. Guo, Z. The Farmland Protection System of China: Implementation Performance Evaluation, Implementation Deviation and Optimization Methods. *J. Zhengzhou Univ.* **2017**, *50*, 64–68.
58. Liu, Y. Study on the Level of China Agricultural Ecological Welfare and Promotion Strategy: Case of Hubei Province. Ph.D. Thesis, Huazhong Agricultural University, Wuhan, China, 2014.

59. Zhang, L. The breaking and reconstruction of nutrient recycling chains in agro-ecosystem in China. *Ecol. Econ.* **2006**, *2*, 103–105.
60. Peng, J.; Tian, L.; Liu, Y.; Zhao, M.; Hu, Y.; Wu, J. Ecosystem services response to urbanization in metropolitan areas: Thresholds identification. *Sci. Total Environ.* **2017**, *607–608*, 706–714. [CrossRef] [PubMed]
61. Francis, C.A.; Hansen, T.E.; Fox, A.A.; Hesje, P.J.; Nelson, H.E.; Lawseth, A.E.; English, A. Farmland conversion to non-agricultural uses in the US and Canada: Current impacts and concerns for the future. *Int. J. Agric. Sustain.* **2012**, *10*, 8–24. [CrossRef]
62. Chen, Z.; Zhang, A.; Song, M.; Zhang, Z. Measuring external costs of rural–urban land conversion: An empirical study in Wuhan, China. *Acta Ecol. Sin.* **2016**, *36*, 30–35. [CrossRef]
63. Glenk, K.; Colombo, S. Modelling outcome-related risk in choice experiments. *Aust. J. Agric. Resour. Econ.* **2013**, *57*, 559–578. [CrossRef]
64. Garcia, X. The value of rehabilitating urban rivers: The Yarqon River (Israel). *J. Environ. Econ. Policy* **2014**, *3*, 323–339. [CrossRef]
65. Baró, F.; Palomo, I.; Zulian, G.; Vizcaino, P.; Haase, D.; Gómez-Baggethun, E. Mapping ecosystem service capacity, flow and demand for landscape and urban planning: A case study in the Barcelona metropolitan region. *Land Use Policy* **2016**, *57*, 405–417. [CrossRef]
66. Hart, J.F. Urban Encroachment on Rural Areas. *Geogr. Rev.* **1976**, *66*, 1–17. [CrossRef]
67. Swinton, S.M.; Lupi, F.; Robertson, G.P.; Hamilton, S.K. Ecosystem services and agriculture: Cultivating agricultural ecosystems for diverse benefits. *Ecol. Econ.* **2007**, *64*, 245–252. [CrossRef]
68. Yu, K. Three proposals for preventing the potential damage by building the new countryside and protecting the local cultural landscape and the industrial heritage. *Chin. Landsc. Archit.* **2006**, *8*, 8–12.
69. Li, H.; Zhang, A. Ecological compensation boosted ecological protection and human well-being improvement. *Acta Ecol. Sin.* **2013**, *33*, 1065–1070.
70. Sherren, K.; Verstraten, C. What Can Photo-Elicitation Tell Us About How Maritime Farmers Perceive Wetlands as Climate Changes? *Wetlands* **2012**, *33*, 65–81. [CrossRef]
71. Scheaffer, R.L.; Mendenhall, W. *Elementary Survey Sampling*, 6th Revised ed.; Brooks Cole: Boston, MA, USA, 2005; p. 486.
72. Cheli, B.; Lemmi, A. A totally fuzzy and relative approach to the multidimensional analysis of poverty. *Econ. Notes* **1995**, *24*, 115–134.
73. Gao, J.; Qiao, R. Analysis on variation in farmers welfare after rural-urban land conversion. *China Popul. Resour. Environ.* **2011**, *21*, 99–105.
74. Schirmer, J.; Berry, H.L.; O'Brien, L.V. Healthier land, healthier farmers: Considering the potential of natural resource management as a place-focused farmer health intervention. *Health Place* **2013**, *24*, 97–109. [CrossRef] [PubMed]
75. Smith, C.L.; Clay, P.M. Measuring subjective and objective well-being: Analyses from five marine commercial fisheries. *Human Organ.* **2010**, *69*, 158–169. [CrossRef]
76. Qi, J.; Yang, Z. Global climate changes and human well-being and adaptability. *Acad. Mon.* **2014**, *46*, 21–26.
77. Sen, A.K. *Commodities and Capabilities*; North-Holland: Amsterdam, The Ntherlands, 1985; pp. 1–104.
78. Nussbaum, M.; Sen, A.K. *Capability and Well-Being*; Clarendon Press: Oxford, UK, 1993.
79. Sen, A. *Development as Freedom*; Oxford University Press: New York, NY, USA, 1999.
80. Erickson, J.J.; Martinengo, G.; Hill, E.J. Putting work and family experiences in context: Differences by family life stage. *Human Relat.* **2010**, *63*, 955–979. [CrossRef]
81. Ding, S.; Zhang, Y.; Ma, Z. Research on changes of livelihood capabilities of rural households encountered by land acquisition: Based on improvement of sustainable livelihood approach. *Issues Agric. Econ.* **2016**, *6*, 25–34.
82. Skandrani, Z.; Daniel, L.; Jacquelin, L.; Leboucher, G.; Bovet, D.; Prevot, A.C. On Public Influence on People's Interactions with Ordinary Biodiversity. *PLoS ONE* **2015**, *10*, e0130215. [CrossRef] [PubMed]
83. Peel, D.; Berry, H.L.; Schirmer, J. Farm exit intention and wellbeing: A study of Australian farmers. *J. Rural Stud.* **2016**, *47*, 41–51. [CrossRef]

84. Pan, Y.; Marshall, S.; Maltby, L. Prioritising ecosystem services in Chinese rural and urban communities. *Ecosyst. Serv.* **2016**, *21*, 1–5. [CrossRef]
85. Jiang, C.; Jin, J.; Li, L. Non-market valuation of cultivated land protection using cvm: A case study of Wenling City. *Resour. Sci.* **2011**, *33*, 1955–1961.

© 2018 by the authors. Licensee MDPI, Basel, Switzerland. This article is an open access article distributed under the terms and conditions of the Creative Commons Attribution (CC BY) license (http://creativecommons.org/licenses/by/4.0/).

Article

Estimation of Carbon Dioxide Emissions Generated by Building and Traffic in Taichung City

Chou-Tsang Chang and Tzu-Ping Lin *

Department of Architecture, National Cheng Kung University, 1 University Road, Tainan 701, Taiwan; archcct@hotmail.com
* Correspondence: lin678@gmail.com

Received: 23 October 2017; Accepted: 27 December 2017; Published: 5 January 2018

Abstract: The emissions of carbon dioxide generated by urban traffic is generally reflected by urban size. In order to discuss the traffic volume generated in developed buildings and road crossings in a single urban block, with the metropolitan area in Taichung, Taiwan as an example, this study calculates the mutual relationship between the carbon dioxide generated by the traffic volume and building development scale, in order to research energy consumption and relevance. In this research, the entire-day traffic volume of an important road crossing is subject to statistical analysis to obtain the prediction formula of total passenger car units in the main road crossing within 24 h. Then, the total CO_2 emissions generated by the traffic volume in the entire year is calculated according to the investigation data of peak traffic hours within 16 blocks and the influential factors of the development scale of 95 buildings are counted. Finally, this research found that there is a passenger car unit of 4.72 generated in each square meter of land in the urban block every day, 0.99 in each square meter of floor area in the building and the average annual total CO_2 emissions of each passenger car unit is 41.4 $kgCO_2$/yr. In addition, the basic information of an integrated road system and traffic volume is used to present a readable urban traffic hot map, which can calculate a distribution map of passenger car units within one day in Taichung. This research unit can be used to forecast the development scale of various buildings in future urban blocks, in order to provide an effective approach to estimate the carbon dioxide generated by the traffic volume.

Keywords: CO_2 emissions; transport; urban block; urban design

---

1. Introduction

According to the statistical data of the International Energy Agency (IEA), the global CO_2 emissions increased by 88.7% from 1971 to 2004 and according to the energy consumption structure of the transportation department in Taiwan, the energy consumption of road transportation accounts for maximally 90% of the total energy consumption of the transportation department [1]. Prior to the Revolution, the content of CO_2 in the atmosphere was about 280 ppm and this concentration increased to 403.3 ppm in 2016 [2].

At present, the energy consumption of the transportation department and CO_2 emissions increase continuously and there are many factors influencing urban energy consumption [3]. The location of the housing and its size are the dominant factors determining energy use and greenhouse gas emissions [4]; however, in the future, research should be conducted regarding the relation between the large proportion of energy consumption in urban transportation and the development of buildings. In addition to the fact that urban areas have great influence due to land use control, zoning and building design scale, it is required to carry out overall research and analysis of the CO_2 emissions generated by traffic demand, in order to effectively and completely know the overall urban energy consumption.

1.1. Method of Research on the Energy Consumption of Urban Building Groups

Of the various energy consumptions in urban areas, electricity accounts for the most and urban energy consumption can be calculated from the urban electric energy consumption model or energy use intensity, in order to obtain the urban energy consumption amount.

There are many researches in Taiwan, which are used to count and investigate the power consumption regarding the usage category, scale, or total power consumption of various buildings, in order to discuss the future power use mode or energy consumption situation of various buildings and provide forecasts and simulations in the future. In terms of power use analysis of various buildings, the building shell energy consumption ENVLOAD simple algorithm [5] is taken as the 12 variable factors to simplify climate and building design and effectively evaluate the annual air conditioning load and usage as the energy saving design method of evaluating the building shell.

In another research, the total carbon dioxide generated by a building is calculated through the life cycle assessment (LCA) [6], where the fossil fuels and electricity consumed are calculated to obtain the total amount of building material, the total amount of electricity used by the buildings, in order to calculate the complex variable regression model of CO_2 emissions of RC buildings. In the current building specifications of Taiwan [5], the emissions of CO_2 in RC buildings are 331 kg/m^2, as obtained through the assessment of CO_2 reduction indicators in the green building evaluation system.

In addition, urban energy is analyzed with the total energy consumption relation of various buildings in the urban block and the mutual change relation with the scale and use form patterns of various buildings. For various buildings, the "dynamic Energy Use Intensity (EUI) indicator method" [7] is employed to calculate the energy consumption density standard EUI (kwh/(m^2·yr)) of various classified spaces and power consumption can be calculated after the summary. By calculating carbon emissions from buildings according to electricity emission factor released by the Bureau of Energy and energy usage intensity [8]; in this manner, the carbon dioxide emitted during the use of various buildings can be calculated.

Regarding the research literature of the total power consumption of all buildings in a block of a residential area [9], the site area is used as a single variable, where the floor area ratio, block shape and location, business factor, road and park area proportion, etc., are included as the variables, in order to forecast and master all power usage behaviors in the residential area. After conversion of the power consumption in the residential area, the annual power consumption is equivalent to the emissions of about 650 mt CO_2.

As stated above, there have been relevant researches for the simulation and forecast mode of the relation between a single building and the total power consumption and CO_2 emissions of a block and such achievements should be further used in the future.

1.2. Research Method of the Carbon Dioxide of Traffic Systems

In the exiting transportation planning forecast mode, regarding the analysis and forecast mode of travel demands, the overall procedure planning of traditional transportation planning is usually adopted, which is classified into trip generation, trip distribution, model split and traffic assignment processes [10] and after calculation by sequence, the trip generation in an area is forecasted or calculated for traffic planning.

Additionally, regarding the assignment of static traffic volume, it is assumed that the traffic flow and traveling time within a road section do not change with the time [11], thus, the user equilibrium and system equilibrium [12] of traffic assignments, as proposed by Wardrop [13], are applied for further simulation in order to calculate the traffic assignment mode under a mixed traffic flow. Then, the design reference of the highway capacity is considered to calculate and distribute the expected design vehicle speeds to complete the overall traffic planning of urban transportation planning.

The proposal and execution of a traffic project can promote urban land use and will influence land price in the future [14]. Regarding the influence of urban form on travel demand [15], the potential variables and measured variables of three aspects, "urban form", "travel demand" and "control factors" are used for inclusive empirical analysis on the influential relationship. The research results show that, in the urban form characteristics, while higher development density will increase the trip generation rate, it can reduce the opportunity of selecting private vehicles; meaning the higher equilibrium of mixed land use will reduce the trip generation rate but will increase the selection of private vehicles.

Analysis of the relation of local temperature to the land use, which residential areas, traffic areas and greenhouse agricultural areas all contributed to an increase in local temperatures [16].

In addition, in the research of the relation between measured CO_2 concentration and land use [17], the land use type is classified into buildings and built-up areas with impervious pavement and the urban environments are free city spaces [18]. Then the road service level and highway capacity of the original design are used to calculate the traffic volume of various roads and "vehicle mileage emission factor method" is used to calculate the CO_2 emissions. Therefore, the transportation system must estimate the passenger car units (PCU) per hour from different road service levels, or calculate the total CO_2 emissions according to the statistical data of the usage rate of different vehicle types, or with the calculated parameters, such as different vehicles, mileage, fuels and emission coefficient.

Urban spatial and statistical data for metropolitan Tainan in southwestern Taiwan are used to explore inside and outside of the CO_2 system of the city and estimate the amount of CO_2 emissions from road traffic. Therefore, CO_2 emissions are concentrated in over-urbanized areas, where the population density is higher than 5000 people/km^2 [19].

Road engineering designs, such as different road slopes, road widths, speed limitations and pavement types will also influence vehicle speed, traffic flow and fuel consumption rate [20]. Regarding urban roads with two-way 4 lanes without central dividing strip, the result of analysis on the influence of road width on vehicle speed shows that, during the off-peak period, with every reduction of 0.3 m for the lane width, the average vehicle speed will be reduced by 0.97 km/h [21].

According to the research of Ardekani and Sumitsawan (2010) on roads in Texas, upon comparison of fuel consumption rate changes of vehicles traveling on flexible payment materials, the result shows that rigid pavement can reduce the required fuel amount by 3–17%. Regarding CO_2 and fuel consumption tests aimed at special vehicles, the statistics show that slope has a present correlation trend with carbon dioxide, where the larger the slope, the greater the carbon dioxide [22].

As stated above, in the calculation of internationally emitted greenhouse gases, the "IPCC Guidelines for National Greenhouse Gas Inventories" rules are usually used and the estimation of CO_2 emissions by the transportation department can be classified into large-range estimation by considering the fuel type, fuel consumption and carbon emissions of various fuels, where priority calculate is the total amount of fuel, then calculate the types of the fuel consumption rate, namely the Top-down approach; while small-range estimation calculations are based on the Bottom-up principle, according to the activity strength with a fuel consumption rate and this applies to the assessment of the small-range traffic management strategy [23].

Whether applying the Top-down principle or Bottom-up principle, the total CO_2 emissions generated by traffic is calculated by multiplying the type of vehicles (such as passenger car, truck and bus) by the fuel type (diesel, gasoline, etc.) and then by the emission factor for the usage of unit fuel or the emission coefficient of unit mileage [17].

1.3. Existing Carbon Dioxide Issue

In this research, based on existing transportation planning and execution procedures and methods of urban planning, the following research topics are proposed upon review.

Issue 1: Transportation analysis was mainly based on large-scale networks, lacking the information from the urban blocks

Regarding the calculation method of existing transportation planning, the regional total traffic volume can be estimated only by large investigations of the urban road networks and numerical data of road facilities. Moreover, the investigation and statistical method of the existing urban transportation traffic volume is to conduct model splits and traffic assignments on the vehicle flows in various regions, in order to form large-scale transportation results in urban areas.

However, follow-up urban planning or regular overall inspections adopt zoning and cooperate with traffic road system planning to form various basic divided urban block units. Therefore, measured investigation on road vehicle flow is salutatory and there is a lack of research on the mutual influence between the usage characteristics of urban block units, building scales, road system, etc.

Issue 2: There is no effective permitted data for the buildings to integrate the overall forecast of urban energy consumption

During an application for a building permit, if it is required to assess the environmental impact or consider urban design, the traffic volume in that region must be evaluated and measured. However, in the current document application, under the condition of no integrated application, it is impossible to immediately and properly use the relevant legally considered data above, which reduces the added value after consideration, or requires further application by the government or folk practitioners.

Therefore, in the application of environmental impact assessment or urban design review, in addition to the stated measured traffic volume data for the region, it is possible to integrate and analyze traffic data and building permits, in order to evaluate the application in the overall urban development.

Issue 3: The design of transportation planning lacks mapping of urban traffic environment characteristics

In transportation system planning, set the model split according to the data collection and measured investigation, forecast trip generation and trip distribution in the future and then evaluate the optimal feasible plan. In the follow-up design of a road system, the road geometry, road slope, running speed, etc., are used for planning and alignment.

The transportation system above is planned according to the completion of the road geometrical design and only the numerical design and calculation are taken as the main achievements, which lack a map evaluation mode of integrating the urban form and transportation trunk into complete readability [24] and urban traffic hot spots.

1.4. Research Purpose and Improvement Countermeasures

In this research, the small-scale application mode is used to analyze the urban energy consumption of urban areas, thus, with a single urban block as the basic unit to estimate traffic volume, the research purpose and countermeasures are proposed as follows:

To completely analyze the overall energy consumption of entire urban areas and taking the urban block as the basic analysis unit, in combination with the vehicle flow generated at the junction of the road system, as the basic application unit, develop and analyze the overall energy consumption correlation of the urban areas and establish the forecast model.

To effectively integrate the legal application deliberation of the documentation of a building permit and be beneficial to the follow-up modification of urban planning, this research applies and analyzes the report of the urban design review in order to research the influential factors of block development and building design, conduct in-depth analysis of the mutual relation with the traffic carbon dioxide and evaluate the total CO_2 emissions generated by urban development in the future.

This study used the basic urban block units formed by an urban road system, where relevant road systems and road engineering are simplified into road design factors and upon this research, the simple forecast model of urban roads and urban planning configurations are analyzed and the basic information of the traffic volume is presented to obtain a readable traffic volume map of urban areas. In addition, the research results will be conveniently and rapidly used in the follow-up revisions of urban planning, or the periodical overall review of urban planning.

2. Research Method

The research area is located in Taichung, which is in the middle of Taiwan, with a land area of 2214.9 km^2 and a total of 29 administrative regions. The research scope is mainly the 8 administrative regions in the Taichung metropolitan area, with a land area of 163.4 km², population of 1,121,128 (March 2015) and population density of 6612 persons per square kilometer.

The Taichung metropolitan area is a metropolitan form of commercial service and commercial mixing, the west is an industrial zone and the east is the Dakeng Scenic area (Figure 1).

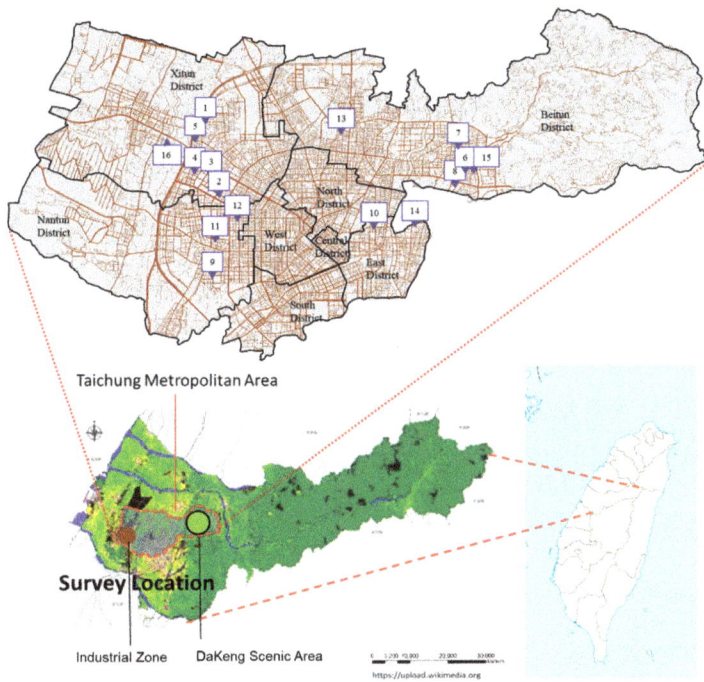

Figure 1. Location of Taichung city (Taiwan).

Due to the forecast mode of traffic volume, the investigation range of the traditional transportation planning method must often contain the regional characteristics formed by dozens of urban blocks and then, the measured traffic volume is used, thus, it belongs to large-range investigation and research. Regarding the mutual relation between land use control and building scale in urban areas, it is difficult to conduct in-depth research and analysis.

This research is aimed at the number of vehicles at four junctions in the urban blocks of Taichung's metropolitan area, as well as the applied development cases of various buildings in the urban blocks, in order to further calculate the total CO_2 emissions generated by the traffic volume and investigate and analyze the building development scale factors.

2.1. Estimation Model of Total Traffic Volume in the Urban Blocks

Regarding the main urban road crossings in this research, the number of vehicles within 24 h is measured and investigated. During an application for a building permit, the environmental impact evaluation case should be dominated by law. Due to the assessment contents of environmental impact, for the measuring of traffic volume generated by motorcycles, passenger cars and buses, it is required to calculate the passenger car unit (PCU) at the road crossings, as specified in the Taiwan Highway Capacity Manual [25], in order to obtain the traffic volume of the entire day as the basis of environmental impact assessment and design considerations.

First, this research analyzed the total number of vehicles of five main road crossings and then analyzed the forecasted regression formula of dependent variables during the morning peak and evening peak hours with the measured vehicle flow data of 07:30–08:30 in the morning and 17:30–18:30 in the evening.

In this research, the traffic investigation data of morning peak and evening peak hours in the urban design review cases are substituted into the forecasted regression formula of total number of vehicles in the five main road crossings, in order to respectively calculate the passenger car units in each junction of the block within one day.

Regarding the four road crossings in an urban block, the measured data of traffic volume are taken as the case and on the basis of practical cases, there is an extreme lack of hard-won cases, which are difficult to obtain. Therefore, in this research, hundreds of cases of urban design reviews over the years (2011–2013) are screened to determine the application of the site area of the blocks and those with road crossing investigation data are the preferred cases. Then, those with traffic volume in three crossings are taken as the minor cases.

Next, passenger car unit (PCU) at the road intersection in the city block was estimated and the Geographic Information System for metropolitan Taichung was used to calculate the length of each side road in the block as the travel distance of automobiles. The amount of fuel used by an automobile to travel an entire road was estimated on the basis of the travel distance per liter of fuel of the automobile in the city (9200 m), which was calculated using the research data published by the Industrial Technology Research Institute [26]. The carbon emission coefficient of gasoline was accordingly estimated to be 2.26 kgCO$_2$/L [27]. The carbon emission coefficient of gasoline was multiplied by the total PCU and the travel distance of the automobile was subsequently estimated. In this way, the total CO$_2$ emissions from traffic were estimated using Equation (1).

$$\sum_{t=1}^{n} PCE_i = N_i \times FE \times EC \times D_i \qquad (1)$$

where, PCE_i is the CO$_2$ emissions generated by each passenger car unit (kgCO$_2$/PCU); N_i is the passenger car units (PCU) for each of the road, FE is the fuel consumption rate (0.000109 L/m), EC is the CO$_2$ emissions coefficient (2.26 kgCO$_2$/L), D_i is the length of the vehicle driving in the block (m). There are 64 data of the total traffic volume in each section, as well as the total CO$_2$ emissions generated by the vehicles.

2.2. Statistical Model of Building Development Factors

The building design case should be subject to the urban land use zoning and provisions of the Building Act to be convenient for the building design. Therefore, in this research, legal buildings are used as the main research targets. Upon reference to the "building cadastral mapping" system of the competent government building authority (Figure 2), after inquiry of construction permits in all buildings, the building permit data are referred and the design value is logged in.

Due to insufficient official government budget and in order to accelerate obtaining the public facility reserved land of urban areas, provisions on encouraging floor area for those who have obtained public facilities have been established, that is to say, after private practitioners obtain public facilities and give them to the government free of charge, the equivalent building floor area is obtained through private transformation and moved to the future development site area, in order to increase the actual floor area ratio of buildings in the future.

In addition, in order to award the private practitioners, the government increases the parking space to solve the urgent parking demand in urban areas; those private practitioners who have increased the parking space obtain the award of increased floor area. In addition, due to the insufficiency of green parkland space in urban areas, for example, a site area is reserved with a square up to a certain scale by law, which is connected for pedestrians and is opened to the public for passing or recreation; private practitioners also can obtain the award of increasing the floor area of the open space.

Figure 2. Study case reference to the "building permit data mapping" system of the competent government building authority.

Therefore, in order to reduce building development costs and consider increasing the value of building development products, private practitioners usually need to increase the building development scale and will often apply for the legal provisions applicable to the above three measures for awarding floor area. Thus, in addition to incorporating the general building plan of influential factors in this research, floor award area factors are also considered, in order that the research forecast result is pragmatic and broader in application.

As stated above, in this research, regarding the major analysis method of traffic volume and block-related factors, in addition to the site area, building coverage ratio, floor area ratio, etc. (Table 1), as stated in the building permit, the thirty-four factors of various buildings, green coverage areas, green coverage rate, etc. of the entire block are summarized. Therefore, the land use zoning in this research case is dominated by the residential area, for a total of 16 blocks, the building purpose is dominated by residential use; however, in some buildings, the first floor is used as shops, for a total of 95 buildings.

Table 1. Statistical for block-related factors.

Case	The Total Area of an Urban Block (m^2)	Site Area (m^2)	Building Development Land Ratio (%)
1	10,291.7	9971.5	96.9%
2	16,679.2	11,773.1	70.6%
3	13,554.5	10,047.6	74.1%
4	9747.7	6145.4	63.0%
5	3822.4	2944.5	77.0%
6	16,143.3	14,387.3	89.1%
7	6606.4	4843.9	73.3%
8	16,400.0	9739.2	59.4%
9	6856.0	6145.4	89.6%
10	3522.3	3127.7	88.8%
11	20,261.5	16,617.5	82.0%
12	7548.9	7548.9	100.0%
13	9813.0	8432.7	85.9%
14	9831.8	3767.8	38.3%
15	9264.0	9264.0	100.0%
16	26,498.8	10,548.6	39.8%

This research constructed a map of the total carbon budget of traffic, buildings and parks in metropolitan Taichung, central Taiwan [19]. Carbon emissions from buildings in the city were estimated according to electricity usage [28,29], e-Question results, a common building type in Taiwan (i.e., residential and commercial mixed-use building) and the energy consumption of low-story and high-story apartments, office buildings, hospitals, educational institutions and other building types. Therefore, the total CO_2 emissions from buildings in Taichung were measured according to 15 different building uses, EUI, land types as specified during urban planning and presented in grids of 100 × 100 m and the total floor area, with the estimation results shown in Table 2.

The forms of energy used in a building vary according to how the energy is used. For example, some residential buildings use fuel gas for cooking and heating bath water, whereas others use electricity in kitchens, industrial buildings have diverse use of energy, while other types of buildings use mainly electricity. Therefore, CO_2 emissions from buildings in grids were estimated as follows:

$$CO_2e_{Building} = [EU_{elec} + EU_{gas}] \times C_{CO_2} \quad (2)$$

$$EU_{elec} = \sum_{t=1}^{n} EUI_i \times TFA_i \quad (3)$$

where EU_{elec} (kWh/yr·grid) is the usage of electricity; EU_{gas} (kWh/yr·grid) is the usage of fuel gas (which is estimated to consume 0.23 times more domestic energy than does electricity, according to the CO_2 emissions data for residential buildings provided in [6]) and C_{CO_2} is the amount of CO_2 emissions per kWh of electricity generated in Taiwan in 2014. On the basis of the amount of CO_2 emissions per kWh of electricity generated in the year, the EEF was estimated to be 0.521 $kgCO_2$/kWh.

Table 2. Carbon budget coefficients of buildings and green spaces.

	Land use	CO_2 (kgCO$_2$ m^{-2} yr^{-1})	References
Building carbon dioxide emission	Residential I (House)	15.29	Lin [28]
	Residential II (Apartment without elevator)	13.09	Lin [28]
	Residential III (Apartment with elevator)	18.62	Lin [28]
	Commercial	111.33	Lin [28]
	University/High school	30.74	Lin [28]
	Junior high school	28.66	Lin [28]
	Elementary school	21.88	Lin [28]
	Traffic Station	184.96	Lin [28]
	Post office/Government agencies	63.04	Lin [28]
	Stadium	97.43	Lin [28]
	Industrial area	105.76	Lin [28]
	Hospital	136.03	Chen [29]
	Religious buildings	80.76	Lin [28]
	Traditional Market	42.71	Chen [29]
	Landfill	44.81	Lin [28]
Carbon dioxide absorption	Water	−0.02	Lin [28]
	Park	−2.24	Huang [30], Lin [28]
	Farmland	−4.59	Huang [30]

The CO_2 absorption of different land types (i.e., parks, soils and water bodies, as determined during urban planning) in metropolitan Taichung was analyzed and the carbon sink in each grid was estimated using the following equation:

$$CO_{2Sink} = \sum_{t=1}^{n} CS_i \times A_i \quad (4)$$

where CO_{2Sink} (−kgCO$_2$/yr·grid) is the total amount of CO_2 absorbed in each grid, CS_i (−kgCO$_2$/m^2) is the CO_2 absorption coefficient of each land type and A_i (m^2) is the area of each land type.

Upon counting the design values of various buildings in the urban block, as well as the total number of traffic volume in each road section, as well as the respective calculations of the total CO_2 emissions generated by fuels, multiple regression analysis is applied on the relations of various relevant factors of overall urban energy consumption, in order to obtain the research results of the influence of changes in various dependent variables on the corresponding variables.

3. Results

3.1. Result of the Relation between Urban Blocks and Traffic CO_2 Emissions

In this research, regarding the forecast analysis model of the total traffic volume in each road crossing in the urban block, after counting the total number of vehicles within 24 h in the road crossing, linear regression analysis is conducted according to the morning peak and evening peak data (Figure 3).

In this research, the forecast of total traffic volume in the road crossings is obtained as shown in Equation (5) as follows:

$$y = 2261.52 + 2.36y_1 + 10.18y_2 \qquad R^2 = 0.99 \tag{5}$$

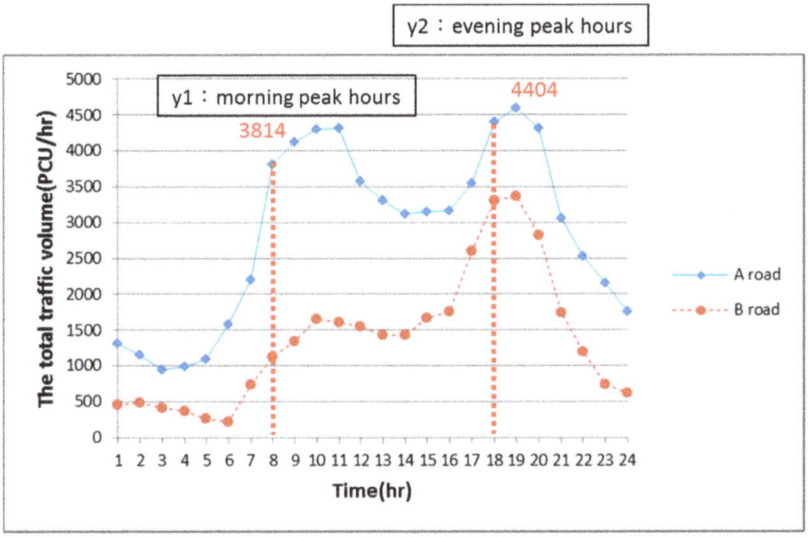

Figure 3. Relational diagram of the traffic volume at the morning peak and evening peak.

This formula is the passenger car unit (PCU/day) generated at the road crossings in an entire day, where y_1 is the morning peak hours (PCU/h) and y_2 is the evening peak hours (PCU/h).

In this research, the total area of an urban block, the total floor area of buildings, the total traffic volume generated at the four road crossings of the block and the total CO_2 emissions generated by the total number of vehicles (Table 3), are analyzed and counted (Figure 4) and the research results on the distribution of the total CO_2 emissions generated by total number of vehicles in each block is as follows (Figure 5):

Regarding the influential relation between the development scale of various buildings and the car traffic volume, through the analysis result in this research, each square meter area in the urban block generates a passenger car unit (PUC/day) of 4.72 every day. Each square meter of floor area of the building generates a passenger car unit (PCU/day) of 0.99 every day. Each passenger car unit (PCU) generates total CO_2 emissions of 41.4 $kgCO_2$/yr throughout the year.

Table 3. Statistical of the total traffic volume generated and the total CO_2 emissions generated by the total number of vehicles.

Case	The Total Area of an Urban Block (m^2)	The Total Floor Area (m^2)	The Total Traffic Volume (PCU/Day)	The Annual CO_2 Emissions by the Total Vehicles (kgCO_2/yr)
1	10,291.7	106,516.1	48,015	699,090.5
2	16,679.2	130,662.3	92,276	1,295,753.5
3	13,554.5	75,925.2	48,654	616,086.0
4	9747.7	26,305.0	47,321	455,057.5
5	3822.4	12,675.2	31,922	215,804.7
6	16,143.3	107,978.7	27,933	395,774.9
7	6606.4	42,656.7	10,945	98,181.4
8	16,400.0	47,283.0	94,676	1,511,754.3
9	6856.0	47,341.0	28,063	249,743.9
10	3522.3	18,860.2	81,547	616,230.2
11	20,261.5	62,957.3	63,484	1,028,001.9
12	7548.9	63,631.0	89,963	854,514.5
13	9813.0	44,720.1	24,158	245,017.3
14	9831.8	16,838.6	45,465	571,277.4
15	9264.0	38,424.3	16,007	160,771.2
16	26,498.8	40,730.6	130,807	2,441,902.1

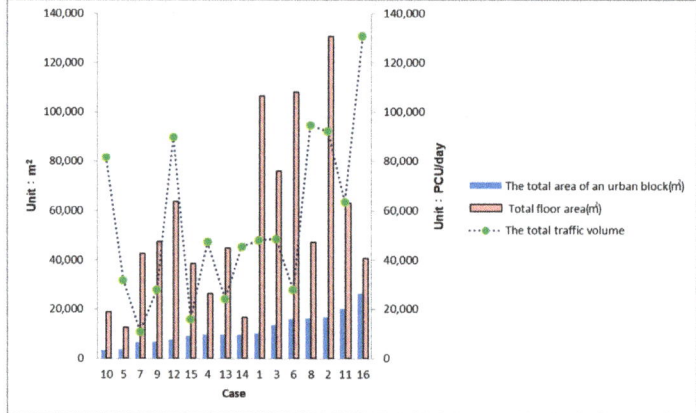

Figure 4. Relational diagram of street blocks and total traffic volume.

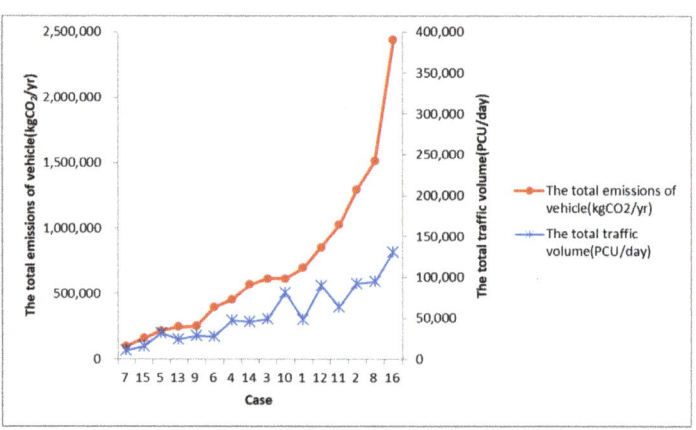

Figure 5. Relational diagram of total emissions and total traffic volume.

3.2. Results of Relation between an Urban Building Complex and Overall Traffic Energy Consumption

This study conducts in-depth analysis of the relation between traffic volume and building design scale. Regarding building design in this research, due to the applicable factors of award decrees of various floor areas, the ultimate design of the building scale is influenced, thus, this research incorporates the total area of the reward of building capacity with parking, total area of the reward of building capacity with open space and the urban building capacity transfer with floor area (Figure 6) and the result is shown as follows (Table 4).

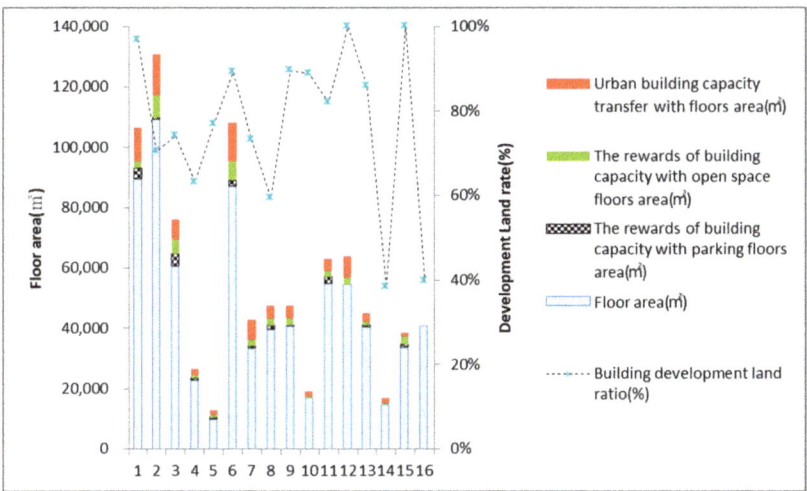

Figure 6. Relational diagram of applicable factors of award decrees of various floor areas.

In this research, an urban block is taken as the basic unit, the total traffic volume generated at the four road crossings of the block is analyzed in order to forecast the mutual relation and efficiently and conveniently estimate the overall urban energy consumption. The research result is as follows:

The relation between the total annual CO_2 emissions of urban blocks generated by vehicles and the total area of an urban blocks (Figure 7) is as follows:

$$y = 79.5 \times TA - 212{,}719 \qquad R^2 = 0.64 \qquad (6)$$

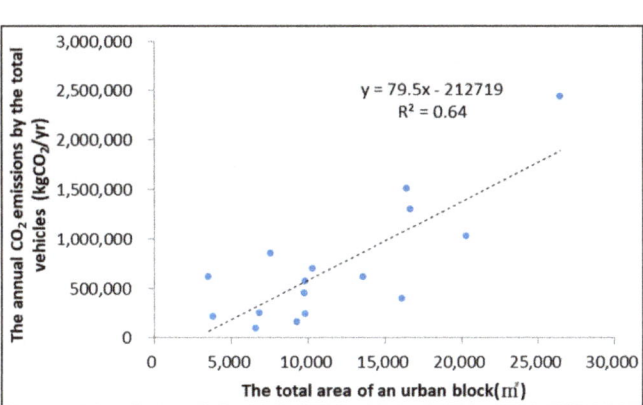

Figure 7. Relational diagram between the total area of an urban block and the total CO_2 emissions all year round.

Table 4. Statistical for the relation between traffic volume and building design scale in urban block.

Case	The Total Area of an Urban Block (m²)	Site Area (m²)	Building Area (m²)	The Total Floor Area (m²)	The Rewards of Building Capacity with Parking Floors Area (m²)	The Rewards of Building Capacity with Open Space Floors Area (m²)	Urban Building Capacity Transfer with Floors Area (m²)	The Annual CO_2 Emissions by the Total Vehicles (kgCO_2/yr)
1	10,291.7	9971.53	5172.38	106,516.05	3566.66	2369.12	11,168.07	699,090.5
2	16,679.2	11,773.10	4905.56	130,662.31	533.33	7684.54	13,197.35	1,295,753.5
3	13,554.5	10,047.57	4631.52	75,925.21	4047.45	4753.48	6642.00	616,086.0
4	9747.7	6145.43	3503.77	26,305.03	833.33	1062.63	1928.18	455,057.5
5	3822.4	2944.46	809.45	12,675.17	433.33	903.44	1445.31	215,804.7
6	16,143.3	14,387.31	6831.72	107,978.68	1913.33	6401.51	12,615.63	395,774.9
7	6606.4	4843.89	2461.70	42,656.69	515.15	2320.36	6542.02	98,181.4
8	16,400.0	9739.19	4677.38	47,283.02	1350.00	2268.40	4203.50	1,511,754.3
9	6856.0	6145.35	2962.79	47,340.98	250.00	2470.84	4039.36	249,743.9
10	3522.3	3127.67	1510.91	18,860.16	0	483.23	1570.45	616,230.2
11	20,261.5	16,617.52	7571.38	62,957.32	2120.00	2115.73	4004.74	1,028,001.9
12	7548.9	7548.92	4252.06	63,631.02	0	2298.50	6895.75	854,514.5
13	9813.0	8432.73	4366.38	44,720.06	406.67	1115.10	2750.15	245,017.3
14	9831.8	3767.80	1706.17	16,838.64	0	571.17	1705.20	571,277.4
15	9264.0	9263.96	4527.37	38,424.30	963.33	2676.77	1255.83	160,771.2
16	26,498.8	10,548.63	4741.82	40,730.55	0	0	0	2,441,902.1

This formula is the total annual CO_2 emissions of vehicles in the blocks ($kgCO_2/yr$) and TA is the total area of an urban blocks (m^2).

The relation between the total annual CO_2 emissions of urban blocks generated by vehicles and the total floor area of buildings (Figure 8) is as follows:

$$y = 3.3 \times FA_{TA} + 532{,}167 \qquad R^2 = 0.04 \tag{7}$$

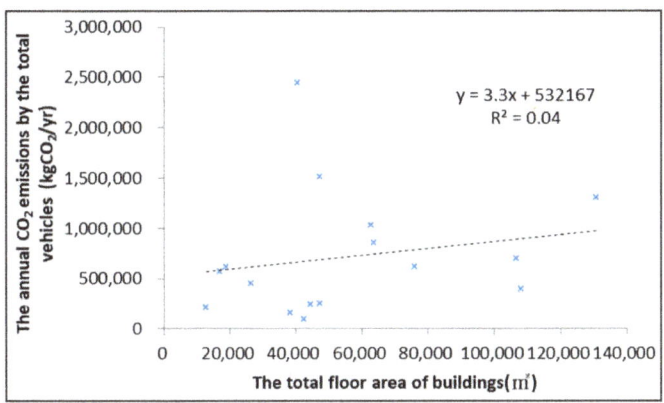

Figure 8. Relational diagram between the total floor area of buildings and the total CO_2 emissions all year round.

This formula is the total annual CO_2 emissions of vehicles in the urban blocks ($kgCO_2/yr$) and FA_{TA} is the total floor area of buildings (m^2).

After analysis of the 16 blocks, this research aimed at the seven important factors influencing building planning and the total CO_2 emissions of vehicles and the multiple regression forecast formula of the total annual CO_2 emissions of vehicles in the blocks is concluded as follows:

$$\begin{aligned}TVC = {} & 14{,}359.71 + (109.94 \times TA) - (60.92 \times SA) - (88.17 \times BA) + (15.36 \times FA_{TA}) - (65.30 \times FA_P) \\ & - (118.30 \times FA_O) - (39.60 \times FA_T) \qquad R^2 = 0.84\end{aligned} \tag{8}$$

where, TVC is the annual CO_2 emissions generated by the total vehicles ($kgCO_2/yr$), TA is the total area of an urban blocks (m^2), SA is the site area (m^2), BA is the building area (m^2), FA_{TA} is the total floor area of buildings (m^2), FA_P is the total area of the reward of building capacity with parking (m^2), FA_O is the total area of the reward of building capacity with open space (m^2) and FA_T is urban building capacity transfer with floor area (m^2). In this research, the relation between seven important design factors of the building design and the daily passenger car unit is further analyzed and the forecast multiple regression formula of the total passenger car units of the blocks within one day is concluded as follows:

$$\begin{aligned}TV = {} & 30{,}065 + (4.72 \times TA) - (4.99 \times SA) - (2.19 \times BA) + (1.09 \times FA_{TA}) - (6.47 \times FA_P) \\ & - (6.38 \times FA_O) - (3.39 \times FA_T) \qquad R^2 = 0.57\end{aligned} \tag{9}$$

where, TV (total vehicles) is the passenger car unit per day (PCU/day) and other symbols are the same as Equation (8).

Regarding the total floor area, new households and actual construction and development rate of blocks of building designs in an urban block, the forecast regression formula is concluded as follows:

$$TV = 94{,}042 - (59{,}234.2 \times BE) + (0.86 \times FA_{TA}) - (181.65 \times HH) \qquad R^2 = 0.66 \tag{10}$$

where, TV is the passenger car unit per day (PCU/day) and BE is the building development land ratio, which is the ratio of applied land for a building permit divided by the total block area, FA_{TA} is the total floor area of buildings (m^2) and HH is the number of households.

Regarding the relation with the total traffic volume, as generated by the urban traffic system within one day, the forecast regression formula in this research is concluded as follows:

$$TV_i = 8873.61 - (116.28 \times R_{di}) + (7.51 \times R_{ai}) \qquad R^2 = 0.83 \qquad (11)$$

where, TV_i is the passenger car unit per day (PCU/day) generated on road i, R_{di} is the length of the road (m) at block i and R_{ai} is the area of road (m^2) at block i.

3.3. Result of the Distribution Map of One-Day Vehicle Volume in Urban Areas

According to the "Code for design of urban roads and auxiliary engineering" (1999) and the general practical application, the method to plan a lane usually depends on lane width. If the road is 15 m wide, planning is dominated by car lanes and motorcycle vehicle mixed lanes; if the road width is more than 20 m, the division of two lanes is taken as the principle.

In this research, in order to analyze the road width, as well as the mutual relation between the length of each road in the block and the amount of traveling vehicles generated in the road throughout the day, three road widths are classified (Type A: 8~12 m; Type B: 15~20 m; Type C: 23~60 m). This research found that the relation between the passenger car unit and road length in the three road widths is, the wider the road is, the longer the road and larger the traffic volume will be (Figure 9), thus, there is non-positive correlation for urban roads with a width less than 12 m.

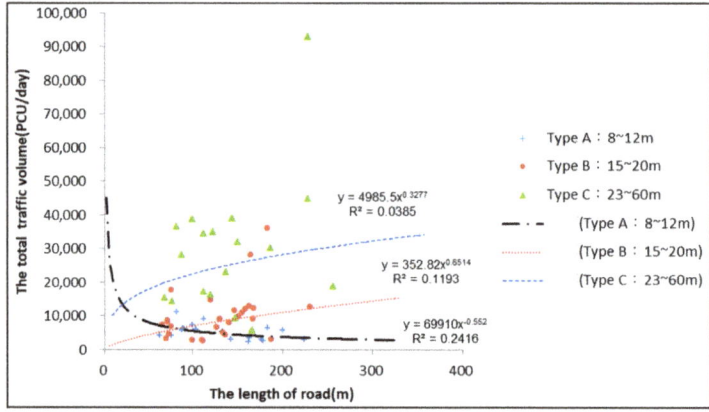

Figure 9. Relational diagram between the length of each road in the block and the amount of traveling vehicles generated in the road by three road widths.

This research takes the Geographic Information System data of Taichung in a scale of 100 × 100 m mesh, the urban planning road by law is taken as the main explanation target and the research result of Equation (11) is used to calculate the road length and road area; after calculation and layer analysis, the distribution map of passenger car units in Taichung within one day is presented (Figure 10) and amount of CO_2 emissions from traffic (Figure 11) and the urban traffic environmental characteristic map can effectively provide the basic information to judge traffic volume.

This figure presents a large distribution of traffic volume in the Taichung metropolitan area, which is dominated by linear distribution in the main external road, while a great characteristic is formed with the traffic flow on the ring road. In addition, for roads with higher traffic volume, a single hot spot is generated at the main road crossing.

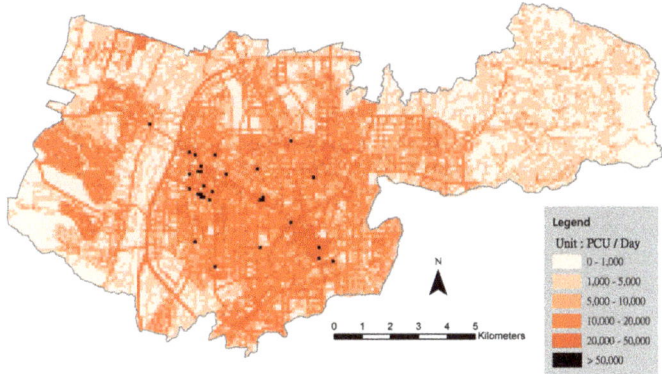

Figure 10. The distribution map of passenger car units in Taichung metropolitan area within one day.

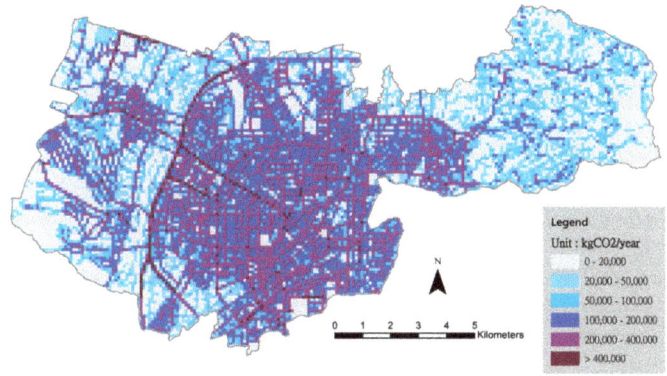

Figure 11. Amount of CO_2 emissions from traffic in Taichung metropolitan area.

Finally, the total CO_2 emissions from traffic and buildings in each 100 × 100 m grid in Taichung were estimated according to the type of land specified and the effects of the city's carbon sinks (e.g., rivers and parks) on that estimated volume of CO_2 emissions were analyzed. On the basis of these results, a carbon budget map for the city was constructed (Figure 12).

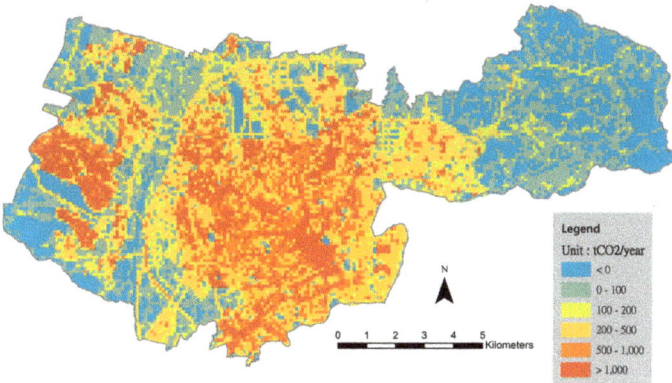

Figure 12. Annual total CO_2 emissions of Taichung metropolitan area.

4. Discussion

Due to climate change and the influence of the reduction of CO_2 emissions, in this research regarding the method and purpose of forecasting traffic volume and estimations of CO_2 emissions, the discussion and model application limitations are proposed as follows:

Daily traffic volumes were measured largely at thoroughfares; little such data were collected at typical roads. To increase the availability of traffic volume data at different types of roads, future studies can collect this data continuously and measure the traffic volume of both thoroughfares and common roads on a daily basis and analyze these data to improve the accuracy of prediction models. This research summed up hourly traffic volumes measured at thoroughfares and conducted a regression analysis of maximum traffic volumes at thoroughfares during morning and evening peak hours. Therefore, the CO_2 emissions from road traffic predicted by this research were characteristically large.

Regarding urban road travel distance, due to different forms of vehicles, different carbon emissions are generated; in terms of calculation, due to simplification and no consideration of types of fuels, driving speeds, road slopes and other fuel consumption influences, the research result belongs to the result of a larger forecast value. In the four road crossings in a single block in an urban area; regarding the case of measured data of existing traffic volume, practical data are difficult to obtain, thus, this research suggested that in order to increase the accuracy of the estimation formula, in the future, researchers can directly measure the road crossing traffic volume at the four road crossings of a single urban block and expand it for use.

The fuel usage and efficiency and CO_2 emissions of vehicles in use vary according to their types and road infrastructure planning. Therefore, future studies that analyze the amount of carbon emissions in an entire city can account for vehicle types and road infrastructure plans to inform their estimation of carbon emissions from road traffic. To estimate the amount of carbon emissions from buildings, future researchers can measure, on the basis of local EUI, the amount of CO_2 emissions from electricity used by different types of buildings. The carbon-sink effects of constructed landscapes such as trees or green spaces on the amount of CO_2 emissions from different types of buildings can also be measured to explicate how these effects on individual buildings help reduce CO_2 emissions on the scale of city blocks.

5. Conclusions

By simplifying the transportation demand mode in this research, the forecast formula of the total traffic volume at important road crossings throughout the day in Taichung is obtained and the relation between building complex development factors and traffic volume generated by the overall urban blocks is analyzed.

Regarding the relation between the development scale of various buildings and the influence of vehicle traffic flows, this research calculates the passenger car units generated by each square meter of area every day, as well as the result of the total annual CO_2 emissions. This research aimed at the relation between seven important design factors influencing building design, as well as the total traffic CO_2 emissions, in order to further obtain the multiple regression forecast estimation model of the total annual carbon dioxide of vehicles in the blocks.

Furthermore, it is possible to propose the integration of practical deliberation case information of building permits, such as urban planning laws and the Building Act, thus, when an architect designs the development factors, or when urban planners regularly conduct overall inspections of urban planning, it is possible to easily and effectively plan and evaluate urban areas and rapidly calculate the traffic volume in the future. Moreover, after proposing the cooperation of the conversion Geographic Information System information in urban areas, it is possible to clearly obtain the basic information of the traffic volume and present a map of urban traffic environmental characteristics.

Author Contributions: The author would like to thank the Urban Development Bureau of Taichung City Government in all the building permit data. Thanks to Building Permit Management Division for their document support.

Conflicts of Interest: The authors declare no conflicts of interest.

References

1. Liao, L.C. *The Assessment of the Energy Saving and the Carbon Dioxide Reduction Strategies for Road Transportation Sector*; Taipei University Institute of Natural Resourse and Environment Management: Taipei City, Taiwan, 2009; pp. 1–10.
2. Li, S.S. Method in Estimating CO_2 Emissions from Gasoline Vehicles in Taiwan. Master's Thesis, National Cheng Kung University, Tainan, Taiwan, 2007.
3. Henning, N.; Dagmar, H.; Martin, L.; Heidi, W. Environmental impact assessment of urban land use transitions—A context-sensitive approach. *Land Use Policy* **2009**, *26*, 233–248.
4. Fuller, R.J.; Crawford, R.H. Impact of past and future residential housing development patterns on energy demand and related emissions. *J. Hous. Built Environ.* **2011**, *26*, 165–183. [CrossRef]
5. Lin, H.T. *Green Building in Hot-Humid Climate*; Chan's Arch-Publishing Co., Ltd.: Taipei City, Taiwan, 2003.
6. Chang, Y.S. Life Cycle Assessment on the Reduction of Carbon Dioxide Emission of Buildings. Ph.D. Thesis, Department of Architecture, National Cheng Kung University, Tainan, Taiwan, June 2002; pp. 67–136.
7. Lin, H.T. *Footprint of Architecture Evaluation Theory*; Chan's Arch-Publishing Co., Ltd.: Taipei City, Taiwan, 2014; pp. 67–164.
8. Gan, J.J. A Study on the Electricity Consumption Prediction Method of City Blocks. Master's Thesis, Department of Architecture, National Cheng Kung University, Tainan, Taiwan, 2009; pp. 1–11.
9. Wang, R.J. A Study on Electricity Consumption Analysis of Residential Area. Ph.D. Thesis, Department of Architecture, National Cheng Kung University, Tainan, Taiwan, May 2005; pp. 107–126.
10. Po, C.S. Analysis of Energy Saving and Carbon Reduction by the Mode Choice and Trip Assignment Combined Model. Master's Thesis, Department of Civil and Ecological Engineering, I-Shou University, Kaohsiung, Taiwan, January 2011; pp. 8–40.
11. Hu, T.Y. Micro-simulation Based Dynamic Traffic Assignment. *Transp. Plan. J.* **2001**, *30*, 1–32.
12. Chen, J.N. A Study of Micro-simulation Based Dynamic Traffic Assignment Model under Mixed Traffic Flow Environment: The Principle of User Equilibrium. Master's Thesis, Department of Transportation Engineering and Management, Feng Chia University, Taichung, Taiwan, 2005; pp. 5–37.
13. Wardrop, J.G. *Some Theoretical Aspects of Road Traffic Research*; Institution of Civil Engineers: London, UK, 1952; pp. 325–378.
14. Doron, L. Land use for transport projects: Estimating land value. *Land Use Policy* **2015**, *42*, 594–601.
15. Yang, A.T. Structuralized Analysis of Urban Form Impacts on Travel Demand. *J. Chin. Inst. Transp.* **2006**, *18*, 391–416.
16. Jeong, Y.; Lee, G.; Kim, S. Analysis of the Relation of Local Temperature to the Natural Environment, Land Use and Land Coverage of Neighborhoods. *J. Asian Archit. Build. Eng.* **2015**, *14*, 33–40. [CrossRef]
17. Dai, G.F. The Research on the Interaction between Urban Land Use and Carbon Dioxide Concentration. Master's Thesis, Department of Urban Planning, National Cheng Kung University, Tainan, Taiwan, June 2010.
18. Hopkins, A.S.; Schellnhuber, H.J.; Pomaz, V.L. Urbanised territories as a specific component of the Global Carbon Cycle. *Ecol. Model.* **2004**, *173*, 295–312. [CrossRef]
19. Lin, T.P.; Lin, F.Y.; Wu, P.R. Multiscale analysis and reduction measures of urban carbon dioxide budget based on building energy consumption. *Energy Build.* **2017**, *153*, 356–367. [CrossRef]
20. She, P.C. A Study of the Influence of Highway Design Factors on Vehicle Carbon Emissions in Taiwan. Master's Thesis, Institute of Construction Engineering and Management, National Central University, Taoyuan, Taiwan, 2012; pp. 37–58.
21. Heimbach, C.L.; Cribbins, P.D.; Chang, M.S. Some partial consequences of reduced traffic lane widths on urban arterials. *Transp. Res. Rec.* **1983**, *923*, 69–72.
22. Chen, C.L.; Zhuang, Z.W.; Jiang, Y.H. Research energy consumption and pollution emissions by vehicle. In Proceedings of the Combustion Institute of R.O.C. 18th Symposium, Taiwan, 29 March 2008.

23. Huang, Y.K.; Tsao, S.M. An analysis of carbon dioxide emissions in transportation sector. *Urban Traffic Q.* **2003**, *18*, 1–14.
24. Ren, C.; Wu, E.N. *Urban Climatic Map—An Information Tool for Sustainable Urban Planning*; China Building Industry Publishing: Beijing, China, 2012; pp. 11–30.
25. Lin, F.B. *Taiwan Highway Capacity Manual*; Institute of Transportation, MOTC: Taipei City, Taiwan, 2011.
26. Bureau of Energy, Ministry of Economic Affairs, Industrial Technology Research Institute Printed, Vehicle fuel Consumption Guidelines, February 2010. Available online: https://www.moeaboe.gov.tw/ecw/populace/content/wfrmStatistics.aspx?type=5&menu_id=1303. (accessed on 25 December 2017).
27. Jiang, M.F. Using 3D Remote Sensing Data to Analyze the CO_2 Balance in Vicinity of Road-Case Study of Highway No. 84 at Taiwan. Master's Thesis, Department of Resources Engineering, National Cheng Kung University, Tainan, Taiwan, 1999; pp. 6–31.
28. Lin, H.T. *Building Carbon Footprint*, 2nd ed.; Chan's Arch-Publishing Co., Ltd.: Taipei City, Taiwan, 2015.
29. Chen, J.H.; Lin, H.T. The Classification Model of Energy Use Intensity Based on Building Function Types. Master's Thesis, Department of Architecture, National Cheng Kung University, Tainan, Taiwan, June 2009.
30. Huang, P.H.; Huang, S.L. Metabolism Approach for Studying the Relationship Between Urbanization and CO_2 Emission and Sequestration, first edition. Master's Thesis, National Taipei University, Taipei, Taiwan, June 2014.

© 2018 by the authors. Licensee MDPI, Basel, Switzerland. This article is an open access article distributed under the terms and conditions of the Creative Commons Attribution (CC BY) license (http://creativecommons.org/licenses/by/4.0/).

Article

An Integrated Carbon Policy-Based Interactive Strategy for Carbon Reduction and Economic Development in a Construction Material Supply Chain

Liming Zhang *, Wei Yang, Yuan Yuan and Rui Zhou

Business School, Sichuan University, Chengdu 610064, China; yang_benjun@163.com (W.Y.); yuanyuan1129@scu.edu.cn (Y.Y.); ruizhou_283@163.com (R.Z.)
* Correspondence: zhangliming@scu.edu.cn

Received: 17 October 2017; Accepted: 13 November 2017; Published: 18 November 2017

Abstract: Carbon emissions from the construction material industry have become of increasing concern due to increasingly urbanization and extensive infrastructure. Faced with serious atmospheric deterioration, governments have been seeking to reduce carbon emissions, with carbon trading and carbon taxes being considered the most effective regulatory policies. Over time, there has been a global consensus that integrated carbon trading/carbon tax policies are more effective in reducing carbon emissions. However, in an integrated carbon reduction policy framework, balancing the relationship between emission reductions and low-carbon benefits has been found to be a critical issue for governments and enterprises in both theoretical research and carbon emission reduction practices. As few papers have sought to address these issues, this paper seeks to reach a trade-off between economic development and environmental protection involving various stakeholders: regional governments which aim to maximize social benefits, and producers who seek economic profit maximization. An iterative interactive algorithmic method with fuzzy random variables (FRVs) is proposed to determine the satisfactory equilibrium between these decision-makers. This methodology is then applied to a real-world case to demonstrate its practicality and efficiency.

Keywords: construction materials; green supply chain; integrated carbon policy; interactive strategy; low carbon

1. Introduction

The "low carbon" concept was introduced at the World Climate Change Conference in Copenhagen, Denmark, 2009, after which low carbon economies became the major focus in many countries, leading to the development of the green supply chain (GSC) [1]. As one of the industries with the highest carbon emissions, the construction sector accounts for over one-third of global carbon dioxide emissions [2–5]. In addition to the carbon emissions from the daily operation of buildings, China has been undertaking many urban construction projects [6], which has led to a tremendous rise in construction carbon emissions [7]. In particular, as one of the six largest energy-consuming industries in China, the construction material industry represents 9% of the total energy consumption and 6% of total electricity consumption in China [8]. It is also a pillar industry in China since its added value makes up about 1% of the gross domestic product (GDP) each year [9]. The construction material industry has great potential with respect to energy conservation and carbon dioxide emission reduction, which could be of great significance to the achievement of total energy consumption control and transformation of low-carbon development. Enterprises, as the basic elements in the supply chain (SC), are required to take responsibility for the environmental performance of the supply chain participants [10]. As enterprises in supply chains are closely related, all are simultaneously affected by any carbon emission regulations. Therefore, SC enterprises must jointly adjust their operating and production plans to effectively achieve individual environmentally friendly performance [1]. Several advantages of the GSC have been identified, such as a positive corporate image, improved efficiency, and innovative leadership [1], all of which have encouraged more decision-makers to embrace GSC management (GSCM). When carbon emission regulations are imposed in a marketplace, scientifically designed environmental plans can enhance innovation, reduce total production costs, and highlight enterprise value [11]. Therefore, for each GSC member, low-carbon operations can represent a valuable, non-substitutable advantage [12]. Further, GSCM is a means for reducing potential losses from poor carbon emission performance that can intensify regulatory pressures [13], damage an enterprise's image, attract government fines, and lead to customer boycotts or order cancellations [14,15].

At the same time, the focus on the protection of benefits in environmentally-related construction issues has grown, becoming a primary norm for the development of socioeconomic policies [16–18]. Therefore, further environmental policies and institutional acts on this topic are urgently required for greener approaches in the area of construction engineering [19]. Governments around the world have promulgated various policies to reduce carbon emissions, with carbon trading and carbon taxes considered the most effective policy schemes for reducing carbon emissions [20,21]. Carbon trading, which is a mainstay in emission trading programs, is specifically aimed at reducing carbon emissions [22]. In the carbon trading market, enterprises which want more than their allocated carbon emissions can purchase rights to emit more, and firms who do not require their allocated carbon emissions can sell their carbon emission rights to other enterprises [22]. There has been a sharp rise in the number of carbon emission trading schemes in recent decades. For example, in 2005, 374 million t of equivalent carbon dioxide were exchanged, but by 2011, the carbon trading volume had risen to 10.28 billion CO_2 t, with the global carbon trading market valued at 176 billion US dollars [23]. Carbon taxes, which are a type of Pigovian tax [24], are a potentially cost-effective method for reducing greenhouse gas emissions [25]. Many European countries, such as the Netherlands, Sweden, Finland, and Norway, implemented carbon taxes many decades ago [26]. However, the Chinese government only introduced a carbon tax around 2013, which has severely affected the domestic market in China [27].

As stated above, most scholars have tended to study GSC from government or enterprise perspectives; however, while there have been many studies on carbon trading and carbon tax, many have only focused on the impact of a single policy on the macroeconomic development of carbon emission reductions, and the mutual relationships between supply chain enterprise operations and government policies have been ignored. In addition, there has been a lack of research on the performance of integrated carbon trading and carbon tax policies. To address this research gap, this paper explored the government and SC producer carbon reduction problems associated with integrated carbon trading and carbon tax policies. The government initially determines the annual free carbon emission allowances for the producer based on the average carbon emission level of the industry and its historical emission data. To control total carbon emissions and reduce the adverse impact of carbon emission reduction, the government, whose objective is to maximize social welfare, imposes a carbon tax on the producers. Under the dual constraints of emission allowance and carbon tax, the SC producers must be allocated sufficient carbon emissions to satisfy their daily operations. As the SC producers cannot exceed emission allowance limitations, they must either trade any remaining carbon emission allowances on the carbon trading market or directly purchase additional allowances to meet their emissions requirements. However, now carbon tax and the consumer's low-carbon preference must be taken into serious consideration, while at the same time considering the carbon tax and consumer preference for low-carbon operations. Producers can achieve emission reductions by flexibly combining emission reduction investments and emission rights purchasing. Finally, to maximize their own profits, the SC producers must weigh up the emission costs and benefits under different strategy combinations to determine the final emission level and the associated product prices.

Because of the multiple decision-makers and the complex interactions, bi-level mathematical programming is proposed as it can accurately describe the interests of the decision-makers. Bi-level models have been widely applied in SC management [28–32]. For example, Ghosh and Shah [28] developed a bi-level supplier/manufacturer SC to examine SC coordination issues under a carbon emission policy. Song and Leng [29] included the carbon emission factors into a single-cycle newsboy model to examine the influences of different carbon emission policies on producers' orders. Choi [30] examined the impact of a carbon footprint tax on bi-level fashion SC systems, and the importance of the carbon footprint tax on SC fashion management. Du et al. [31] considered an emission-dependent SC to examine an emission-dependent manufacturer and an emission permit supplier under a cap-and-trade system. Jaber et al. [32] researched bi-level manufacturer/retailer SC game processes and coordination mechanisms under carbon cap-and-trade conditions.

These studies have inspired researchers with novel management insights into government carbon emission regulations and GSC operations; however, the carbon emission regulation parameters have been generally regarded as exogenous variables, with the governments not being involved in the decision-making processes. Therefore, the main contributions of this paper are as follows. First, the integration of carbon reduction policies and their relationships within GSCs are explored. Second, optimal decision results are theoretically derived through the development of a bi-level optimization model, in which the leader, which has a social welfare maximization objective, determines the carbon tax and emission allowance allocations, and the following producers, which have a profit maximization objective, determine their production output and sales quantities. Third, it is shown that the sustainable GSC development and a trade-off between environmental protection and economic development can be achieved by employing the proposed methodology.

The remainder of this paper is organized as follows. Section 2 gives the research and problem statement, including the research background and the decision-making relationship analysis. In Section 3, a methodology, including a bi-level mathematical model and an interactive algorithm, is established as an abstraction of the real problem. To confirm the generality of the methodology, a general case, results, and some further discussions are given in Section 4. Finally, Section 5 gives the conclusions and suggestions for future studies.

2. Research and Problem Statement

In an integrated carbon policy-based carbon emission reduction problem, there are various decision-makers: the government as the leading decision-maker and the GSC producers as the following decision-makers who act based on the government's decisions. Both parties have individual contradictory carbon reduction targets. The government seeks to effectively reduce overall carbon emissions, while safeguarding the economic interests of the producers, with the aim of stimulating participation and enthusiasm for emission reduction, and maximizing total social welfare, while the producers seek to obtain as high a carbon emission allowance as possible to reduce their emission costs and maximize profits. Both of parties have individual but interacting decision-making variables; the government's decision-making variables are the free carbon emission allowances and the carbon tax; while the producer's variables are production and sales quantities. It is assumed that the GSC producers have an equal market position and each independently trades their carbon emission allowances on the open market. As the producers' profits are considered when the government sets the carbon emission reductions targets, the decisions made by the producers not only determine their own objectives but also influence the government's goals. Therefore, the government's decisions also need to consider the influence of the producers' responses to its own goals. Therefore, the carbon emission reduction problem in this paper is a dynamic optimization decision-making process, within which the government needs to monitor the carbon trading market, assess the effectiveness of the carbon tax level, and improve their emission reduction strategies based on the responses from the market and the producers.

The above analysis has shown that the carbon emission reduction decision-making process is an interactive decision-making mechanism comprised of the government as the leading decision-maker responsible for the overall carbon reduction plan and control, and the producers as the following decision-makers who have independent decision-making rights in terms of their own carbon reduction goals. As the government is in the lead decision-making position, it has the advantage of moving first. From this description, this integrated emission reduction decision-making policy problem has the same characteristics and mechanisms as a general bi-level decision-making problem, which is similar to the hierarchical decision leader–followers Stackelberg game, in which the leader is more powerful and the follower reacts rationally to the leader's decisions [33,34]. Therefore, the bi-level Stackelberg game can be used to examine this level of this government/producers carbon emission reduction decision-making relationship.

This problem can be abstracted as a bi-level mathematical model for calculation, which, along with the hierarchical structure makes it a complex problem, as shown in the concept model in Figure 1.

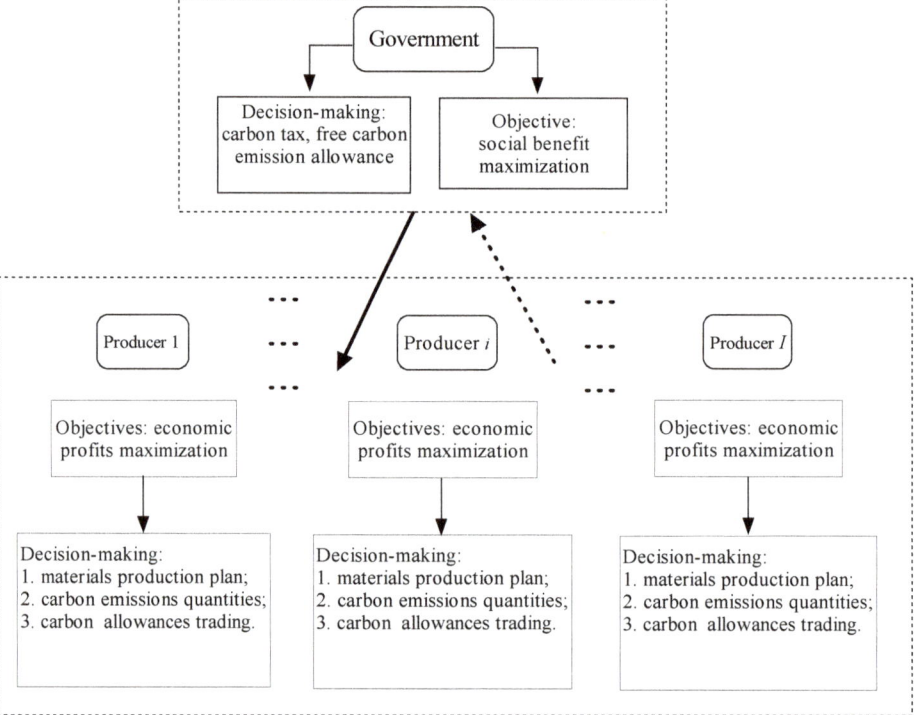

Figure 1. Model of the bi-level decision-making mechanism in carbon emission reduction.

3. Methodology

A bi-level mathematical model and a corresponding solution approach are proposed in this section.

3.1. Bi-Level Programming

Assumptions, notations, objective functions and constraints of an integrated emission reduction decision-making policy problem are introduced in this section.

3.1.1. Assumptions

Before formulating the model, the following assumptions are given:

A_1 This integrated carbon emission policy problem is a single-period decision-making problem; therefore, a static optimization problem is assumed.
A_2 The government is responsible for the initial free carbon emission allowance allocation and the producers freely transact with other producers in the CET market.
A_3 None of the producers can individually meet all the projects' requirements [35].
A_4 Each decision-maker fully understands the objective functions and inherent constraints, and behaves rationally [36].

3.1.2. Notations

Sets:
$\mathbb{I}$: Set of producers, $\mathbb{I} = \{1, 2, \cdots, I\}$.
$\mathbb{J}$: Set of projects, $\mathbb{J} = \{1, 2, \cdots, J\}$.

Indices:
i: Index for the construction material producers, $i \in \mathbb{I}$.
j: Index for the projects, $j \in \mathbb{J}$.

Decision variables:
α_i: Free carbon emission allowance for construction material producer i.
γ: Carbon tax rate for construction material production.
q_i: Total construction materials produced by producer i.
q_{ij}: Total construction materials purchased from producer i for project j.

Certain parameters:
Cap: Actual free carbon allowance allocation for the construction material industry in the last production period.
γ^l: Lower bounds for the unit carbon tax.
γ^u: Upper bounds for the unit carbon tax.
CE_i^u: Carbon emissions produced by producer i with no emission reduction measure.
CE_i^l: Carbon emissions produced by producer i after the carbon reduction efforts.
PC_i: Unit production cost for construction material producer i.
IC_i: Unit inventory cost for construction material producer i.
P_{ij}: Price of a unit of construction material k from producer i for project j.
GP_i: Production carbon emission coefficient for producer i.
PM_i^l: Lower bounds for the production capacity for producer i.
PM_i^u: Upper bounds for the production capacity for producer i.
SC_i^{max}: Maximum storage capacity of producer i.
β: Market price of carbon emission trading.
λ: Sale taxes an enterprise pays for a unit of construction material.

Uncertain parameters:
$\tilde{D}_j$: Total demand of project j for construction materials.

3.1.3. Model Formulation

This section gives a detailed description of the global model including the model of the government and the model of the producers.

Objective 1: The government's social benefit objective. To balance the environmental protection and economic development of the construction material industry, the government must control carbon emissions while also considering the social benefits. However, the social benefits are the primary objective, which are made up of three parts; sales taxes on materials, revenue from carbon emission trading, and the carbon tax revenue. Let λ be the unit sales tax for the construction material. The total sales taxes for the materials are therefore $\lambda \sum_{i=1}^{I} \sum_{j=1}^{J} q_{ij}$. It is assumed that β is the carbon emission trading price. Therefore, the total carbon emission trading revenue is $\beta \sum_{i=1}^{I} (GP_i q_i - \alpha_i)$. Under the carbon tax policy, construction material producers have paid taxes for their carbon emissions. Let γ be the unit carbon tax for producing the construction materials. The total carbon tax revenue is therefore $\gamma \sum_{i=1}^{I} GP_i q_i$, and the overall social benefit for the government is:

$$\max W = \lambda \sum_{i=1}^{I} \sum_{j=1}^{J} q_{ij} + \beta \sum_{i=1}^{I} (GP_i q_i - \alpha_i) + \gamma \sum_{i=1}^{I} GP_i q_i. \quad (1)$$

Constraint 1: Free carbon emission allocation constraint. To limit the carbon emissions, the government initially allocates free allowances to the producers. To protect the construction material industry, the government cannot allocate a carbon emission allowance beyond a producer's capacity. Therefore, there exists an upper bound, $\leq CE_i^m$, for producer i, which represents the maximum carbon emissions emitted when producer i is under full-load production; this constraint is denoted as $\alpha_i \leq CE_i^m$. However, to guarantee each producer's basic rights, the government must allocate a carbon emission allowance that ensures that the producer can produce at capacity. Therefore, there exists a lower bound, CE_i^l, for producer i, which is the minimum carbon emission allowance allocation needed to maintain basic operations. This constraint is denoted as $CE_i^l \leq \alpha_i$. To ensure both sides are fully considered, the producers' restrictions when the government makes decisions are:

$$CE_i^l \leq \alpha_i \leq CE_i^m. \quad (2)$$

Constraint 2: Industry free carbon emission allowance allocation constraint. As the government must guarantee the atmospheric environment, they may alter their intentions to control the carbon emissions of the whole industry within an acceptable range, which cannot surpass the actual free carbon allowance allocation given to the construction material industry in the last production period,

$$\sum_{i=1}^{I} \alpha_i \leq Cap. \quad (3)$$

Constraint 3: Carbon tax constraint. The formulated unit carbon taxes must be within the minimum and maximum carbon tax limitation bounds, which can be expressed as:

$$\gamma^l \leq \gamma \leq \gamma^u. \quad (4)$$

Constraint 4: Demand constraint. As construction materials are required to ensure the project meets its construction deadlines, the producers must satisfy the demand for each type of project material. However, because of the inherent complexity and uncertainty in construction technology as well as the fluctuating demand, accurate data for the material supply level is difficult to obtain. Therefore, this demand is dealt with using an expected value operator. The material quantities provided to each project, therefore, must satisfy the respective demands, namely,

$$\sum_{i=1}^{I} q_{ij} \geq E\left[\tilde{D}_j\right]. \quad (5)$$

Objective 2: Producer's profit objective. With the integrated carbon policies, each producer, as an independent decision-maker, seeks to maximize his individual profit, which is the difference between total revenue and total cost. Total revenue comes from construction material sales $\sum_{j=1}^{J} P_{ij} q_{ij}$, while total costs are made up of material production costs $PC_i q_i$, inventory costs $IC_i \left(q_i - \sum_{j=1}^{J} q_{ij}\right)$, sales taxes $\lambda \sum_{j=1}^{J} q_{ij}$, CET costs $\beta (GP_i q_i - \alpha_i)$, and carbon taxes $\gamma GP_i q_i$. Therefore, the profit function is:

$$\max P_i = \sum_{j=1}^{J} P_{ij} q_{ij} - PC_i q_i - IC_i \left(q_i - \sum_{j=1}^{J} q_{ij}\right) - \lambda \sum_{j=1}^{J} q_{ij} - \beta (GP_i q_i - \alpha_i) - \gamma GP_i q_i. \quad (6)$$

Constraint 5: Producers' carbon emissions constraint. The amount of carbon emissions for each producer cannot exceed the emissions without a reduction measure but must not be lower than the amount of emissions with the greatest carbon reduction efforts, that is,

$$CE_i^l \leq GP_i q_i \leq CE_i^m. \tag{7}$$

Constraint 6: Production capacity constraint. When providing construction materials for multiple projects, the material production q_i of producer i must be within a specified range between the maximum and minimum production capacity. Therefore, the production capacity constraint is:

$$PM_i^l \leq q_i \leq PM_i^u. \tag{8}$$

Constraint 7: Inventory constraint. Each producer owns a warehouse for temporarily storing construction materials that are not yet sold. The construction material inventory level of producer i cannot exceed the storage capacity, namely

$$0 \leq q_i - \sum_{j=1}^{J} q_{ij} \leq SC_i^{\max}. \tag{9}$$

3.1.4. Global Model

To sum up, the global model is built as in Equation (10). The decision-makers impact on each other as the government's decisions (α_i, γ) affect the construction material producers' decisions (q_i, q_{ij}). The government attempts to expand the social benefit by reducing total carbon emissions, however, each construction material producer seeks profit maximization. At the same time, the producers' actions (q_i, q_{ij}) affect the government's subsequent actions (α_i, γ). Consequently, all decision-makers seek satisfactory solutions based on their respective optimization targets. At the beginning, based on previous information and its own objectives, the government determines the initial carbon emission allowance allocations, the decisions for which are then is delivered to the construction material producers. Each producer, as a follower, sets its own plan in view of the government's decisions, the market demands, and their own technological conditions. The producers' plans are then submitted to the government, after which the government adjusts its initial decisions in consideration of each producer's emission performance, and an improved plan is then sent to the producers. Therefore, the government influences the behavior of the construction material producers without completely controlling their actions, and the construction material producers' behavior affects the government's decisions. Relationships between each construction material producer are also assumed to be non-cooperative, as each producer makes decisions independently and without collaboration [37]. Each producer, therefore, has an optimization problem, in which the other producers' decisions are regarded as the certain parameters. Therefore, the problem can be expressed mathematically in a bi-level programming model. In summary, the global model is:

$$\max W = \lambda \sum_{i=1}^{I}\sum_{j=1}^{J} q_{ij} + \beta \sum_{i=1}^{I} (GP_i q_i - \alpha_i) + \gamma \sum_{i=1}^{I} GP_i q_i \qquad (10)$$

$$\text{s.t.} \begin{cases} CE_i^l \leq \alpha_i \leq CE_i^m \\ \sum_{i=1}^{I} \alpha_i \leq Cap \\ \gamma^l \leq \gamma \leq \gamma^u \\ \sum_{i=1}^{I} q_{ij} \geq E\left[\tilde{D}_j\right] \\ \max P_i = \sum_{j=1}^{J} P_{ij} q_{ij} - PC_i q_i - IC_i \left(q_i - \sum_{j=1}^{J} q_{ij}\right) - \\ \qquad \lambda \sum_{j=1}^{J} q_{ij} - \beta(GP_i q_i - \alpha_i) - \gamma GP_i q_i \\ \text{s.t.} \begin{cases} CE_i^l \leq GP_i q_i \leq CE_i^m \\ PM_i^l \leq q_i \leq PM_i^u \\ 0 \leq q_i - \sum_{j=1}^{J} q_{ij} \leq SC_i^{\max} \end{cases} \\ \alpha_i, \gamma, q_i, q_{ij} \geq 0 \\ i \in \mathbb{I}, j \in \mathbb{J} \end{cases}$$

3.2. Iterative Interactive Algorithm

As is well known, bi-level programming optimization is a non-deterministic polynomial (NP) hard problem even in its most common formulation [38–40]. The decision variables of the government's upper-level mathematical model and the producers' lower-level mathematical models are therefore nested in each decision-maker's objective function and constraints of the model (10), for which an iterative interactive algorithm based on evolutionary game theory between the two decision-makers is established to deal with the complex model (10) interaction. The iterative interactive algorithm solves both the upper and lower optimization problems in the bi-level programming mathematical model. At first, all constraints in the government's upper-level optimization model are set using Matlab R2013a (MathWorks, Natick, MA, USA) and a feasible zone of the upper-level optimization model is built. Then a vector (α_i, γ) is generated, which is the initial solution to the upper-level optimization model, after which vector (α_i, γ) is put into the lower-level optimization model as the constant, and the model is transformed into a single-level programming model for (q_i, q_{ij}). By employing the mathematical toolbox in Matlab R2013a, an initial solution to the producers' optimization model, (q_i, q_{ij}), is obtained, which is fed back to the upper-level optimization model, and the model is also transformed into a single-level programming model for (α_i, γ) for the government. The mathematical toolbox in Matlab R2013a is employed again to obtain an improved solution, (α_i, γ). If the government is satisfied with the improved solution, the computation is terminated. If not, the new solution (α_i, γ) is again imported into the lower-level optimization model and the solution to the model calculated, thus generating a new vector for (q_i, q_{ij}), which is then conveyed to the upper-level model again. As an interactive mechanism in this bi-level decision-making structure, this process is repeated several times until an overall satisfactory solution to both the upper and lower level optimization models is obtained, which is the final solution to the proposed bi-level optimization model. The procedures for this solution approach are shown in Figure 2.

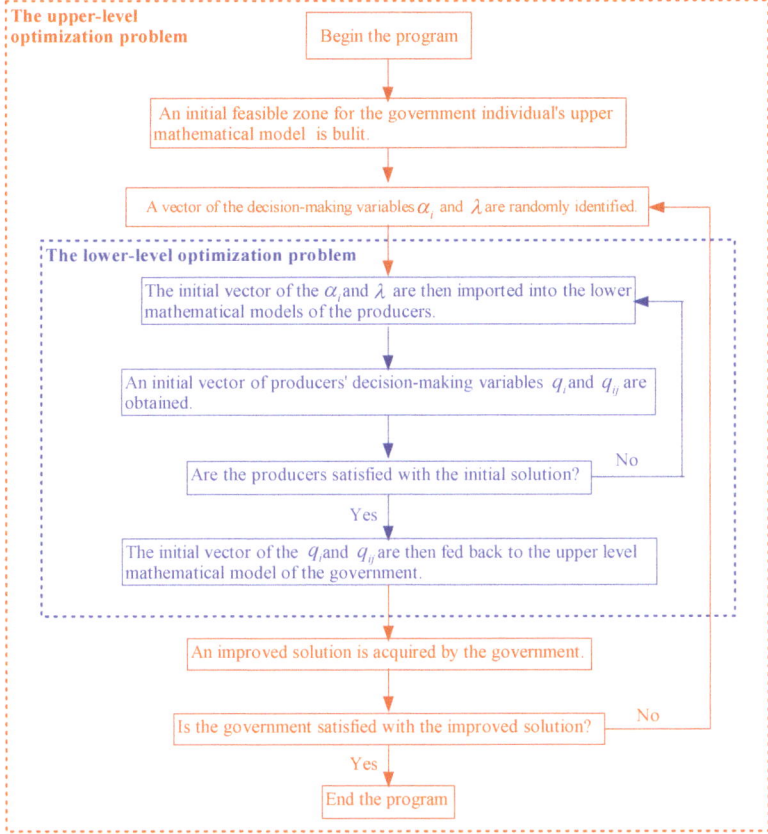

Figure 2. Framework for the iterative interactive algorithm.

4. Case Study

In this section, the construction material industry in City X is employed as a practical case to probe into the integrated carbon emission reduction policy and to demonstrate the practicality of the proposed optimization methodology.

4.1. Case Description

Industry is the dominant consumer of energy and producer of CO_2 emissions in China. China's urbanization process has undoubtedly promoted infrastructure construction, which has stimulated demand for cement, ceramics, glass, and other construction materials [41]. The emission reduction potential is of great significance to the achievement of the total carbon reduction control. City X, the main region for cement production in Shanxi province, supplies the Shanxi province with 20% of its cement demands. The cement industry plays a key role in the economic development of City X, however, emissions from material production are the primary source of carbon emissions, and represent a serious menace to local air quality. In terms of air quality, City X has been judged as one of the worst cities in the Shanxi province. Thus, reducing carbon emissions in construction material industry is regarded as the primary goal for the local government in the next production period.

4.2. Data Presentation

In City X, three main cement producers, referred to as A, B, and C, are engaged in the production and supply of cement for three key projects, referred to as I, II, and III. The carbon trading price was set at 60 CNY/t based on the average carbon allowance trading price at City X's Emissions Trading Institution. The lower and upper bounds for the carbon tax rate were 10 CNY/t and 30 CNY/t, respectively, based on the recommendations from National Development and Reform Commission experts. The unit sales tax for cement was 50 CNY/t, and the total carbon emissions in the last planning period Cap were about 3.2×10^6 t CO_2. Information for producers, such as the emission allowances, production capacity bounds, the carbon emission coefficient, production and inventory unit cost, maximum storage capacity, and unit prices, are listed in Table 1. From Table 1, it is clear that the production costs were slightly higher for Producer B, but fewer emissions were generated due to the advanced, more efficient machinery and manufacturing technology. The cement demands for the three key projects are listed in Table 2.

Table 1. Basic information for the cement producers.

Parameters	Producer A	Producer B	Producer C
Lower bounds for carbon emissions (10^5 t)	6.3	5.4	6.2
Upper bounds for carbon emissions (10^5 t)	14.7	13.3	14.3
Lower bounds for production capacity (10^4 t)	140	138	135
Upper bounds for production capacity (10^4 t)	180	175	180
Carbon emission coefficient (kg CO_2/t)	630	608	622
Unit production costs (CNY/t)	52	46	54
Unit inventory costs (CNY/t)	24	27	26
Maximum storage capacity (10^4 t)	54	52	55
Unit price of cement for Project I (CNY/t)	290	305	296
Unit price of cement for Project II (CNY/t)	292	310	305
Unit price of cement for Project III (CNY/t)	280	300	286

Table 2. Project demands for cement.

	Project I	Project II	Project III
Cement (10^4 t)	$(135, \mathcal{N}(160, 25), 197)$	$(146, \mathcal{N}(172, 18), 196)$	$(126, \mathcal{N}(155, 20), 173)$

4.3. Results Analysis

By importing the collected data into the proposed optimization model (10) and running the iterative interactive algorithm on Matlab R2013a, the results for the proposed model were determined, as shown in Table 3. Satisfactory solutions were obtained for both the government and the producers, in which the social benefits for the government were estimated at $W = 3.08 \times 10^8$ CNY, and the profits for Producers A, B, and C were respectively $P_A = 2.56 \times 10^8$ CNY, $P_B = 3.17 \times 10^8$ CNY, and $P_C = 3.24 \times 10^8$ CNY. The total carbon emission allowance for the construction material industry for the government was 29.5×10^5 t CO_2, of which 9×10^5 t CO_2, 9.5×10^5 t CO_2, and 11×10^5 t CO_2 were allocated to Producers A, B, and C, respectively. The optimal carbon emissions for Producers A, B, and C were 9.26×10^5 t CO_2, 9.72×10^5 t CO_2, and 11.20×10^5 t CO_2, respectively, and extra carbon emission allowances were purchased from the government.

Table 3. Satisfactory solution.

Decision-Makers	γ (CNY/t)	α_i (10^5 t CO_2)	q_i (10^4 t)	q_{ij} (10^4 t) I	II	III	Social Benefits (Profits) (10^8 CNY)
Government	20	-	-	-	-	-	3.08
Producer A	-	9	147	55	50	42	2.56
Producer B	-	9.5	160	55	55	50	3.17
Producer C	-	11	180	50	67	63	3.24

4.4. Carbon Tax Analysis

In this section, different values for the carbon tax were set to verify its influence on the overall and individual emissions and economic performances, in which the γ ranged from 10 to 30 CNY/t in intervals of 5 CNY/t. Figures 3 and 4 show the influence of the different values for the carbon tax on industry carbon emissions, social benefits, and profits for the government and Producers A, B and C.

In Figure 3, as the carbon tax increased, the total carbon emissions in the construction industry continued to decrease, and when $\gamma \geq 20$ CNY/t, the total emissions were less than the government's annual emission objectives (i.e., 29.5×10^5 t CO_2). This indicated that the a carbon tax policy could be a good carbon emission reduction method. From the producers' perspective, with an increase in the carbon tax, the carbon emissions of Producers A and C decreased; however, the carbon emissions of Producer B increased, indicating that Producer B was a lower carbon emission enterprise than the other two producers. When the government imposed stricter carbon tax regulation, consumers tended to purchase materials from Producer B to avoid the carbon tax being passed on. Therefore, Producer B was able to produce a greater number of materials than previously, thereby generating greater carbon emissions.

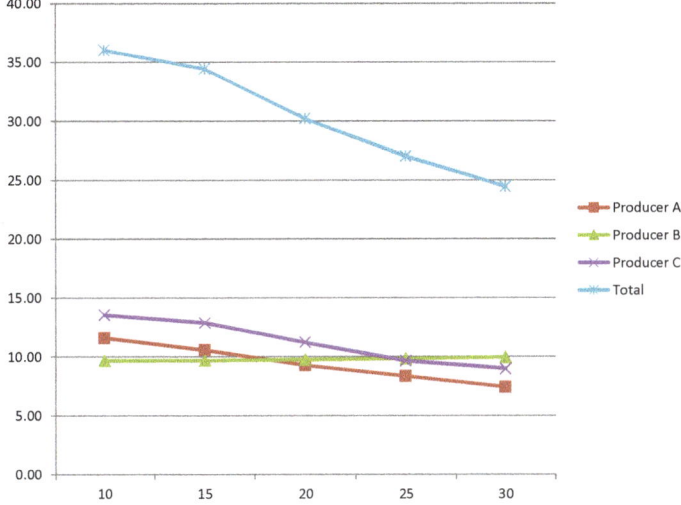

Figure 3. Influence of different carbon tax values on total carbon emissions (unit: 10^5 t CO_2).

In Figure 4, as the carbon tax grew, the government's social benefits and Producer B's profits increased, and the growth rates of both when $\gamma \in [20, 30]$ were obviously larger than when $\gamma \in [10, 20]$. Because the increased carbon tax revenue raised social benefits, the promoted projects purchased more materials from Producer B. However, under the pressure of a higher carbon tax, Producers A and C had to reduce their production output, leading to a decrease in profits. From the producers' perspective, when faced with different carbon tax changes, enterprises with lower carbon emissions would be more favored by the government and the market, and enterprises with higher carbon emissions would need to improve machinery and invest in cleaner manufacturing technology. From the industry perspective, total profits were relatively unchanged, which indicated that the carbon tax policy had little negative impact on the economic development of the construction material industry.

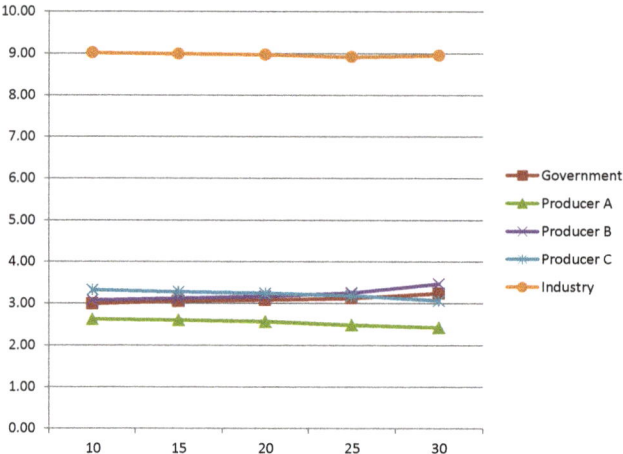

Figure 4. Influence of different carbon tax values on social benefits and profits (unit: 10^8 CNY).

4.5. Comparative Analysis in Different Decision-Making Environments

The computational results were acquired by employing some fuzzy random variables (FRVs) under a fuzzy random environment; however, the results may alter in different decision-making environments. To measure the robustness of the methodology when some parameters from model (10) had some perturbations, a sensitivity analysis was conducted to demonstrate the effect on the solutions under different decision-making environments. First, model (10) under a fuzzy random environment was compared to a model under a certain environment, in which the mean values for the FRVs were kept to eliminate uncertainty in the determined environment. For example, for the demands for Project I (i.e., $\tilde{\bar{D}}_I = (135, \mathcal{N}(160, 25), 197)$), the adopted value was 166, which was the mean value from 135 and 197. The computational results are shown in Table 4. Second, model (10) under a fuzzy random environment was also compared to a model in a fuzzy environment, in which in the fuzzy environment the fuzziness in the FRVs was retained, but the stochastic nature was neglected. To obtain useful data, Gaussian distributions were removed and the expectations were reserved. For instance, the demands of Project I in the fuzzy environment were denoted as $(135, 160, 197)$. The expected method was also used to convert the objective functions into equivalent crisp functions. The comparative results are also shown in Table 4.

The results in Table 4 indicated that the solutions in the determined environment had greater deviations than those in the fuzzy random environment, indicating that model (10) under a fuzzy random environment was able to provide more reliable references for the decision-makers. It was also found that the results under the fuzzy random environment were also better than in a fuzzy environment. Therefore, the proposed methodology was found to be robust in solving a carbon emission reduction problem for the construction material industry for the government, and the use of the fuzzy random environment was better able to match actual circumstances based on the post hoc analysis.

Table 4. Results comparison for the government in model (10) under different environments.

Objective	Unit: 10^8 CNY	Fuzzy Random Environment	Certain Environment	Fuzzy Environment
	Best	3.34	3.38	3.32
W	Average	2.95	2.86	2.94
	Worst	2.65	2.45	2.52

5. Conclusions

Carbon emission reduction is important to the sustainable development of the construction material industry. Most construction material activities, such as material producer selection, material production plans, dynamic inventories, and assignment problems, are interrelated, which means that contradictions between the government and construction material enterprises are unavoidable. Previous research has demonstrated that to achieve sustainable development in the construction material industry, these contradictions need to be reduced or fully resolved [2,4,36,42].

However, research on carbon emission reduction problems in the construction material supply chain has been restricted to a simple overall GSC system, or only a single carbon emission policy was assumed. Therefore, there were several difficulties that remained unresolved. Firstly, the hierarchical decision-making relationships between the government and the producers were not considered; however, as there are multiple stakeholders within this structure, contradictions must be considered. Secondly, linearization assumptions and simplifications were often employed to ensure model tractability, which led to loss of generality in the mathematical models. Thirdly, uncertainty and complications in the decision-making environment were not considered when dealing with the collected data.

To overcome these difficulties, this paper proposed an integrated methodology for carbon emission reduction problems under an integrated carbon reduction policy. The integrated methodology combined a bi-level mathematical model and an iterative interactive algorithm. The bi-level mathematical model was formulated to handle the leader–followers contradictions and competition between the stakeholders, and FRVs were used to reflect the inherent uncertainties in the problem, all of which made the bi-level mathematical model more complex but better related to the practical environment. An iterative interactive algorithm was designed to solve the non-linear bi-level mathematical model. Then, the proposed methodology was applied to a real-world case, the results from which clearly showed that satisfactory solutions for both the government and the producers could be obtained, and a suitable trade-off reached between economic development and environmental protection. Solution analysis, study of the impact of carbon tax, and comparative analysis in different decision-making environments were conducted to illustrate the applicability and robustness of the proposed methodology.

Acknowledgments: This research was supported by Social Science Research Project in Sichuan Province (Grant No. 17GL070), The Research Center for Systems Science & Enterprise Development, Key Research of Social Sciences Baseof Sichuan Province (Grant No. Xq17C03) and National Natural Science Foundation of China (71702118).

Author Contributions: Liming Zhang and Wei Yang conceived the idea and designed the structure of this paper; Liming Zhang and Wei Yang collected related study literature; Yuan Yuan and Rui Zhou contributed data; Liming Zhang analyzed the data; Wei Yang wrote the paper.

Conflicts of Interest: The authors declare no conflict of interest.

Abbreviations

The following abbreviations are used in this manuscript:

FRV   fuzzy random variable
GDP   gross domestic product
GSC   green supply chain
SC   supply chain

References

1. Wang, C.; Wang, W.; Huang, R. Supply chain enterprise operations and government carbon tax decisions considering carbon emissions. *J. Clean. Prod.* **2017**, *152*, 271–280.
2. Chau, C.K.; Hui, W.K.; Ng, W.Y.; Powell, G. Assessment of CO_2 emissions reduction in high-rise concrete office buildings using different material use options. *Resour. Conserv. Recycl.* **2012**, *61*, 22–34.

3. Atmaca, A.; Atmaca, N. Life cycle energy and carbon dioxide emissions assessment of two residential buildings in Gaziantep, Turkey. *Energy Build.* **2015**, *102*, 417–431.
4. Herrera, J.C.; Chamorro, C.R.; Martín, M.C. Experimental analysis of performance, greenhouse gas emissions and economic parameters for two cooling systems in a public administration building. *Energy Build.* **2015**, *108*, 145–155.
5. Shi, Q.; Chen, J.; Shen, L. Driving factors of the changes in the carbon emissions in the Chinese construction industry. *J. Clean. Prod.* **2017**, *166*, 615–627.
6. Zhang, X.; Wang, F. Life-cycle assessment and control measures for carbon emissions of typical buildings in China. *Build. Environ.* **2015**, *86*, 89–97.
7. Zhang, Z.; Wang, B. Research on the life-cycle CO_2 emission of China's construction sector. *Energy Build.* **2015**, *112*, 244–255.
8. Jiang, Z.; Lin, B. China's energy demand and its characteristics in the industrialization and urbanization process. *Energy Policy* **2012**, *49*, 608–615.
9. Dhakal, S. Urban energy use and carbon emissions from cities in China and policy implications. *Energy Policy* **2009**, *37*, 4208–4219.
10. Seuring, S.; Müller, M. From a literature review to a conceptual framework for sustainable supply chain management. *J. Clean. Prod.* **2008**, *16*, 1699–1710.
11. Porter, M.E.; Van der Linde, C. Green and Competitive: Ending the Stalemate. *Harv. Bus. Rev.* **1995**, *28*, 128–129.
12. Hollos, D.; Blome, C.; Foerstl, K. Does sustainable supplier co-operation affect performance? Examining implications for the triple bottom line. *Int. J. Prod. Res.* **2012**, *50*, 2968–2986.
13. Reid, E.M.; Toffel, M.W. Responding to public and private politics: corporate disclosure of climate change strategies. *Strat. Manag. J.* **2009**, *30*, 1157–1178.
14. Bansal, P.; Clelland, I. Talking Trash: Legitimacy, Impression Management, and Unsystematic Risk in the Context of the Natural Environment. *Acad. Manag. J.* **2004**, *47*, 93–103.
15. Hajmohammad, S.; Vachon, S. Mitigation, Avoidance, or Acceptance? Managing Supplier Sustainability Risk. *J. Supply Chain Manag.* **2016**, *52*, 48–65.
16. Miccoli, S.; Finucci, F.; Murro, R. A monetary measure of inclusive goods: The concept of deliberative appraisal in the context of urban agriculture. *Sustainability* **2014**, *6*, 9007–9026.
17. Miccoli, S.; Finucci, F.; Murro, R. Assessing Project Quality: A Multidimensional Approach. *Adv. Mater. Res.* **2014**, *1030*, 2519–2522.
18. Miccoli, S.; Finucci, F.; Murro, R. Criteria and Procedures for Regional Environmental Regeneration: A European Strategic Project. *Appl. Mech. Mater.* **2014**, *675*, 401–405.
19. Moretti, L.; Caro, S. Critical analysis of the Life Cycle Assessment of the Italian cement industry. *J. Clean. Prod.* **2017**, *152*, 198–210.
20. Sorrell, S.; Sijm, J. Carbon trading in the policy mix. *Oxf. Rev. Econ. Policy* **2003**, *19*, 420–437.
21. Weisbach, D.A.; Metcalf, G.E. The Design of a Carbon Tax. *Soc. Sci. Res. Netw. Electron. J.* **2009**, *33*, 499–556.
22. Carbon Emission Trading, 2017. Available online: https://en.wikipedia.org/wiki/Carbon_emission_trading (accessed on 11 November 2017).
23. Perdan, S.; Azapagic, A. Carbon trading: Current schemes and future developments. *Energy Policy* **2011**, *39*, 6040–6054.
24. Intergovernmental Panel on Climate Change. *Climate Change 2007 Synthesis Report*; Intergovernmental Panel on Climate Change: Geneva, Switzerland, 2007.
25. Klemmensen, B.; Pedersen, S.; Rydén, L.; Dirckinck-Holmfeld, K.R.; Marklund, A. *Environmental Policy: Legal and Economic Instruments*; Baltic University Press: Uppsala, Swdden, 2007.
26. Baranzini, A.; Goldemberg, J.; Speck, S. A future for carbon taxes. *Ecol. Econ.* **2000**, *32*, 395–412.
27. Fang, G.; Tian, L.; Fu, M.; Sun, M. The impacts of carbon tax on energy intensity and economic growth—A dynamic evolution analysis on the case of China. *Appl. Energy* **2013**, *110*, 17–28.
28. Ghosh, D.; Shah, J. A comparative analysis of greening policies across supply chain structures. *Int. J. Prod. Econ.* **2012**, *135*, 568–583.
29. Song, J.; Leng, M. *Analysis of the Single-Period Problem under Carbon Emissions Policies*; Springer: New York, NY, USA, 2012; pp. 297–313.
30. Choi, T.M. Carbon footprint tax on fashion supply chain systems. *Int. J. Adv. Manuf. Technol.* **2013**, *68*, 835–847.

31. Du, S.; Zhu, L.; Liang, L.; Ma, F. Emission-dependent supply chain and environment-policy-making in the 'cap-and-trade' system. *Energy Policy* **2013**, *57*, 61–67.
32. Jaber, M.Y.; Glock, C.H.; Saadany, A.M.A.E. Supply chain coordination with emissions reduction incentives. *Int. J. Prod. Res.* **2013**, *51*, 69–82.
33. Von Stackelberg, H. *The Theory of the Market Economy*; Oxford University Press: Oxford, UK, 1952.
34. Gibbons, R. *Game Theory for Applied Economists*; Princeton University Press: Princeton, NJ, USA, 1992.
35. Xu, J.; Zhao, S. Noncooperative Game-Based Equilibrium Strategy to Address the Conflict between a Construction Company and Selected Suppliers. *J. Constr. Eng. Manag.* **2017**, *143*, 04017051.
36. Xu, J.; Yang, X.; Tao, Z. A tripartite equilibrium for carbon emission allowance allocation in the power-supply industry. *Energy Policy* **2015**, *82*, 62–80.
37. Nash, J. Non-cooperative games. *Ann. Math.* **1951**, *54*, 286–295.
38. Ben-Ayed, O.; Boyce, D.E.; Iii, C.E.B. A general bilevel linear programming formulation of the network design problem. *Transp. Res. Part B* **1988**, *22*, 311–318.
39. Bard, J.F. *Practical Bilevel Optimization*; Springer: New York, NY, USA, 1998; pp. 144–146.
40. Colson, B.; Marcotte, P.; Savard, G. An overview of bilevel optimization. *Ann. Oper. Res.* **2007**, *153*, 235–256.
41. Lin, B.; Ouyang, X. Energy demand in China: Comparison of characteristics between the US and China in rapid urbanization stage. *Energy Convers. Manag.* **2014**, *79*, 128–139.
42. Hong, J.; Shen, G.Q.; Feng, Y.; Lau, S.T.; Mao, C. Greenhouse gas emissions during the construction phase of a building: A case study in China. *J. Clean. Prod.* **2015**, *103*, 249–259.

 © 2017 by the authors. Licensee MDPI, Basel, Switzerland. This article is an open access article distributed under the terms and conditions of the Creative Commons Attribution (CC BY) license (http://creativecommons.org/licenses/by/4.0/).

Concept Paper

Environmental Activation of Inner Space Components in Sustainable Interior Design

Magdalena Celadyn

Academy of Fine Arts in Krakow, Faculty of Interior Design, pl. Matejki 13, 31-157 Krakow, Poland; mceladyn@asp.krakow.pl; Tel.: +48-667-899-959

Received: 21 May 2018; Accepted: 7 June 2018; Published: 11 June 2018

Abstract: Implementation of environmental responsibility issues into the interior design methodology considers many aspects of the design process, but analyzes them separately. These include building materials' and products' specifications based on the assessment of their parameters impact on the users of indoor environments, or resource management within an ecological efficiency context. This concept paper concentrates on the analysis of an environmental activation of inner space components, identified by the author as the holistic and systemic design model, which is to empower the foundation of a contemporary sustainable interior design model. The proposed design scheme is supposed to assure the environmental effectiveness of interiors and their structure, as well as complementing functional components. The contributions of interiors completed in accordance with this concept can refer to the enhancement of the performance of building mechanical systems and the improvement in the indoor environment quality parameters. They can be achieved with the appropriate environmental activation-oriented structural, technical, and material solutions, applied to the selected inner space components. The theoretical scheme presented should become the basis for further investigations and studies to establish the comprehensive methodology design framework assuring the integrative role of interior design in the creation of a sustainable near environment.

Keywords: environmentally responsible interior design; sustainable interior design; environmental activation of interior elements; indoor environment quality

1. Introduction

The environmental responsibility of the interior design profession has been explored by researchers, architecture critics, and academics since the 1990s [1–3]. The increasing recognition of environmentally-sustainable design [1,4–6] has imposed on designers the necessity for a comprehensive and informed approach toward the interior design process. The notion of environmental responsibility in interior design can be interpreted as comprising the issues of: (1) an object's ecological effectiveness, with regard to the minimization of its negative impact on the natural environment; (2) the economic consequences and implications of the building spaces' energy performance; and (3) complementing social system's considerations related to the inner space quality parameters and their influence on the occupants' psychological and physical comfort. The constant interconnectedness and interdependence of these three systems is a major factor affecting the stability of the human ecosystem model and should be the subject of continuous investigation of environmentally-responsible interior designers [1] when searching for the optimization of the functionality and quality of inner spaces [4].

Although the term 'green design' is interchangeably used with ecological, environmentally-responsible, or sustainable design [3,7], it applies to the micro-scale of the interior [1]. The interior is defined as part of the built environment being in direct mediation with the space occupants and constituting the nearest area for their activities. It directly influences their health and well-being, and stimulates their behavioral patterns [8], as created in accordance with the sustainability paradigm. The interior designers accomplishing this model are to respect the environmental implications, considering the multiple life cycle consequences of completed objects and, therefore, addressing global environmental problems [4]. The position of interior designers in the process of creation of the so-defined environmentally-oriented built environment is still not precisely and comprehensively determined [1] and appreciated. The possible range of their interventions into the indoor environment design is mainly based on the appropriate specifications of introduced building materials, products, equipment, and appliances. These are made with the introduction of the comparative and whole-life-material-cycle environmental preference method (EPM) [9], consideration of the embodied energy (EE) measures, inclusion of the life cycle assessment (LCA) method associating the energy and material flows with potential environmental impacts [10], and the multi-criteria environmental evaluation schemes into the interior design process. All are necessary for finding informed and knowledgeable interior design solutions.

All these issues reflect the environmentally-oriented consciousness of designers, although they are methodically verified under reliable research-based assessment schemes and, when being considered separately as subsequent sustainable goals to be achieved, lack a cohesive vision designed to establish and endorse their environment-oriented effectiveness.

The main objective of the presented study is to define the role and technical, as well as formal, opportunities for the specific interior components' environmental activation, which would enable the identification of sustainable goals for interiors and the successful achievement of sustainability requirements in architectural and interior design. The intention of the article is to propose an integrated interior design framework, enabling the development of a perspective demonstrating the range of the impact of interior design on the built environment performance, and the development of the holistic approach to interior design, as postulated for the built environments' design [11,12].

The author discusses the possible implications of the recommended comprehensive model on the improvement in the closed spaces' quality parameters through the coordinated design process. Subsequent sections identify the term of the interior design's environmental contextualization with regard to the accomplishment of the sustainability paradigm. They also provide an overview of possible effects obtained in the case of properly-conceived interior components which are oriented towards the ecological effectiveness in the creation of the sustainable built environment. The analysis of the effects of environmentally conscious interior design, presented in this article, is restricted to its substantial features which are the question of the means of improvement in the indoor environment quality parameters, achieved through the cohesive and evidence-based design of the interior [13], with emphasis on the role of interior components in the optimization of sustainability goals.

2. A Model for Environmental Contextualization of Interior Components

Design strategies that are supposed to be applied to the environmentally-conscious interior design involve meeting the demand for the sustainable use of environment and resources. This postulate combines demands for the reduction in material resources and the enhancement of the indoor environment quality, achieved due to the research findings derived from peer-reviewed journals or conducted observations and surveys, which constitute a basis for design solutions concerning the built environment [3].

The achievement of sustainable goals in interior design requires the redefinition of its position, and the provision of the contribution of interior design to shaping the quality of built environments. This has to be augmented by the advanced environmental contextualization [14,15] consequently imposed on the interior and its components' creation (Figure 1).

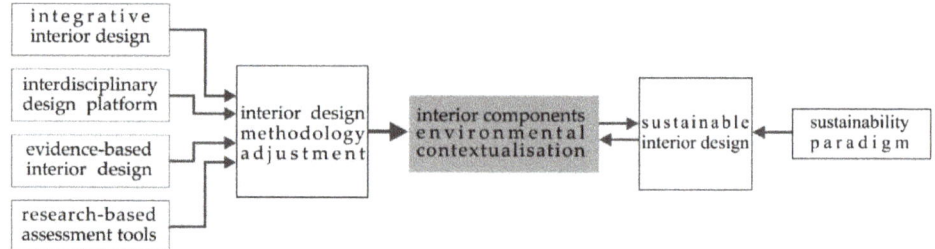

Figure 1. Interior components' multifaceted environmental contextualization as a function of the interior design methodology adjustment. Source: author's drawing.

The environmental context in the creation of internal spaces, by adding to this process a multidimensional perspective within which they can be analyzed and explored, requires the introduction of several adjustments concerning the conventionally-applied architecture design methodology. These corrections should be oriented on holistic thinking and interdisciplinary communication [1] marked with the inclusion of different specialists into the decision-making process; among them are interior designers, working as partners and co-learners [16] on the basis of integrative design teams. Intensive workshop sessions, as the interdisciplinary communication platform for professionals involved into the design process, enable the verification of proposals based on evidence provided by the participants in search of the optimization of buildings' performance. These eco-charrettes, focused on the design sustainability-compliance [6], allow the interior designers engaged in the multi-disciplinary integrated design process, to identify the area of their possible intervention in search for the healthy built environment and to contribute their knowledge to informing the design.

Applying research-based design tools, as the instruments assuring systemic and comprehensive pondering of the interior design postulates as detailed in the whole building certification system guidelines, may enable the interior designers to identify the sustainability goals to be achieved in their projects. Multi-criteria evaluation schemes based on the parametric assessment dedicated to the whole building, or its inner spaces' energy saving-, pro-ecological- and social-oriented solutions evaluation (e.g., Leadership in Energy and Environment Design Green Building Rating System for Interior Design and Construction LEED ID + C certification scheme established by the USGBC in 2004), have become a contemporary professional architectural design tool for defining and measuring the sustainability of green buildings [17]. It assists in the prevention of greenwashing effect [1]. The interior certification systems are considered by many researchers and practitioners as formal assessment schemes ensuring the rigorous approach leading to the achievement of sustainability features [4,6]. They are also the reliable and systemic verification tools for the compliance with the sustainability demands of indoor environment and are designed to define the quantitative and qualitative criteria for the main indoor environment evaluation categories (e.g., WELL Building Standard, a certification system conceived in 2014 and administrated by the International WELL Building Institute). They seem to be specifically recommended for interior design practitioners, enabling them a systematic approach to fundamental sustainability demands.

3. A Model for Environmental Activation of Interior Components

The effectiveness of environmental contextualization claimed by the author assigned to the designed inner space as an entity, and to its structural and complemented elements, relies on the simultaneous and equivalent consideration of several involved interior areas with regard to their impact on economic, ecological, and social systems. These design features, in terms of their role in the space structure, include the interior layout respecting the building orientation and climatic requirements, building materials specification based on the environmental preference methods, and the inner space multifunctional forming and supplementing components with emphasis on their technical and formal solutions. The interior components' design concept should be oriented toward their shaping considering an active response to the question of constant interaction of natural and man-made systems, with the provisions of its consequences as to the inner space quality endorsement and the building's performance.

Two aspects seem to mostly determine the components' environmentally-oriented features, and the effectiveness of the idea of environmental contextualization, as major sustainable interior design imperatives. These are the multi-functionality of interior components combined with the adaptability of inner spaces to accommodate different activities, along with the resources management framework. They might facilitate the fulfillment of environmental postulates in practice.

Elements of inner space, as indicated by researchers and academics for decades now, should not be created by designers in a traditional way, including functional, formal, and aesthetic contexts, but should be conceived by interior designers in a more complex [8] and environmentally-oriented context. This process of the creation of interior components should rely on their consideration as those conventionally formed with the sustainability features [3] (p. 267).

The typologies of components constituting the inner space, as proposed by critics and researchers [5,18,19], usually identify the groups according to their basic and auxiliary functions to be fulfilled. The structural forming of indoor components, as proposed in the author's classification [5], comprise: external walls determined as enclosures separating the inner space from the natural environment actively responding to the changing climate conditions and usually accompanied by various technical devices or natural finishing on the inner side; partitions and inner space dividers of various dimensions and finishes; raised floors; and suspended ceilings. Supplementing or completing interior components include: furniture, furnishings, equipment, and fixtures assigned to a separate category [5], enabling the proper usage of spaces in accordance with the exigencies.

The methods of structural forming of internal elements' and their integration with the building components, should be the consequence of designers' considerations of the interconnectedness of both natural and man-made environments, as well as predictions of possible consequences of their mutual interrelationship. As there are both direct and indirect relations between interiors and the environment [8] (p. 49), they should be specified, analyzed with scientific means, and reflected in the adjusted integrated interior design methodology.

The interior components' design strategy should identify the measures undertaken that respond to the users' needs, and positively affect their health and comfort, as well as minimize the objects' environmental load through the reduction in energy consumption by building systems. The strategies enabling the fulfillment of these, should combine the rational management of material resources, functional efficiency and formal diversity of components, their active inclusion into the overall integrated and indoor environment high-quality as supplementary means, and energy concepts.

Environmental Activation of Components

The presented model of multi-functionality of inner space components to be explored in the sustainable interior design framework, can be developed in order to identify its leading role in the consequent and comprehensive multi-faceted environmental activation of interior components. This design concept for complex components activation can be regarded as a valuable interior designers' contribution to the comprehensive accomplishment of sustainability imperatives.

A successful implementation of this model based on the comprehensive and environmentally-conscious interior design, and complying with the sustainability paradigm, requires designers to take into account the results of scientific studies in the field of technical disciplines including building physics, and climate engineering. The model can enable the integration of closed spaces in the built and natural environments obtained through: (1) endorsement of building systems as a substantial factor in the optimization of building's performance and related savings in energy consumption; (2) resource conservation as the result of rational management of building materials and products;

and (3) improvement in indoor environment quality parameters (Figure 2). The latter aspect of the suggested design model, discussed in further sections, concentrates on the presentation of the components' creation methods that are introduced into interior design as passive modes may influence and positively stimulate the interior's quality parameters.

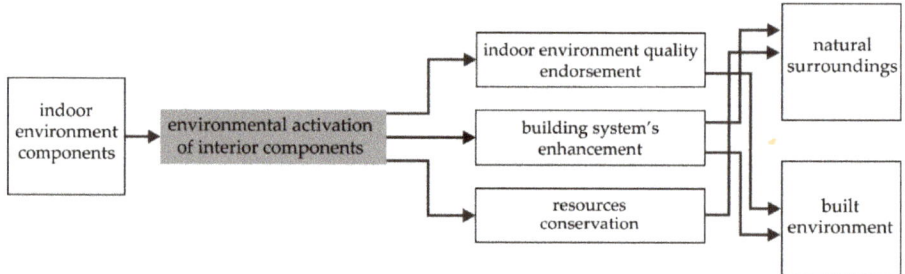

Figure 2. Design concept for environmental activation of interior components Source: author's drawing.

The improvement in lowering of the environmental impact of buildings on natural surroundings and inner spaces depends on the application of adequate technological and technical measures, as well as on the introduction of supportive passive methods. The latter may become a valuable contribution to the transition of interior design toward the accomplishment of sustainability paradigm, where the ecological, economic, and societal areas of human activities, are to remain in balance. The indicated requirements, being essential for the execution of sustainability issues in architectural design, can be met through the complex activation of inner space elements focused on the environmental responsibility.

Proposals for the definition of environmental activation of interior elements [14,15] are formulated as specific means of stimulation of these components to their action, performed and based on their' comprehensive and multi-criteria evaluation. This design method should play a substantial role in the interior decision-making process, for its multidimensional perspectives, identifying the functional, spatial, and temporal contexts. The main purpose of the environmental activation of interior elements, proposed in this paper, is to reduce the dependency of indoor environment quality on mechanical building systems, as well as to enhance the systems' performance. The recommended components' activation can be achieved by the properly analyzed setting of interior elements in inner spaces according to the users' individual requirements and organizational demands, by the elements' formal and structural forming, as well as their purposeful functional and formal integration with the main building structural components or building services.

The main objectives, which may be assigned to the activation concept, remain in accordance with a triple bottom line, which addresses the designers' responsibility in meeting the demands for the sustainability of economic, ecologic, and societal aspects of architectural design. These aims include the envisaged structurally, formally, and functionally combined solutions that should be considered as supportive building systems supplementing the passive means enabling:

- Enhancement of heating, ventilation, air conditioning (HVAC), and building lighting systems' effectiveness, assuring the increase in building performance and the decrease in energy and water consumption, as well as the reduction in operational costs;
- Reduction in the emission of toxic substances released in the production of building materials, and in the course of maintenance and conservation of buildings; and
- Optimization of the indoor environment quality (IEQ) parameters, related to the inner air characteristics, daylighting control, and acoustic conditions, in relation to the occupants' health and well-being.

The analysis of the selected parameters, presented in the following sections, indicates the possibilities of innovative spatial and technical methods to positively stimulate their optimization. The daylight transmission and its redistribution in interiors' regulation of solar thermal energy gains preventing the spaces from overheating, combine the demand for reduction in material resources with the enhancement of indoor environment quality through the applied structural solutions.

4. Components' Activation for the Optimization of Indoor Environment Quality

The complexity of indoor environment quality issues warrants the integrated approach to the design of buildings [20] (p. 430), which are functional-spatial-formal entities. This relates also to their inner spaces and, specifically, to these conceived in the course of refurbishments or adaptation of existing objects. The closer analysis of this problem being substantial for the interior evaluation proves, that there are many considerations within the Indoor Environment Quality category for the work of an interior designer [3] (p. 261) which should be identified while carrying out an environmentally-responsible design. Interior designers, along with other professionals involved in the design process, may contribute to the expected properties of the main components of integrated Indoor Environment Quality design [21]. These features combine the inner air quality with thermal, lighting, daylighting, and acoustic parameters, providing users with the health environment [22] having an impact on the occupants' health, physical and psychological comfort, well-being and productivity [1].

These objectives can be realized by interior designers with the applied different measures based on the determined sustainable goals, concerning the cautious selection of building materials, and the innovative and comprehensive approach to the formation of inner structural elements oriented toward their functional integration with the enhanced building components and systems (e.g., inner walls designed as thermal masses and means of reduction in the consumption of heating energy).

4.1. Indoor Air Quality

The indoor environmental and air quality is strictly health-related [23] (p. 242) and, thus, its significance should be recognized by interior designers while preparing the specifications of products and building materials to be introduced into an inner space, as the occupants' exposition to the pollutants contained in products or furnishings may cause several diseases (e.g., cardiovascular and respiratory diseases, multiple chemical sensitivity (MCS)). The selection procedures should be based on the credible information provided by independent institutions in the result of research-based measurement and certification processes concerning the occupants' health-affecting attributes. This would enable the provision of designers with the knowledge on building biology [10], which reveals the interrelationship between humans and their living environment.

The designers' ability to assessing impacts (...) leads to managing the risks [23] (p. 247) and, thus, to verifying of the consequences of their presence and effects on interiors features. Prior to the final decisions regarding the selection of constituting component products, the designers should provide the verification of possible influences of their introduction on the users' health, well-being, and comfort. The carefully selected building construction materials, structural components' solutions, as well as finishes can improve the indoor air quality. This refers to the following concerns:

- Regulation of the amount of pollutants in closed spaces and supported air purification;
- Reduction in solar heat gains and the enhancement of the monitoring of inner temperature, achieved by the individual users' control on thermal conditions and the way of spaces usage (e.g., presence of inner solar protection equipment manually adjustable);
- Reduction in daily fluctuations of inner air temperature by the free cooling effect (e.g., introduction of unfinished solid inner walls as passive thermal stores for the heating or cooling purposes and being exposed to solar heat [4,24], suspended ceilings, inner side finishes of external walls or space forming components, and partitions filled with integrated phase change materials PCM);

- Regulation of the radiant temperature and the usage of solid internal walls' thermal capacities with different finishes (e.g., porous surfaces of selected materials easy in maintenance, and increasing the heat absorption; removable wall finishing devices or hangings responding to seasonal climate changes and enabling the solid external walls to act as thermal masses in winter; and dark finishes of components situated in close proximity to glazed envelopes as thermal store and light finishes reflecting the heat [22]);
- Control of the relative humidity level of inner air (e.g., the supply of an adequate amount of water for growing plants and impeding the formation of mold in poorly-ventilated areas, 'bioclimatic' openwork inner walls made with materials of high permeability, components, or wall covering made with wood fiber panes).

These exemplary passive design methods should be recommended as environmentally-responsible complementing means of maintaining the high-quality indoor air with the related reduction in the number of technologically-advanced and energy-consuming heating, ventilating, and air conditioning systems. Living walls, recommended for their air purifying- and conditioning potential, and vastly implemented in interiors, can become the essential passive means of maintaining good air quality, through the stabilization of the ambient temperature and indoor humidity level. The consequent installation of space dividers in the form of active green walls can contribute to the decrease in inner temperature, and play a complex role of airborne chemical filters, carbon dioxide absorbers, and the source of oxygen released during the daytime into the inner space. The indoor air quality can be further enhanced by the inclusion of living moss walls playing the role of internal buffers and helping in the removal of concentrated volatile organic compounds (VOC) and other chemical pollutants from the air and, thus, allowing the reduction in demanded ventilation rates. Plants broadly introduced into closed spaces assure for the users a connection to nature and affect their well-being, responding to the demand for the inclusion of the biophilic concept into the interior design model.

The 'bioclimatic' openwork inner walls and space dividers, made with gypsum, light concrete or ceramic tiles, may be seen as another valuable solution, which contributes to the thermal microclimate conditions and inner air relative humidity. These climate-stimulating structures in modernized and adapted buildings can be completed with building materials recovered and reused in new spatial contexts. Due to their abilities to exert a positive impact on the indoor environmental quality, they are examples of innovative techniques allowing the accomplishment of sustainability principles in interior design, through the concept respecting equally valuable design strategies.

4.2. Daylighting

The lighting scheme developed for interiors should include a vast amount of daylighting, since the use of natural light within the building and its interior is not only fundamental for the reduction in energy demand [5] but also for the minimization of the negative influence of seasonal affective disorder (SAD) which is a result of deficiency of daylight. Its abundance usually positively affects occupants' health and well-being with respect to the circadian stimulation and its positive influence on physical activity and perceptual effectiveness [25].

Apart from the technical devices assuring the users' individual control of the intensity of light at workplaces, the daylight-level and occupancy sensors or dimming devices keeping the electric energy demands on a reasonable level, there are many other possible solutions. The effective use of daylight can be a valuable contribution of interior designers frequently enhanced by consultations with lighting engineers. They may advance the occupants' well-being, ensuring their increased productivity and satisfaction.

These are supposed to ensure good daylighting levels on the floorplates and to provide a good amount of daylight in accordance with the enforced regulations, codes and standards. They have to be appropriate for space purposes, their functional requirements, and should enable the achievement of a high-quality visual environment. The main positive effects of designers' intervention into the methods of components' introduction with regard to the optimization of daylighting combine:

- Minimization of the negative glare effect by the introduction of some functional components (e.g., manually controlled internal solar control blinds, louvers, supplementing devices eventually mounted externally, screens modulating the brightness of glazed surfaces, other components in the proximity of openings with highly reflective finishes);
- Minimization of the solar thermal heat gain (e.g., blinds, manually-controlled curtains);
- Ambient daylighting in fully-glazed spaces obtained with the structures suspended from the glazed roof support, translucent perforated fabrics [6], or framing structures made with lightweight bright finished materials mounted to the glazed envelope;
- Maximization of the quantity of incoming daylight by its reflection (e.g., window reveals finished with light-colored surfaces or reflective materials, like safety glass or metal sheets, reflective surfaces of internal components adjacent to openings [3], and enclosures separating spaces situated in a distance from glazed walls with their upper parts made with transparent or translucent panels);
- Reduction in the distracting contrast on the surfaces of working area caused by direct sunlight (e.g., fabrics mounted to the ceiling next to windows [6];
- Penetration of distracting daylight (e.g., introduction of clerestory windows with the complementing reflective finishing materials of sloped ceilings to optimize their performance [1,3]);
- Redirection and transmission of daylight, as well as an increase in the illuminance of daylighting in the overall lighting supply (e.g., technical devices like reflective optical light tubes providing under-lit spaces with the demanded daylighting quantity, translucent inner light shelves integrated with the structure of the building's south side glazed envelope [1,8,24], or adjustable reflective panels suspended under the ceiling and anidolic integrated ceilings (AIC) [24,26]); and
- Diffusion of daylight (e.g., wall cladding with reflective materials supplementing the semi-translucent diffusing panes suspended from the glazed roofs to obtain an effective daylight distribution [27]).

The methods commonly introduced into inner spaces which enable the control of the amount of daylight through the use of reflected light, combine the installation of blinds, curtains, louvers configured depending on the window orientation, or more technically-advanced means including opaque and translucent inner light shelves provided with reflective finishes of suspended ceilings. The additional structures like passive-daylight devices introduced by designers with respect to the existing interior layout, building orientation, and the volume of the glazing of envelopes (e.g., translucent polymer mobile optic diffusors suspended from the skylight and integrated with passive solar optic systems PSO), may effectively participate in the redirection of the daylight and its transmission into the parts situated at a distance from the glazed external walls [28]. Other implemented components made with glass fiber reinforced gypsum and painted white may enable the control of the amount of incoming daylight and the reduction of the light contrast, or a negative glare effect inside. The above-mentioned possible effects of the installation of internal components as passive-daylight means, define the range of the interior designers' formal contribution to the daylighting quality through the supportive, innovative and complex design approach, as well as by the environmental implications of assigned components' activation. The daylighting concept, for its effectiveness, should be conceived in accordance with the site conditions and activity-related illuminance requirements, measured and verified with computer simulations and integrated with the artificial lighting concept from the pre-design planning phase [3].

4.3. Acoustic Conditions

The rising demands among users for the controlled noise levels and the improvement in acoustical conditions of interiors, as well as the postulates of necessary acoustic privacy, are other interior design issues that have to be carefully considered by the environmentally-conscious designer. As the surveys conducted on the contribution of different factors to the distraction of workplaces confirm [6,29], noise remains the one most complained about. The acoustical comfort is one of the substantial features in the occupants' assessment of interior quality, influencing their health and wellbeing. On the other hand, the internal acoustic quality still neither has been recognized by designers as a primary issue in sustainable design [22], nor as a substantial assessment criterion of the indoor environment quality. Interior designers can carry out a balanced acoustical design of inner spaces and obtain a 'non-intrusive' speech privacy level [6,29], including the elimination of uncontrolled transmitting of distractive internal noises, absorbing sound or its masking. The efficiency of these and other applied solutions is assured by the improved space planning, through the selection of proper building materials and adequate formal and structural characteristics of internal components.

The design methods and techniques directed toward the improvement in acoustical comfort consist of:

- Reduction of the internal noise transmission (e.g., a functionally-resolved interior layout and spatial zoning, ceiling height enclosures complemented with transparent upper panes to enable light transmission, broad introduction of the sound proofing building materials constituting the layers of enclosures, ceiling finishes, or functional interior components); and
- Absorption of sound (e.g., acoustic absorber panels mounted directly to the ceiling, parallel to the glazed walls, and complemented with a finishing layer adjusted to the interior's spatial and formal concept [24]).

The accomplishment of good acoustic conditions, through the introduction of sound-absorbing building materials, may provide the interior design with substantial quality parameters, as well as to enhance the space unique formal and stylistic values. The sound-absorbing finishing layer of the composite partition wall made with demolished parts reclaimed from building bricks or ceramic tiles, and other reused materials, can separate intensively-used circulation areas from adjacent spaces. The exposed rough texture of the cut hollow bricks forming the inner wall layer, being a sound-dissipating and sound diffusing multi-faceted space divider, may be seen as an innovative means of control of the sound level, and the reduction in the reverberation time.

In addition, the building material of the finishing layer, massively incorporated as a sound absorber, participates in the process of modification of the level of inner air relative humidity, being one of the substantial considerations related to comfortable inner thermal conditions. Therefore, the choice of reclaimed building materials, made on the basis of an analysis of their physical and chemical parameters, enables the fulfillment of yet another sustainable design demand, regarding the optimization of indoor environment parameters and interior microclimate characteristics, as essential for the users' comfort. The introduction of building materials reclaimed on site and implemented into the objects' structure, as presented in the example of the possible installation of interior components, allows to achieve compliance with the requirement for the reduction in the amount of demolition wastes [16], as well as the effective waste resources' management.

5. Conclusions

This concept paper provides a proposal regarding the modification of interiors and their structure, as well as complementing the component design methodology enabling the compliance with the environmental sustainability paradigm. It demonstrates the multi-dimensional environmental approach in the creation of closed spaces as the basic concept for a sustainable interior design process.

The environmental activation of interior components, as the substantial element of the design method, results in multi-faceted benefits addressing the optimization of buildings and their indoor environment performance, reduction of their negative impact on the natural environment, and the establishing of real interconnectedness between manmade and natural environments.

The benefits of components' environmental activation, as an interior design imperative enabling the consideration of sustainability issues and the achievement of sustainable goals, in terms of the optimization of building materials and product consumption include:

- Efficient management of interior components related to their multi-functionality;
- Resource management optimization due to the components' adaptability and applied structural and technical solutions; and
- Reclaiming of dismantled or removed components and their parts from refurbished spaces and their implementation into new structures as a design imperative to be considered within the environmentally-responsible design process.

The social benefits of the applied design method oriented on active response of components to their environmental interaction combine:

- Broad inclusion of stakeholders, including professionals of different specialties, owners, and end-users into the decision-making process;
- Responsible usage of the indoor environment treated as a scenery for the accommodation of different human activities, as well as a complex building product requiring rationality in its use affecting its performance;
- Increase in the environmental consciousness of the interiors' occupants through the understanding of interior component roles in shaping the quality of the indoor environment and their impact on natural surroundings; and
- Establishing a developed knowledge-based identity of occupants with their inner spaces in the process of experienced results of the environments' interconnectedness.

The components' environmental activation, as presented in the analysis, affects the performance of spaces within the following aspects:

- Enhancement of the building's artificial light system achieved with the optimization of daylight management through the implementation of multi-functional inner devices integrated with building components;
- Enhancement of ventilation systems through the properly executed space layout and space dividers' configuration; and
- Endorsement of heating/cooling, as well as air conditioning systems, with the assignment of additional functions to the components.

The qualitative assessment of the above-mentioned positive results derived from the enclosure of environmental activation concept into the interior design methodology indicate that this innovative cohesive approach toward the indoor environment and its components' creation may significantly affect the position of interior design discipline in the integrative sustainable design process. The examples of a range of structural and formal interventions by interior designers undertaken toward the effective compliance with the sustainability requirements presented above, prove that the concept of interior components' environmental activation, preceded by their multi-functionality model and multi-faceted environmental contextualization, can become a fundamental design method for the planning of inner spaces and the forming of components. This may assure the achievement of sustainable goals in the creation of interiors. The assignment of the essential meaning to the interior components' environmental contextualization in the interior design process, may become crucial in a search for synergy in sustainable interior and architectural design, as well as mechanical systems' designs.

The increasing recognition of environmentally-sustainable interior design [1,12] encourages designers to modify their design methodology in order to comply with the sustainability paradigm. The systemic problem solving applied to interior design, enabling the transition from degenerative architectural design through the transitional phase of sustainable design, assures the achievement of coexistence of natural and man-made environments. This can lead to a model actively supporting the idea of regenerative [30] architectural and interior design.

Conflicts of Interest: The author declares no conflict of interest.

References

1. Jones, L. *Environmentally Responsible Design: Green and Sustainable Design for Interior Designers*; John Wiley & Sons Inc.: Hoboken, NJ, USA, 2008; ISBN 978-0-471-76131-0.
2. Pilatowicz, G. Sustainability in Interior Design. *Sustainability* **2015**, *8*, 101–104. [CrossRef]
3. Winchip, S.M. *Sustainable Design for Interior Environments*, 2nd ed.; Fairchild Books: New York, NY, USA, 2011.
4. Moxon, S. *Sustainability in Interior Design*; Laurence King Publishing: London, UK, 2012.
5. Raymond, S.; Cunliffe, R. *Tomorrow's Office. Creating Effective and Human Interiors*; E & FN Spon: London, UK; Taylor & Francis Group: New York, NY, USA, 2000.
6. Bonda, P.; Sosnowchik, K. *Sustainable Commercial Interiors*, 2nd ed.; John Wiley & Sons: Hoboken, NJ, USA, 2014.
7. Hayles, C.S. Environmentally sustainable interior design: A snapshot of current supply of and demand for green, sustainable or Fair Trade products for interior design practice. *Int. J. Sustain. Built Environ.* **2015**, *4*, 100–108. [CrossRef]
8. Pilatowicz, G. *Eco-Interiors. A Guide to Environmentally Conscious Interior Design*; John Wiley & Sons, Inc.: New York, NY, USA, 1995; ISBN 0-471-04045-2.
9. Anink, D.; Boonstra, C.; Mak, J. *Handbook of Sustainable Building. An Environmental Preference Method for Selection of Materials for Use in Construction and Refurbishment*; James & James (Science Publishers) Ltd.: London, UK, 1998.
10. El Khouli, S.; John, V.; Zeumer, M. *Sustainable Construction Techniques. From Structural Design to Interior Fit-Out: Assessing and Improving the Environmental Impact of Buildings*; Institut fuer internationale Architektur—Dokumentation GmbH & Co. KG: München, Germany, 2015.
11. Vale, B.; Vale, R. *Green Architecture. Design for a Sustainable Future*; Thames and Hudson Ltd.: London, UK, 1996.
12. Kang, M.; Guerin, D.A. The State of Environmentally Sustainable Interior Design Practice. *Am. J. Environ. Sci.* **2009**, *5*, 179–186. [CrossRef]
13. Nussbaumer, L.L. *Evidence Based Design for Interior Designers*; Fairchild Books: New York, NY, USA, 2009.
14. Celadyn, M. The inner space elements in shaping indoor environment quality parameters. In *Integration of Art and Technique in Architecture and Urbanism*; Wydawnictwa Uczelniane Uniwersytetu Technologiczno-Przyrodniczego: Bydgoszcz, Poland, 2017; Volume 5, pp. 41–50; ISBN 978-83-65603-35-7.
15. Celadyn, M. Inner space elements in environmentally responsible interior design education. *World Trans. Eng. Technol. Educ.* **2016**, *14*, 495–499.
16. Reed, B. Integrated design. In *Sustainable Commercial Interiors*; Bonda, P., Sosnowchik, K., Eds.; John Wiley & Sons: Hoboken, NJ, USA, 2007; pp. 28–31.
17. Vallero, D.; Brasier, C. *Sustainable Design. The Science of Sustainability and Green Design*; John Wiley & Sons Inc.: Hoboken, NJ, USA, 2008.
18. Brand, S. *How Buildings Learn. What Happens After They Are Built*; Penguin Books: London, UK, 1994.
19. Duffy, F. *Design for Change. The Architecture of DEGW*; Birkhauser Verlag: Basel, Switzerland; Boston, MA, USA; Berlin, Germany, 1998.
20. Kibert, C.J. *Sustainable Construction: Green Building Design and Delivery*, 4th ed.; John Wiley & Sons, Inc.: Hoboken, NJ, USA, 2016; ISBN 978-1-119-05517-4.
21. Schoof, J. Vintage design or conservation of resources? Re-use and recycling in architecture. *Detail Green* **2015**, *1*, 6–11.
22. Owen, L.J. *A Green Vitruvius. Principles and Practice of Sustainable Architectural Design*; James & James: London, UK, 1999.
23. Keeler, M.; Vaidya, P. *Fundamentals of Integrated Design for Sustainable Building*, 2nd ed.; John Wiley & Sons Inc.: Hoboken, NJ, USA, 2016.
24. Szokolay, S.V. *Introduction to Architectural Science: The Basis of Sustainable Design*; Architectural Press: Oxford, UK, 2010.
25. Schlaffle, E. Aspects of Office Workplace Lighting. In *A Design Manual Office Buildings*; Hascher, R., Jeska, S., Klauck, B., Eds.; Birkhauser: Basel, Switzerland; Boston, MA, USA; Berlin, Germany, 2002.
26. Yeang, K. *Ecodesign. A Manual for Ecological Design*; John Wiley & Sons: London, UK, 2009.
27. Hegger, M.; Fuchs, M.; Stark, T.; Zeumer, M. *Energy Manual. Sustainable Architecture*; Birkhauser Verlag AG: Basel, Switzerland, 2008.

28. Celadyn, M. Daylighting in sustainable design of office interiors. *Arch. Civ. Eng. Environ.* **2016**, *9*, 1–8.
29. Aronoff, S.; Kaplan, A. *Total Workplace Performance. Rethinking the Office Environment*; WDL Publications: Ottawa, ON, Canada, 1995.
30. Reed, B. Regenerative Development and Design: Working with the Whole. In *Sustainable Construction: Green Building Design and Delivery*, 3rd ed.; Kibert, C.J., Ed.; John Wiley & Sons, Inc.: Hoboken, NJ, USA, 2013; pp. 109–111; ISBN 978-0-470-90445-9.

© 2018 by the author. Licensee MDPI, Basel, Switzerland. This article is an open access article distributed under the terms and conditions of the Creative Commons Attribution (CC BY) license (http://creativecommons.org/licenses/by/4.0/).

MDPI
St. Alban-Anlage 66
4052 Basel
Switzerland
Tel. +41 61 683 77 34
Fax +41 61 302 89 18
www.mdpi.com

Sustainability Editorial Office
E-mail: sustainability@mdpi.com
www.mdpi.com/journal/sustainability

www.ingramcontent.com/pod-product-compliance
Lightning Source LLC
LaVergne TN
LVHW071943080526
838202LV00064B/6660